Fly Ash in Concrete
Properties and Performance

Other RILEM Reports available from Taylor & Francis

1 Soiling and Cleaning of Building Façades
Report of Technical Committee 62-SCF
Edited by L.G.W. Verhoef

2 Corrosion of Steel in Concrete
Report of Technical Committee 60-CSC
Edited by P. Schiessl

3 Fracture Mechanics of Concrete Structures: From Theory to Applications
Report of Technical Committee 90-FMA
Edited by L. Elfgren

4 Geomembranes – Identification and Performance Testing
Report of Technical Committee 103-MGH
Edited by A. Rollin and J.M. Rigo

5 Fracture Mechanics Test Methods for Concrete
Report of Technical Committee 89-FMT
Edited by S.P. Shah and A. Carpinteri

6 Recycling of Demolished Concrete and Masonry
Report of Technical Committee 37-DRC
Edited by T.C. Hansen

7 Fly Ash in Concrete: Properties and Performance
Report of Technical Committee 67-FAB
Edited by K. Wesche

Fly Ash in Concrete

Properties and Performance

Report of Technical Committee 67-FAB
Use of Fly Ash in Building

RILEM

(The International Union of Testing and Research Laboratories for Materials and Structures)

Edited by

K. Wesche

CRC Press
Taylor & Francis Group
Boca Raton London New York

CRC Press is an imprint of the
Taylor & Francis Group, an **informa** business
A TAYLOR & FRANCIS BOOK

CRC Press
Taylor & Francis Group
6000 Broken Sound Parkway NW, Suite 300
Boca Raton, FL 33487-2742

First issued in paperback 2019

ISBN-13: 978-0-419-15790-8 (hbk)
ISBN-13: 978-0-367-86396-8 (pbk)
ISBN-13: 978-0-442-31473-6 (USA)

British Library Cataloguing in Publication Data
Fly ash in concrete: Properties and Performance.
I. Wesche, K.
620.1

Library of Congress Cataloging-in-Publication Data
Available

Publisher's Note
The publisher has gone to great lengths to ensure the quality of this eprint but rpoints out that some imperfections in the original may be apparent

Visit the Taylor & Francis Web site at
http://www.taylorandfrancis.com

and the CRC Press Web site at
http://www.crcpress.com

Contents

Preface

RILEM Technical Committee 67-FAB 'Use of Fly Ash in Building' was constituted in September 1981. Its objectives were:

- to produce a state-of-the-art Report documenting current knowledge of the properties of fly ash concrete and of the use of fly ash in building;
- to make recommendations on new or modified test methods relating to the use of fly ash;
- to review research needed in this field and recommend priorities.

There have been four full meetings of the committee:

- 19–22 March, 1982 in Aachen;
- 6–7 June, 1983 in Paris;
- 24–25 September, 1984 in London;
- 22 April, 1986 in Madrid.

Eight Task Groups (TG) were set up to prepare the individual sections of the Report. The Task Group chairmen are listed in the committee list which follows. The results of work in the task groups were reported at the Second International Conference on the Use of Fly Ash, Silica Fume, Slag and Natural Pozzolans in Concrete on 22 April, 1986 in Madrid.

The committee was wound up in October 1986 and an Editorial Group was charged with completion of the Report and the recommendations.

At the time of the Madrid Conference, work in the task groups had reached a point at which work on the final version the state-of-the-art Report could begin. It was agreed that Philip Owens should edit the report. Unfortunately, Mr Owens was compelled to relinquish this task a year later, in April 1987 owing to pressure of work and lack of resources, and for pesonal reasons. Roughly half the text had been completed. After an unsuccessful search for other solutions, I took over the task of editing the report myself. Costs incurred up to that point were paid by the German Research Association (Deutsche Forschungsgemeinschaft). All illustrations were redrawn and standardized at the institute of Building Research (Institut für Bauforschung, ibac) of the University for Technology, Aachen. The standardization of style and terminology and the preparation of the camera-ready manuscript were undertaken by a scientific and technical translation agency in Aachen. The costs of this work were contributed by BauMineral GmbH in Herten, Germany.

This Report collates insights drawn from research results throughout the world up to 1986, providing a foundation for all future research in the fly ash sector and for the assessment of all types of fly ash. Since the report covers the major and most significant areas of all possible research in this field, it will prove an essential aid and source for every researcher working on this topic, now and in the future.

To prepare the Report, all available publications were collected, listed and evaluated. The list of references (Chapter 5), containing 817 individual entries, forms an important element of the report, and will be an indispensable source for further research.

The two sets of final RILEM recommendations on the testing and assessment of fly ash alone (FAB 1) and of fly ash in mortar and concrete (FAB 2) have been included in an Appendix. These recommendations were published in draft form in *Materials and Structures* in July 1989.

My thanks are due to all those who have so kindly sponsored the production of the report, and to all the authors and other colleagues who have contributed to it.

K. Wesche
Aachen,
November 1990

RILEM Technical Committee 67-FAB

K. Wesche (Chairman)
Institut für Bauforschung (ibac), University of Technology, Aachen, Germany

P.L. Owens (Secretary)
Philip L. Owens & Partners Ltd., Tring, Hertfordshire, United Kingdom

P. Schubert (Secretary, Chairman TG 3 and 4.3)
Institut für Bauforschung (ibac), University of Technology, Aachen, Germany.

MEMBERS CONTRIBUTING TO THE STATE-OF-THE-ART-REPORT

M.L. Valero Alonso (Section 1, Chairman TG 1)
Asociación Tecnica Española del Asfalto, Madrid, Spain

Th. Bach (Sections 3.1.2., 3.1.4., 3.1.6)
Dansk Eternit, Aalborg, Denmark

J.M. Bijen (Sections 3.2.5, 3.2.6, 3.6, 3.7, 4.1, 4.2, 4.4, 4.5, 4.10, Chairman TG 5)
INTRON BV, Institute for Material Testing and Environmental Research, Maastricht, Netherlands

H. Grube (Section 3.3)
Forschungsinstitut der Zementindustrie, Düsseldorf, Germany

H. Kukko (Sections 2.2–2.4)
Technical Research Centre, Concrete and Silicate Laboratory, Espoo, Finland

S. Nagataki (Section 3.3)
Tokyo Institute of Technology, Faculty of Engineering, Department of Civil Engineering, Tokyo, Japan

H. Ogha (Section 3.3)
Tokyo Institute of Technology, Faculty of Engineering, Department of Civil Engineering, Tokyo, Japan

P. Schubert (Sections 3.2.1–3.2.4, 3.5)

J.P. Skalny (Sections 3.1.1, 3.1.3, 3.1.5, 3.1.7, 4.3)
W.R. Grace Co., Construction Materials Research, Columbia, Maryland, USA

E. Vazquez (Sections 4.6–4.9)
Universidad Politecnica de Catalunya, Escuela Ing. de Caminos, Canales, Puertos, Barcelona, Spain

W. vom Berg (Section 2.1)
VGB – Vereinigung der Großkraftwerksbetreiber e. V., Essen, Germany

M.A. Ward (Section 3.4, Chairman TG 2.2 and 4.2.)
Department of Civil Engineering, University of Calgary, Calgary, Canada

MEMBERS UNDERTAKING OTHER ACTIVITIES IN THE COMMITTEE

J. Beretka
CSIRO, Division of Building Research, Highett, Victoria, Australia

S. Dazai
Denpatsu Fly Ash Company Ltd., Tokyo, Japan

R.K. Dhir
University of Dundee, Dundee, Scotland, United Kingdom

S. Droljc
ZRMA, Ljubljana, Yugoslavia

P. Dutron
CEMBUREAU, European Cement Association, Brussels, Belgium

R.D. Hooton
Ontario Hydro Research, Toronto, Canada

P.G.K. Knight
Central Electricity Generating Board, London, United Kingdom

J.E. Krüger
Nasionale Bounavorsings – Inst. van die WNNR, Afd. Anorg. Materiale, Pretoria, South Africa

J.D. Matthews (Chairman TG 2.1 and 4.1)
BRE – Building Research Establishment, Garston, Watford, United Kingdom

C.D. Pomeroy
British Cement Association, Slough, United Kingdom

D. Ravina
Technion Israel Institute of Technology, Haifa, Israel

M. Regourd
C.E.B.T.P., Paris, France

H. Scholz
VNDK, Dortmund-Höchsten, Germany

A. Sellevold
Norwegian Building Research Institute, Oslo, Norway

EDITORIAL GROUP

K. Wesche (Chairman)
P. Schubert (Secretary)
M.L. Alonso
J.M. Bijen
W. vom Berg
R. Rankers (Assistance)
Institut für Bauforschung (ibac), University of Technology, Aachen, Germany

---

RILEM
The International Union of Testing and Research Laboratories for Materials and Structures/Réunion Internationale des Laboratoires d'Essais et de Recherches sur les Matériaux et les Constructions.

Secretariat: Ecole Normale Superieure, Pavillon du Crous, 61 av. du Pdt Wilson, 94235 Cachan Cedex, France.

INTRODUCTION

The term "fly ash" is often used to describe any fine particulate material precipitated from the stack gases of industrial furnaces burning solid fuels. The amount of fly ash collected from furnaces on a single site can vary from less than one ton per day to several tons per minute.

The characteristics and properties of different fly ashes depend on the nature of the fuel and the size of furnace used. Pulverization of solid fuels for the large furnaces used in power stations creates an immediate, urgent problem; dry fly ash has to be collected from the stack gases and disposed of quickly and safely. The similarity of some fly ashes to natural pozzolanas of volcanic origin has encouraged the use of fly ash in conjunction with portland cement in concrete making. Not all fly ashes are suitable for this application, however; unstable chemical reactions may have adverse effects on both the hydration process and the ultimate stability of the end product.

Fly ashes generally fall into one of two categories, depending on their origin and their chemical and mineralogical composition /D 19/. Combustion of anthracite or bituminous coal generally produces low-calcium fly ashes; high-calcium fly ashes result from burning lignite or sub-bituminous coal. Both types contain a preponderance of amorphous glass.

In addition, low-calcium fly ashes usually contain quartz, mullite, hematite and magnetite, while high-calcium ashes contain quartz, lime, mullite, gehlenite, anhydrite and cement minerals such as C_3A and C_2S. Both types of fly ash have pozzolanic properties, but high-calcium fly ashes also exhibit cementitious properties. Owing to these differences, the interactions of each of these two types with cement require separate consideration.

It should be noted that, even where fly ash comes from a single source, it may be a highly variable material in terms of both its chemical composition and its physical properties /D 17/. The variations manifest themselves in the reactivities of the fly ash and their effects on hydration and strength development in mortar and concrete. For this reason, general discussions of fly ash cement interactions should emphasize trends rather than quantitative parameters.

Difficulties often arise when the performances of different fly ashes are compared. The interpretation of test results is a frequent cause of unnecessary disputes, since there are considerable differences between the test methods used in various countries. Nevertheless, it should be recognized that where the demand for industrial and domestic energy results in the production of large volumes of fly ash, these should not only be disposed of safely to prevent environmental pollution, but should whereever possible be treated as a valuable resource. There is no doubt that the production of energy from solid fuels will increase on an unprecedented scale during the next 25 years. Some authorities forecast fly ash volumes of more than treble the current world output to some 800×10^6 tons by the year 2010.

1 CHARACTERIZATION OF FLY ASH

J.L. ALONSO and K. WESCHE

1.1 Origin of Coal and Burning Conditions

Coal is a complex, heterogeneous material, in widespread use as an energy source throughout the world. It is the end product of a series of biological and physicochemical processes which have resulted in the wide variety of minable materials currently utilized in industry.

When pulverized coal is burnt to generate electrical power, extremely large quantities of fly ash and bottom ash are produced. Fine grade fly ash has acquired considerable importance in the building materials sector.

Coals are formed in the earth's interior over periods in the order of 300 to 400 million years. Over such long periods, the different kinds of plant material from which coal is formed undergo complex transformations, so that the nature and properties of the great variety of coals we now utilize are dependent on the class of plants which have been transformed and on the depth to which these have been buried. Together with the depth of burial, high temperatures and pressures play an important role in determining coal composition and characteristics. Coal attains its final state in combination with a range of different compounds, and can be sub-divided into various classes or groups such as peat, lignite, sub-bituminous and bituminous coals and anthracite.

The quantity of water present in these different classes of coal decreases in proportion to their ascending rank, ranging from 90 % for peats to 0.5 % for anthracites.

Characterization of coals demands a knowledge of the following parameters:

- moisture,
- ash content,
- volatile matter,
- fixed carbon,
- sulphur content (organic, pyritic and sulphatic sulphur),
- calorific or heating value.

In the case of bituminous coals, the free swelling index (FSI) and Gieseler plasticity are also of importance.

Table 1.1 indicates some of the widely differing fuel specifications for two lignite-fired power plants in the United States of America, including chemical analyses of the coals and their fly ashes /N 22/.

***Table 1.1:** Fuel specifications and chemical analyses for two lignite-fired power stations in the USA*

Parameter	Range of Analysis (%)	
	Power Station 1	Power Station 2
Proximate analysis		
Moisture	27.0 - 35.0	33.6 - 40.0
Ash	24.0 - 29.0	4.3 - 9.5
Ultimate analysis. dry basis		
Carbon	40.3 - 45.4	57.1 - 66.2
Sulphur	2.2 - 2.7	0.6 - 2.5
Ash	34.9 - 41.6	7.2 - 15.8
Oxygen	10.7 - 12.6	15.3 - 27.4
Elemental analysis of ash		
SiO_2	61.1 - 65.1	10.8 - 39.6
CaO	4.0 - 5.5	14.1 - 41.3
MgO	0.5 - 0.8	3.1 - 9.2
Na_2O	2.8 - 3.7	1.0 - 15.4
SO_3	3.3 - 5.9	11.2 - 27.8

Coal is burned in power stations in order to generate the heat required to turn water into steam which can be used to drive steam turbines. The energy of the coal is finally converted into electrical power. In accordance with the ranking noted above, anthracite has the highest and lignite the lowest calorific value of the coals used as power station fuels.

Three different processes are employed for the combustion of pulverized coal in power station boilers /H 32/:

- High temperature combustion: here, combustion occurs at furnace temperatures of some 1 500 - 1 700 °C. The resulting ash melts and falls into water, where it collects in the form of solid, mainly vitreous particles. Only a small quantity of fine particles escapes to electrostatic precipitators in the form of fly ash. Furnaces of this type are generally referred to as slag-tap furnaces.
- Dry combustion: in this case, the pulverized coal is burnt at furnace temperatures of 1 100 to 1 400 °C. Roughly 90 % of the ash collected from the process is in the form of ultra-fine particles retained by electrofilters or precipitators. Since the temperature decreases slowly, the percentage of vitreous particles is low.

- Fluidized-bed combustion: the furnace temperature in the fluidized beds is less than 900 °C, excluding melting. Ashes are irregularly shaped, with a high percentage of crystalline particles. These are not genuine fly ashes, and are of little interest for building material applications.

Table 1.2 indicates fly ash production and utilization in various countries for the years 1977 /F5/, 1984 /R42/ and 1986/87 /M53/.

1.2 Properties of Fly Ash

1.2.1 Definitions and Specifications

Pozzolans are siliceous or siliceous and aluminous materials which, though themselves possessing little or no cementitious value, will, in finely divided form and in the presence of moisture, react chemically with calcium hydroxide at ambient temperature to form compounds with cementitious properties (ASTM Standard C 618-80).

Fly ash is a solid, fine-grained material resulting from the combustion of pulverized coal in power station furnaces. The material is collected in mechanical or electrostatic separators. The term fly ash is not applied to the residue extracted from the bottom of boilers.

Fly ashes capable of reacting with $Ca(OH)_2$ at room temperature can act as pozzolanic materials. Their pozzolanic activity is attributable to the presence of SiO_2 and Al_2O_3 in amorphous form.

Fly ashes may be sub-divided into two categories, according to their origin (ASTM):

- Class F: Fly ash normally produced by burning anthracite or bituminous coal which meets the requirements applicable to this class. Class F fly ash has pozzolanic properties.
- Class C: Fly ash normally produced by burning lignite or sub-bituminous coal which meets the requirements applicable to this class. In addition to pozzolanic properties, Class C fly ash also possesses some cementitious properties. Some Class C fly ashes may have lime contents in excess of 10 %.

Many other forms of classification can be accepted, e.g. classification according to carbon content, SiO_2 reactivity, SiO_2 solubility, pozzolanic activity, etc. *Table 1.3* compiles the standards of different countries in which fly ashes are specified.

Table 1.2: Fly ash production and utilization in various countries

Country	Production (10^6 t/a)			Utilisation (10^6 t/a)			Utilisation in % of Production		
	1977	1984	1986/87	1977	1984	1986/87	1977	1984	1986/87
Australia	5.4	3.5	5.2	0.58	0.25	0.56	11	7	11
Canada	2.6	3.3	3.2	0.71	0.8	1.1	27	24	34
China		35	41		7.2	9.5		21	23
Denmark		1	1.2		0.45	0.9		45	73
France	4.8	5.1	2.2	2.0	1.5	1.3	42	29	58
Germany, F.R.		2.6	2.9		2.0	2.2		77	76
Hungary	5.0		4.1	0.43		1.1	9		26
India		19	39		0.5	1.2		3	3
Japan	2.0	3.7	3.3	0.45	0.5	0.96	23	14	29
Netherlands		0.5	0.74		0.3	0.72		60	97
South Africa		13	13		0.1	0.58		1	4
Spain	5.0		7.4	0.65		1.5	13		20
Sweden		0.1	0.14		0.02	0.08		20	57
UK	13.9	13.8	10.4	5.6	1.3	5.9	40	9	57
USA	61.0	47	38.3	9.1	5	8.0	13	11	18

Table 1.3: Standard specifications for fly ash for use in concrete

Country	Designation of standard		Year
Australia	AS 1129	Fly ash for use in concrete[1]	1971
	AS 1130	Code of practice for use of fly ash in concrete	1971
Austria	ÖNORM B 3319	Fly ash as hydraulic powdered admixture component for cement manufacture	1962
Canada	CAN 3-A23.5-M 82		1982
India	IS 3812		
	Part I	Fly ash for use as pozzolana	1966
	Part II	Fly ash for use as admixture in concrete	1966
	Part III	Fly ash for use as fine aggregate for mortar and con.	1966
	IS 6491	Methods of sampling fly ash	1972
Turkey	Fly ashes for use with Portland cement clinker and Portland cement concrete (TS 639)		
Japan	JIS A6201	Fly ash[1]	1958
		reaffirmed	1967
		reaffirmed	1977
United Kingdom	BS 3892	Pulverised fuel ash for use in concrete[1]	1965
		now under	1982
		revision	1983
USA	ASTM C618	Fly ash and raw of calcined natural pozzolans for use in Portland cement concrete[2]	1971
		reaffirmed	1980
	FEDERAL SS-C-1760/5		1975
	NORTH DAKOTA S.H.D. Sec.818-3		1976
USSR	GOST 6269-63	Binder active mineral additives	1963
Germany, F.R.	Recommendations for testing fly ash		
	DIN 1045	Concrete and Reinforced Concrete	1988

Notes: 1) Methods of sampling and testing are included

2) Methods of sampling and testing are determined in accordance with ASTM C311

1.2.2 Mineralogical Composition

The chemical and mineralogical composition of fly ashes depends upon the characteristics and composition of the coal burned in the power plant. Owing to the rapid cooling of the material, fly ashes are composed chiefly (50-90%) of mineral matter in the form of glassy particles. A small amount of ash occurs in the form of crystals. Unburned coal is collected with the fly ash as particles of carbon, which may constitute up to 16% of the total, depending on the rate and temperature of combustion, the degree of pulverization of the original coal, the fuel/air ratio, the nature of the coal being burned, etc.

Low-angle X-ray diffractometry can be used to ascertain the glass phase.

Infra-red and Mössbauer Spectroscopy, X-ray diffraction and other specialized techniques provide powerful tools for researching the crystalline phases in fly ashes. The most important minerals found in fly ashes from bituminous coal are:

– Magnetite	0.8 - 6.5 %
– Hematite	1.1 - 2.7 %
– Quartz	2.2 - 8.5 %
– Mullite	6.5 - 9.0 %
– Free calcium oxide	up to 3.5 %

Other minerals like wüstite, goethite, pyrite, calcite, anhydrite and periclase range from trace amounts to 2.5 %.

1.2.3 Chemical Composition

Fly ashes are particularly rich in SiO_2, Al_2O_3 and Fe_2O_3, and also contain other oxides such as CaO, MgO, MnO, TiO_2, Na_2O, K_2O, SO_3, etc. Fly Ash with a high content of CaO (15 to 40%) may be regarded as potentially hydraulic and capable of causing unsoundness in mortars and concrete.

Fly ashes can be sub-divided into four groups, depending on the percentage of main compounds (according to a special contribution by S.Droljc)

Type I

SiO_2	> 50 %
$Al_2O_3 + Fe_2O_3$	medium
CaO	< 7 %
other components	small quantities

Type II

SiO_2	35-50 %
Al_2O_3	high
Fe_2O_3	medium
CaO	more than Type I

Type III

SiO_2	< 35 %
CaO	very high
$Al_2O_3 + Fe_2O_3$,	
other components	wide differences, but lower than Type I and Type II

Type IV

SiO_2	very low
CaO	very high
free CaO	
$CaSO_4$	
$Al_2O_3 + Fe_2O_3$	
other components	low

Type I and Type II fly ashes have good pozzolanic activity, while Type III and Type IV fly ashes are almost as inactive as pozzolan. These are inapplicable for use in concrete and may cause unsoundness.

The methods for sampling and testing fly ash for use as a mineral admixture in portland cement concrete are included in Standard Method ASTM C 311 - 77. Chemical analysis must determine:

- Moisture content (105 °C)
- Loss on ignition (1000 °C)
- Silicon dioxide SiO_2
- Aluminium oxide Al_2O_3
- Iron oxide Fe_2O_3
- Calcium oxide CaO
- Magnesium oxide MgO
- Sulphur trioxide SO_3
- Available alkalis Na_2O and K_2O
- Free CaO

Table 1.4 shows the composition of various fly ashes from different types of coals /F 5/.

Table 1.5 shows the composition of fly ashes from different countries.

When a fly ash is burned at about 1000 °C, it suffers a loss of weight through the presence of carbonates, combined water in residual clay minerals, and combustion of free carbon. The oxidation of S and Fe compounds may produce an increase in weight, which must be taken into account in the general balance. The combined effects are termed the **loss on ignition.**

It has been confirmed that carbon is the most important component of ignition loss. The carbon content in fly ashes is decisive in determining the water requirement for mortar and concrete applications. The amount of water necessary to obtain a paste of normal consistency is greater when the carbon content is high. In general, it may be stated that the lower the carbon percentage, the better will be the fly ash. In practice, fly ashes with high carbon content and coarse granulometry will produce low strength concrete, but only at the same workability.

Class F fly ashes may contain a greater amount of carbon than those belonging to Class C.

The carbon contained in fly ash has high porosity and a very large specific surface and is able to absorb significant quantities not only of water, but of organic admixtures in concrete, such as water-reducing agents, air-entraining agents, set-retarders, etc.

The carbon content can be an important parameter for classifying fly ashes into three groups:

	Group A	Group B	Group C
% Carbon	0 - 5	5 - 10	8 -15

Table 1.6 summarizes the chemical requirements for fly ashes in different countries. The values for Spain are included in a Tentative Method (Spanish draft standard UNE).

Table 1.4: Compositional ranges of some ashes from various types of US coals (%)

	Anthracite	Bituminous	Subbituminous	Lignite
SiO_2	47 - 68	7 - 68	17 - 58	6 - 45
Al_2O_3	25 - 43	4 - 39	4 - 35	6 - 23
Fe_2O_3	2 - 10	2 - 44	3 - 19	1 - 18
CaO	0 - 4	1 - 36	2 - 45	15 - 44
MgO	0 - 1	0 - 4	0.5 - 8	3 - 12
Na_2O	-	0 - 3	-	0 - 11
K_2O	-	0 - 4	-	0 - 2
SO_3	0 - 1	0 - 32	3 - 16	6 - 30

1.2.4 Granulometry

The fineness of fly ashes is commonly measured by sieve analysis, which can be performed using dry or wet methods. Other techniques are also utilized. Generally speaking, it is important to know the amount of material retained by 200, 150, 87, and 44/45 micron sieves.

ASTM Designation C 311 - 77 recommends determining the amount of the sample retained when wet-sieved on a No. 325 (45 μm) sieve in accordance with ASTM Method C 430, except that a representative sample of the fly ash or natural pozzolan is substituted for hydraulic cement in the determination.

In general, the amount of fly ash retained on the 80 μm sieve ranges from 6 to 25%, on the 50 μm sieve from about 16 to 40 %, and on the 45 μm sieve from about 3 to 14 % (all mass percentages).

Table 1.5: Compositional ranges of fly ash from different countries (%)

Country	Canada	Denmark	France	Germany	Spain	USA
Numbers of power plants	7	4	4	14		8
Reference	B 68	S 45	A 8	B 47	A 8	A 8
SiO_2	48 - 56	48 - 65	47 - 51	42 - 55	32 - 64	40 - 51
Al_2O_3	22 - 33	26 - 33	26 - 34	24 - 33	21 - 35	17 - 28
Fe_2O_3	4.2 - 11	3.3 - 8.3	6.9 - 8.8	5.4 - 13	5.1 - 26	8.5 - 19
CaO	0.8 - 9.7	2.2 - 7.8	2.3 - 3.3	0.6 - 8.3	1.3 - 20	1.2 - 7.0
MgO	1.9 - 44		1.5 - 2.2	0.6 - 4.3	0.5 - 2.7	0.8 - 1.1
Na_2O	0.3 - 1.8	1.1 - 2.8	2.3 - 6.4	0.2 - 1.3	0.03- 0.7	0.4 - 1.8
K_2O	2.1 - 5.0			1.1 - 5.6	0.4 - 4.0	1.8 - 3.0
SO_3			0.1 - 0.6	0.04- 1.9	0.2 - 4.0	0.3 - 2.8
Loss on Ignition		3.1 - 4.9	0.5 - 4.5	0.8 - 5.8	0.5 - 10	1.2 - 18

Table 1.6: Chemical requirements for fly ashes in different countries

Country	Germany. F. R.	Australia	Austria	Canada		Spain		India	Japan	U. K.	Turkey	URSS	USA	
Standard	DIN	AS	ÖNORM	CAN		UNE		IS	JIS	BS	TS	GOST	ASTM	
No.	1045	1129	B 3319	3-A23 5-M82				3812	A 6201	3892	639	6269	C618	
Type of Fly Ash	-	-	-	C	F	1)	2)			-	-	-	F	C
SiO_2 min. %	-	-	-	-	-	-	-	35	45	-	-	40	-	-
(S+A+F) min. %	-	-	-	-	-	70	70	70	-	-	70	-	70	50
MgO min. %	-	-	-	-	-	5	5	5	-	4	5	-	5	5
SO_3 max. %	4.0	2.5	3.5	5.0	5.0	4	4	3	-	2.5	5	3	5	5
CaO max. %	-	-	-	-	-	-	-	-	-	-	6	-	-	-
LOI max. %	5.0	8.0	7.0	6.0	12.0	12	7	12	5	7	10	10	12	6
Alkalies max. %	-	-	-	-	-	-	-	1.5	-	-	-	-	1.5	1.5
Moisture max. %		1.5	-	3.0	3.0	3	3	-	1	0.5	3	-	3	3

1) in cement
2) in concrete

LOI = Loss on Ignition
S = SiO_2
A = Al_2O_3
F = Fe_2O_3

It has been observed that the grain size of pulverized coal changes during the combustion process, influencing the granulometry of the fly ash, as shown in *Table 1.7.*

Table 1.7: Granulometry of fly ashes and their origin coals

Size (mm)	Sample 1		Sample 2	
	Coal	Fly Ash	Coal	Fly Ash
>0.25	0.8	16.1	55.7	67.3
0.25 - 0.12	9.8	17.8	21.2	25.6
0.12 - 0.09	5.3	11.9	7.0	1.8
0.09 - 0.075	1.6	9.3	3.6	1.5
0.075 - 0.060	7.5	4.3	4.2	2.3
<0.060	75.0	10.6	8.3	1.5

The various national standards specify the maximum residue in % retained on a 45 mm sieve as follows:

West Germany	50
Australia	50
Canada	34
Spain	14
Japan	25
United Kingdom	12.5
USA	34

Optical and scanning electron microscopy of fly ashes have shown that these can vary in size and shape, including fly ashes of spherical, rounded, irregular and angular shape. Spherical and rounded fly ashes vary in size from 0.5 to 200 μm. Fly ashes of irregular and angular shape are usually but not necessarily larger.

The **particle size** distribution of fly ash particles may be defined as the quantification of particles in terms of their size. The term "distribution functions" may also be employed. Such distribution functions combine a number of parameters, such as the size and number of particles, their mass, surface area, volume, chemical composition, etc.

Such distributions are subject to experimental determination. Nonetheless, there is a theoretical basis for the study of size distribution which can be applied to the case of fly ashes.

The specialized literature includes differential and cumulative size distributions, logarithmic particle size distributions, mean and median particle size distributions, lognormal size distributions and general size distributions.

A particle may be defined as a simple, continuous unit of solid (in the case of fly ash) or liquid material of larger than molecular dimensions. In certain cases, a particle may be formed by the agglomeration of a number of small units, a phenomenon commonly encountered in fly ashes.

Microscopic examination of fly ashes reveals a wide variety of particle sizes and shapes. Grain size may vary from 0.2 to 200 μm, most particles being larger than 1 μm in size.

Particles are spherical, irregular or angular, etc., depending on the nature and granulometry of the coal burned and on the combustion conditions in the power plant. If the combustion temperature is low, the mineral ash fails to melt and the final shape is irregular. At high combustion temperatures, the mineral matter in the coal melts, forming hollow spheres referred to as cenospheres, sometimes containing a number of smaller spheres (plerospheres). At a combustion temperature of roughly 1 500 °C, the majority of particles are round-shaped and hollow, with smooth or rough surfaces.

Particle shape is an important parameter affecting a variety of processes, such as particle motion in a fluid medium, the formation of electrostatic charges, light scattering, etc. In the case of fine particles, spherical, cubic, flake, floc, platelet and irregular shapes are the most significant.

Recent SEM studies have confirmed this observation. In a few cases only, the spheres are partially covered with fine spheres or needles. In his authoritative study of the size and shape of fly ashes, Richartz /R3/ observed fly ash particles of some 40 μm diameter containing a small quantity of interior spheres. This confirmed observations by other authors. Further information on particle sampling and analysis may be found in /M54/.

1.2.5 Specific Surface

The specific surface of a material may be defined as the number of units of surface area contained in a unit of mass. The specific surface or fineness of a fly ash as determined by the Blaine method varies from 250 to 550 m^2/kg (2 500 to 5 500 cm^2/g).

Fly ashes collected in electrostatic precipitators range in fineness from 4 000 to 7 000 cm^2/g. Cyclone or mechanically-collected ashes vary between 1 500 and 2 000 cm^2/g. Finally, certain modern electrostatic precipitators collect ashes with a fineness of 12 000 cm^2/g.

Various methods are used to determine the specific surface areas of these materials. The most commonly used is the Blaine method, which is based on the resistance offered by pulverized materials to an air flow. Fly ash samples are prepared according to certain conditions and the external surface of the grains contained in 1 g of the ash is measured. In the case of fly ashes, values for this surface area, designated "Blaine permeability", generally range from 2 500 to 4 000 $cm^2/$. Permeabilities as low as 1 800 cm^2/g or as high as 5 000 cm^2/g may sometimes be encountered.

Another method of determining specific surface is the BET technique, in which nitrogen adsorption isotherms are measured. Data obtained by this means differ from those for the Blaine method. BET specific surface values for ordinary fly ashes are 3 to 4 times greater than the Blaine values, because the BET method measures the totality of voids existing in the surface of grains. Values obtained in this way may be as high as 12 000 cm^2/g.

The study of granulometric curves also provides an indication of fly ash specific surface size. Values are commonly 30 % lower than with the Blaine method, since not all fly ash particles are spherical.

Table 1.8 shows the specific surface of various fly ashes determined by Cabera /B68/ using particle size analysis, the air permeabilimeter (Blaine) and nitrogen adsorption. It will be evident that further research into fly ash surface area is required, in order to verify the relationship between this parameter and the reactivity of fly ashes.

Table 1.8: Specific surface of various fly ashes determined by different test methods

Ash Code	Specific Surface (cm^2/g)			
	Calculated from Particle Size	Air Permeability	N_{dry}	N_{burnt}
A	810	3050	40700	7600
B	970	4130	38200	8700
C	1150	3360	10200	5700
D	920	2090	4800	10000
E	n.d.	1930	47000	11400
F	1020	6710	89000	6700
G	810	3110	65000	8000
H	n.d.	2880	12400	6800
I	800	2540	9700	6600

According to the CEMBUREAU Technical Committee, fly ash must have a specific surface area of not less than 2 700 cm^2/g (Blaine) and a variation no higher than ±500 cm^2/g (Blaine).

Richartz /R43/ determined the Blaine specific surface and the particle size distribution of fly ashes using a laser granulometer and sieve analysis. The results of 20 analyses were as follows:

		Maximum	Average	Minimum
Density	(g/cm^3)	2.66	2.37	2.13
Specific surface (Blaine)	(cm^2/g)	5290	3562	2730

Fig. 1.1 indicates the particle size distribution curves for fly ashes with a high specific surface (top curve), a low specific surface (bottom curve) and an average specific surface. The particle size distribution of portland cement (PZ 35 F) is shown for comparison.

Fly ash particles with grain sizes > 125 μm are very porous. Research on these particles has demonstrated that they are formed when a very large amount of carbon is present; this unburned material is responsible for the high specific surface values obtained in most fly ashes. The large carbon content is associated with a high water requirement in concretes containing fly ash. Fly ash carbon content is also related to the freezing and thawing resistance of concrete. The higher the fly ash carbon content, the lower will be the freezing and thawing resistance of the concrete.

Generally speaking, fly ashes must have granulometries or specific surfaces closely resembling that of portland cement. This is of great importance if physical variations in concrete properties, especially workability, are to be avoided.

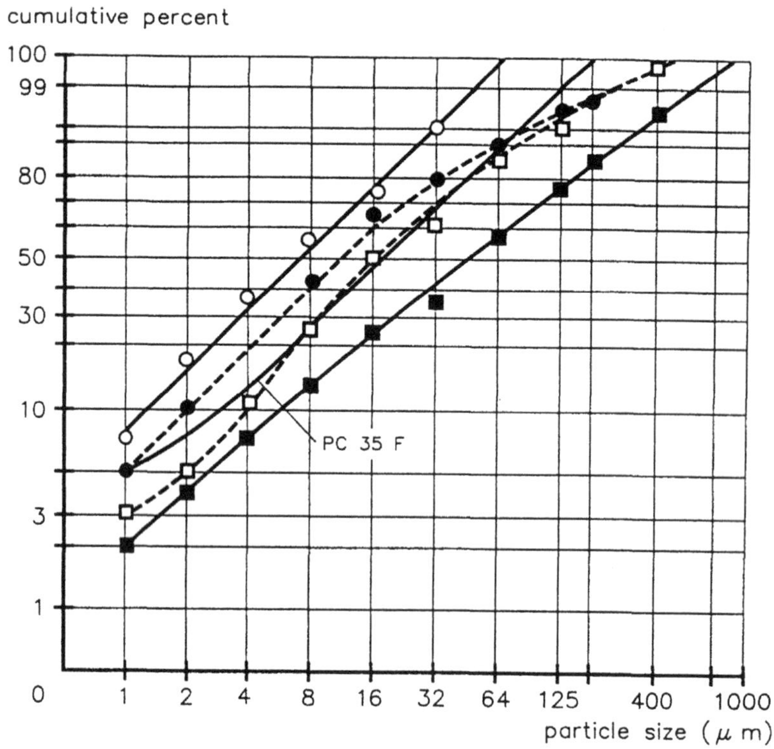

Fig. 1.1: Particle size distribution of various fly ashes

1.2.6 Density and Density Distribution

There are various different concepts for determining density in the field of pulverized materials, especially fly ashes, as is apparent from /J 9/ and ASTM Designation E 12-70 (Reapproved 1981).

ASTM Standard C 188 - 84 deals with specific gravity determination in cements. In cement applications, water must be replaced by an organic liquid. Since fly ashes contain some water-soluble compounds, the use of non-aqueous solvents for determining specific gravity is recommended.

Jarrige /J 9/ found that the mass of 1 dm^3 of dried fly ashes ranged from 0.54 to 0.86 kg for French fly ashes. The apparent mean density of grains is 1.90 to 2.40 kg/dm^3. The Gaussian distribution of the different grains in a sample of fly ash indicates a maximum density value of 2,6 kg/dm^3, with a minimum of 0.5 kg/dm^3. It has been confirmed that the maximum specific gravity value for fly ashes (about 2.98 kg/dm^3) corresponds to the maximum Fe_2O_3 content.

1.2.7 Water Requirement

The amount of water necessary to obtain normal consistency in fly ash concrete varies considerably in accordance with the carbon content. The water absorption is low when the unburned carbon is about 1 %. Conversely, fly ashes with about 10 % free carbon consumes a large quantity of water. Hence, concrete made with fly ashes has a high mechanical strength if the carbon content of the admixture is low.

The ASTM Standard C 618 gives the value 105 % as maximum percent of control for fly ashes of the classes F and C. The same value is included in the USSR Standard, whereas the Japanese Standard indicates a value of 100, and the United Kingdom recommends 95%.

1.2.8 Pozzolanicity

Pozzolanicity is the capacity of certain materials to enter into reaction with CaO or Ca $(OH)_2$ in the presence of water at room temperature, to form solid and water-insoluble masses.

The addition of 20 - 25 % fly ash to portland-clinker has no practical influence on its hydration rate, especially in the first stage of reaction with water. The reaction begins with the solution of cement sulphates, since the rate of solution of anhydrite and hemihydrate in fly ashes is very slow.

Pozzolanic activity is evident from 14 days onwards, especially in the 14 to 150 day period. After 120 days, fly ash particles are practically disintegrated as a result of attack by the $Ca(OH)_2$ produced by the hydration of portland cement. The glass phase of fly ash grains /J9/ is especially affected by this attack.

It is very important to know the "pozzolanic activity index" in the case of portland cement (ASTM C 311 - 77; 1982). Soerensen /S45/ defines the activity index as the weight of cement that can be replaced by one unit weight of pulverized fuel ash (PFA) without altering the concrete compressive strength at a given age.

The pozzolanic activity index with portland cement is:

Pozzolanic activity index = $A/B \cdot 100$

where

A = average compressive strength of test mix cubes.

B = average compressive strength of control mix cubes.

Another index generally calculated for applications involving the use of fly ash in concrete is the pozzolanic activity index with lime. The mortar must contain 1 part hydrated lime and 9 parts of graded standard sand plus a quantity of oven-dry mineral admixture equal to twice the weight of the lime multiplied by a factor obtained by dividing the specific gravity of the mineral admixture by the specific gravity of the lime. The quantity of water must be such as to produce a flow of 110 ±5 %. The index must be indicated by the compressive strength of cylindrical specimens.

There is another standard specification for fly ash and other pozzolans for use with lime in plastic mortars and non-plastic mixtures (ASTM C 593 -76a; reapproved 1981).

The ASTM Standard Specification C618 - 80 requires a compressive strength for fly ash Types F and C

- of at least 75 % by reference to the control mix for portland cement at 28 days
- of at least 5.6 MPa for lime at 7 days.

The effects of fly ash on compressive strength in concrete can also be indicated by means of a comparative test, according to CEN 196 - 1, with three different mixtures.

Pozzolanic activity can be tested according to the Testing Methods of the United Kingdom, India and the USSR /B 69/. This activity is measured at 28 days with specimens containing cement and at 7 days with specimens containing lime.

In the United Kingdom, a test based on accelerated curing with portland cement is used. The method was developed by Lea /C 26/. A minimum of 85 % of the strength of the control mix is required; elsewhere, tests with lime absorption are favoured and a 7 day curing period is adopted. The parameters are 4 MPa at 7 days accelerated curing time, using standard tests.

It has been found that, the finer the fly ash and the lower the carbon content, the greater will be the pozzolanic activity and the greater the contribution to strength in concrete of the same workability /F5/. There is also a relationship between phase composition and reactivity in fly ash. Only the vitreous phase of fly ashes enter into reaction with $Ca(OH)_2$ and water. Heat treatment of fly ashes to 1000 °C followed by rapid cooling increases their pozzolanic activity, while this activity decreases if the cooling process is slow. Chemical reactions between glassy compounds of Si, Al and Fe are induced in order to originate very complex silicates, aluminates and silicoaluminates of calcium. These new compounds are water insoluble and possess very high strength. Crystalline minerals, such as quartz, mullite, hematite, magnetite, etc., do not participate in the calcium hydroxide reaction mechanisms.

The behaviour of fly ashes is similar to that of natural pozzolans and blast furnace slags, which are also predominantly vitreous materials. The highest reactivity corresponds to the lowest particle size.

Many efforts have been made by various authors to correlate the pozzolanic activity, fineness and specific surface area of fly ash particles, so far to no avail.

According to Berry /B 68/, the fraction of fly ash with a particle diameter of about 35 μm was found to be the most appropriate for producing high mechanical strength.

The chemical reactivity of fly ashes is measured by determining the amount of free unreacted $Ca(OH)_2$ remaining in a lime-fly ash mix that has interreacted during a predetermined period. The amount of unreacted calcium hydroxide can be determined by the Franke method.

There are various methods for determining the pozzolanic activity as capacity or aptitude of pozzolans of fly ashes to react with calcium hydroxide at room temperature. Nevertheless, reaction rates under these conditions are very slow, and it has been necessary to operate at temperatures of 40 to 50 °C, in order to accelerate the chemical reactions.

Mention should be made here of a number of methods which have been used to determine the pozzolanic activity of fly ashes.

- Fratini method: this method, developed by Nicola Fratini, is based upon the reaction of fly ash with $Ca(OH)_2$ and subsequent measurement of the Ca^{++} concentration and total alkalinity in liquids in contact with the paste /F 30/.
- Insoluble residue method: studies have been made of the effects of calcinations at 1 000 °C on
  - fly ash alone;
  - mixtures of fly ash + portland cement.

  The decrease or increase in the insoluble residue of materials after this thermal treatment is an indicator of the pozzolanic properties of fly ashes. This method was developed by Guillaume /G 24/.
- Steopoe method: in this procedure, a mixture is made up with pozzolan, calcium hydroxide and water, and specimens measuring $30 \cdot 30 \cdot 5$ mm^3 are prepared on a glass plate. After 3 days of curing under water vapor at room temperature, without the presence of CO_2, the specimens are boiled under water. If the specimens are not disrupted after this treatment, the tested material is considered pozzolanic. Subsequent treatment with HCl and NaOH indicates the insoluble residue. The material is accepted as pozzolanic if its reactive SiO_2 content is at least 10 % /S59/.
- Jambor method: Jambor has created a new method for quick determination of pozzolanic activity /J24/. The procedure is based on studying variations in the development of dissolution heat of a pozzolanic material while it is being dissolved in a diluted mixture of HNO_3 and HF. The insoluble residue obtained in this way approximately represents the percentage of non-active material in the pozzolan.
- Electroconductivity procedure: Leonard /L6/ has described a method of determining the reaction rate of lime with fly ash. An electroconductivity device is used. Research results published by the author yield the same values as those deduced from compressive strength tests. The method enables the reaction rates of a specimen to be determined non-destructively.

Recently, Hubbard /H36/ has proposed a new method based on the relationship of the amorphous component of PFA and the compositional PPI index (based on potash and alumina content). According to the author, all the alumina present in the PFA composition is derived from the impurity of the coal (clay impurity), while the potash content is essentially a function of illite content.

1.2.9 Radioactivity

Most natural materials such as minerals, rocks, coals, etc. possess the property of radioactivity. This phenomenon results from the presence in these materials of very small (trace) quantities of elements whose nuclei disintegrate spontaneously, emitting corpuscular or electromagnetic radiation.

Corpuscular radiation is composed of alpha (α) or beta (β) rays; these are ionized helium atoms or electrons respectively, emerging from the interior of atomic nuclei. Electromagnetic radiation consists of gamma (γ) rays. The velocity of propagation of α and β particles is variable, but that of γ rays is equal to the speed of light, i.e. about 3×10^8 m/s.

The following radioactive properties can be measured:

- Radioactivity is the number of spontaneous disintegrations per unit mass and unit time of a given unstable element. The relevant SI unit is the Bequerel (Bq):
  1 Bq = 1 disintegration /s
- Radiation or exposure is measured in Coulomb/kg (Cb/kg):
  1 Cb = 1 A · s
- The absorbed dose (D) is the quantity of radiation absorbed by a living organism. The unit is Gray (Gy; formerly rad):
  1 Gy = 1 J/kg = 100 rd
- The effective equivalent dose is measured in Sievert (Sv):
  1 Sv = 1 Gy.

According to Beretka /B42/ the natural radioactivity of building materials is usually determined from Ra^{226}, Th^{232} and K^{40} quantities. It has been observed that 98.5 % of the radiological effects of the elements in the uranium series are due to radium and its derivatives. For this reason, we may ignore the contribution of U^{238} and other Ra^{226} precursors. The concentration of radium, thorium and potassium can be determined by gamma ray spectroscopy.

Studies by Mathew, Beretka and other authors indicate techniques for calculating the radium equivalent activity of tested specimens.

The radium equivalent activity is a measure of the sum of the activities of Ra^{226}, Th^{232} and K^{40} in the material specimens. The equation utilized is follows:

$$Ra_{equiv} = A_{Ra} + (1.43\ A_{Th}) + (0.077\ A_K) \tag{1.1}$$

where A_{Ra}, A_{Th}, and A_K are the specific activities of these radioactive elements, expressed in Bq/kg. The generally accepted maximum value for building materials is 370 Bq/kg.

Fly ash contains a certain quantity of K^{40} and the elements of the radioactive series of U^{238} and Th^{232}. These radionuclides give fly ash a radioactivity a few times higher than in the case of ordinary building materials. The activity is of the order of 200 to 750 Bq/kg.

In addition, however, the Rn^{222} gas and its time solid decay products in fly ashes contribute to this activity. Rn^{222} itself is a decay product of U^{238} and Ra^{236}. This isotope can migrate from the interior of building materials into the air, remaining in the atmosphere, and contributing to damage in living organisms.

According to Bijen /B37/ emission from fly ashes is very low, due especially to the dense glassy structure of these residues which prevents most of the radon from escaping.

The U^{238} suffers a decay in six steps, yielding the noble gas radon (Rn^{222}) with a half life of 3.82 days. The intermediate elements formed are Th^{234},Pa^{234}, U^{234}, $Th^{230}$and Ra^{226}.

The Rn^{222} decays, forming the element lead (Pb^{206}) via a series of intermediate elements, such as Pa^{210}, Pb^{214}, Bi^{214}, Pb^{210}, Bi^{210} and Pb^{206}.

Th^{232} decay entails production of Rn^{220}, with a half-life time of 55.6 sec, and results in Pb^{208} as the final and stable element.

There are various limits for indoor radon daughter concentrations in houses. In Sweden, for example, this limit is 75 Bq/m^3. In Germany, limits are imposed on the concentration of uranium and thorium in building materials (370 Bq/kg). In the Netherlands the limit is about 0.3 mSv/a.

Table 1.9 shows the uranium (U) and thorium (Th) contents of fly ashes in three countries.

Table 1.9: Uranium and thorium contents of fly ashes in different countries

Country	U (ppm)	Th (ppm)
USA	10	10
USA	25	35
USA	10	35
USA	6	30
Australia	22	67
Australia	21	62
Poland	11	24

1.2.10 Soundness

In accordance with Brown et al. /B70/, "soundness is the ability of a cement paste, mortar or concrete to withstand internal stresses generated during cement hydration, without cracking". Conversely, "unsoundness" phenomena are usually encountered due to slow hydration of dead-burned CaO and/ or MgO in cement. These reactions occur when the cement paste has hardened. Both hydroxides formed in this process, $Ca(OH)_2$ and/or $Mg(OH)_2$, have an extremely large molecular volume which induces internal stresses eventually leading to expansion of the concrete and, in certain cases, entailing its total disruption.

The use of blended cements (portland cement + fly ash) has been found to be advantageous in reducing expansion phenomena, due probably to the fact that the concrete contains a lower amount of portland cement than concrete without fly ash.

The autoclave expansion test described in ASTM C 151 - 74, is possibly too severe for detecting unsoundness, but is the only method which takes into account MgO. Applied to blended cements, this test indicates that expansion due to the presence of CaO or MgO is smaller than the real expansion that can occur under field conditions. Probably the chemical compounds formed during the hydration process under autoclave conditions are quite different from those encountered in the field.

Further research will be necessary in order to gain a better understanding of the chemical phenomena involved in the CaO and/or MgO hydration processes.

ASTM C 618 defines autoclave expansion or contraction for fly ashes of Types F and C at 0.8 %.

Alonso /A8/ has carried out autoclave expansion tests with 9 types of fly ash. The blended cements contain 30% fly ash in all cases. The measured expansion was between 0.06 and 0.15 %; the corresponding expansion of the plain portland cement was 0.05 %.

1.2.11 Colour

The colour of fly ashes depends on the Fe_2O_3 and carbon content. The most significant factor is the unburned coal content corresponding to loss on ignition. This carbon percentage, ranging from 0.5 to 10 or 12 % in certain cases, is responsible for the "black" or "grey" appearance of some concretes. The presence of large amounts of Fe_2O_3 (brown) in most fly ashes also contributes to the dark colour of concretes. This dark colour in concrete is generally unacceptable, especially in urban buildings, unless the colour is uniformly distributed, since normal concrete without fly ash admixture is light grey in colour.

In certain circumstances, a method of measuring the darkness of fly ashes and of concrete made with fly ashes may be necessary. Research is required to identify suitable techniques for measuring coloration in both concretes and fly ashes, for example by using a reflectometer. A standard method may be achieved by comparing a standardized "white sample" with the tested specimens (cf. ASTM Standard E 306-84).

1.2.12 Moisture

The moisture content of fly ashes depends on the way in which these materials are stored after leaving the filter or precipitator.

Fly ashes are usually stored in stockpiles near the power station. The moisture content of fly ashes taken from these stockpiles is generally high. By contrast, fly ashes collected directly from power station cyclones or filters generally have a low moisture content.

The main international standards for fly ash indicate the following values for the maximum permitted amount of water in %:

Australia	1.5
Austria	no limit
Canada	3.0
Germany F.R.	no limit
India	no limit
Japan	1.0
Spain	3.0
Turkey	3.0
United Kingdom	0.5
USA	3.0
USSR	no limit

In fact, if fly ashes are to be used in cement making, the moisture percent is very important, since fly ash and clinker have to be milled in the factory to obtain blended cements. From an economic viewpoint, water should not be introduced to clinker mills. The moisture content of fly ashes added to concrete does not, however, constitute a disadvantage provided the percentage of water is known.

Moisture content also affects the handling properties of fly ash in silos, lorries, trucks and tankers. It should be remembered that a low moisture content means that fly ashes can be handled as a fluid.

Finally, the moisture of fly ashes is of considerable significance in calculating transport costs and final process, since transporting the extra weights entailed by high water contents is always uneconomic.

2 FRESH MORTAR AND CONCRETE WITH FLY ASH

W. von BERG and H. KUKKO

2.1 Properties of Freshly Mixed Mortar and Concrete (Prepared by W. von Berg)

One of the most important aspects of the use of fly ash in concrete is the fact that, in general, the use of fly ash markedly improves the properties of freshly mixed concrete.

In the available documentation, there is broad agreement as to the effect of fly ash and the importance of the decisive influencing parameters. Opinions still differ, however, as to the physical causes of the observed effects. The following observations based on practical experience with concrete containing fly ash are reported:

- Replacing cement by fly ash reduces the water demand of the concrete /B 10, B 46, B 47, B 55, E 2,F 2, G 5, G 16. K 37, K 38, L 23, L 33, S 7, V 2, V 3, V 5, W 4/.
- The use of fly ash improves concrete pumping or in some cases is a necessary prerequisite for it /G 9, H 21, W 5, J 23, K 37/.
- The workability and, especially, the compactability, flowability, and plasticity of concrete are generally improved /B 1, B 55, G 9, J 1, K 38, L 8, L 27, M 3, R 15, W 5/.
- The work required to cast and compact concrete is reduced /L 23, J 23/, there is less risk of surface shrink holes /B 1, H 21, C 33/.
- Agglomeration capacity is improved and the problem of de-mixing is consequently alleviated /G 9, K 39/.
- Water segregation (bleeding) is reduced /B 55, G 9, H 21, V 5/.

Reasons reported for the **reduced water demand** of fly ash containing concrete (with reduced percentages of cement) and for the improved properties of freshly mixed concrete are both spherical shape and plain surface (ball bearing effect) and also improved grain size composition in the range of the finest particles (filling effect) and gravitational forces respectively.

The reduction of the water demand of mortar and concrete through substitution of fly ash for cement yields a liquefied consistency with a constant water content and an increasing exchange amount. *Fig. 2.1* shows the increase of the spread (flow table test) of ISO mortar with varying water-cement values as a function of fly-ash percentages in relation to cement + fly ash (according to investigations by Venuat /V 2). It should be noted that the ground fly ash investigated in this case had a plasticizing effect. The curves show a more or less distinct optimum for this effect with fly-ash percentages ranging from 20 to 70 %. The position of this optimum is dependent on the properties of both the fly ash and the cement.

The results of the mortar tests are not directly transferable to concrete since the influencing parameters include both the percentage of cement + fly ash and the properties of the aggregate. In concrete tests /B 47/, the water demand of cement/fly-ash mixture was a nearly linear function of the mix proportion (*Fig. 2.2*).

The water demand of certain type of fly ash and its rheological efficiency in concrete are determined mainly by its fineness, its grain composition and shape, and its ignition loss. Lime-containing fly ash may additionally be affected by the lime content.

With an increasing **ignition loss** the water demand of fly ash increases /W4/ yielding a reduction of the relative slump (*Fig. 2.3*) . The relative slump rel s is determined by the slump of fly ash containing mortar (sf) with f/c = 0.25 in relation to the slump of cement mortar without fly ash (sc) :

$$\text{rel s} = \text{sf/sc} \cdot 100 \text{ in } \% \tag{2.1}$$

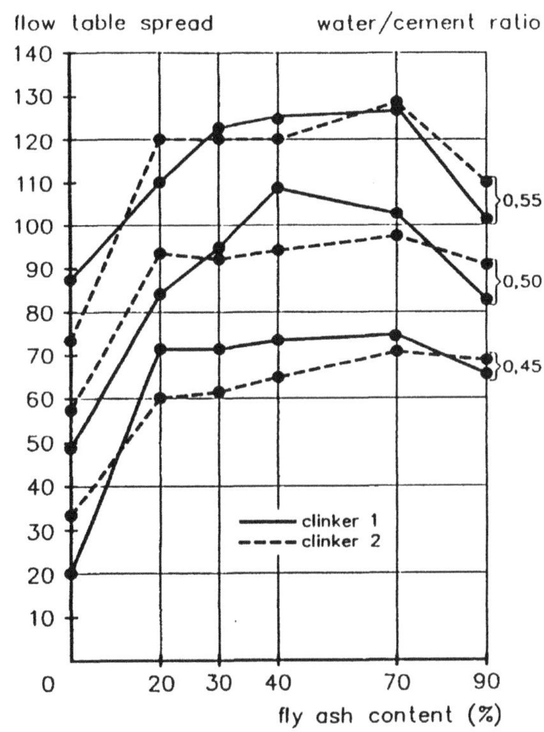

Fig. 2.1: Influence of fly-ash content on consistency /V 2/

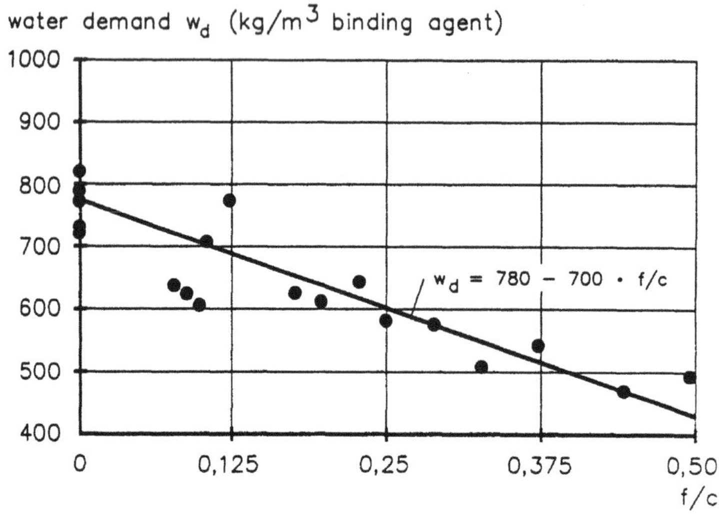

Fig 2.2: Water demand of the binding agent in kg per m^3 binding agent (according to Kluge) as a function of the f/c ratio; test results for a specific cement/fly-ash combination /B 47/

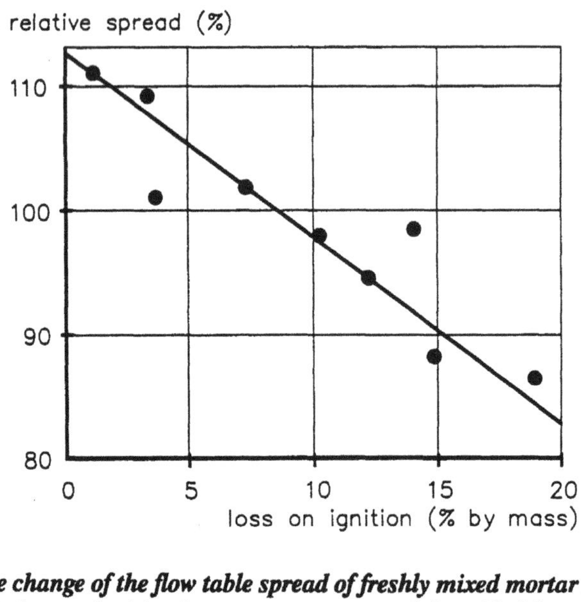

Fig.2.3: Relative change of the flow table spread of freshly mixed mortar as a function of the loss on ignition in an oxygen stream of the added fly ash in relation to the pure cement mortar /W 4/

In the case shown, the water demand of the fly ash tested was lower than that of the cement used up to an ignition loss of 8 % by mass and higher at a higher ignition loss. Lewandowski /L 27/ reports that the reduction of the water demand of concrete with a constant spread of 42 cm is distinctly greater for fly ash with an ignition loss of 3.6 % by mass than for an ignition loss of 9.3% by mass (*Fig. 2.4*).

Fig. 2.5 shows the effect of **grain shape** on the water demand of cement paste having a standard consistency according to Vicat in which 30 % of the cement has been replaced by fly ash. A reduction of the water demand is accordingly likely if about 70% of the fly-ash particles adopt an approximately spherical grain shape.

The **fineness** of the fly ash also has a decisive influence on the water demand of fly-ash mortars and concretes. The water demand generally decreases with increasing ash fineness where the cement is replaced by fly ash /E 2/ . Scholz /S 7/ attributes this to better grading of the grain composition of the cement/fly-ash mix.

According to /W 11/, the relative spread increases with the quantity of grains < 0.04 mm (*Fig. 2.6*). It is also evident from the graph that other material properties are significant, apart from the grain size of < 0.04 mm.

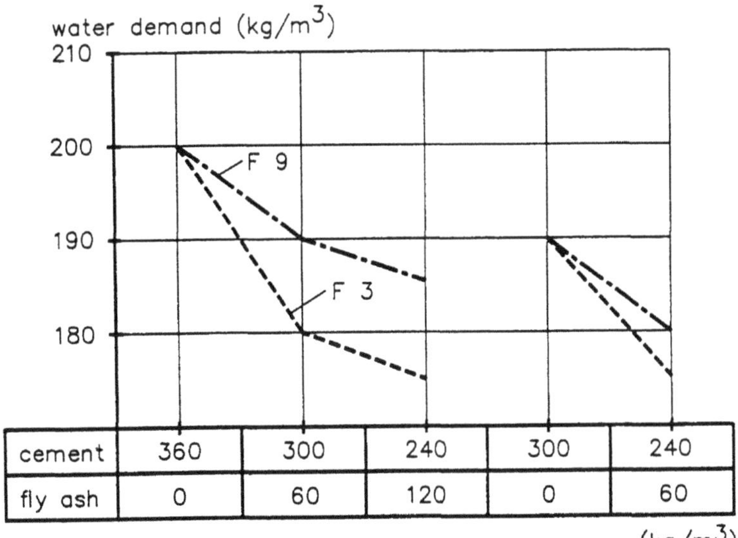

Fig. 2.4: Reduction of water demand of fresh concrete with spread a = 42 cm due to substitution of fly ash for portland cement Z 35 F: ash with loss on ignition of 3.6 % (F 3) and 9.3 % (F 9) /L 27/

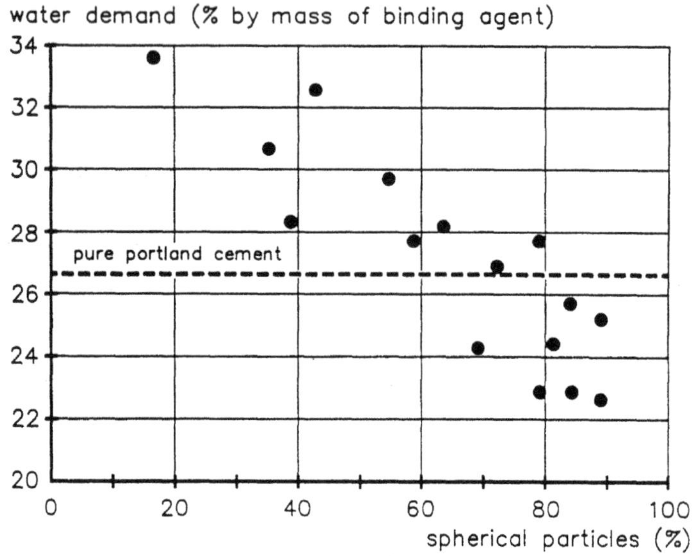

Fig.2.5: Influence of spherical particles of fly ash on the water requirement of standard paste: portland cement (70 %) — fly ash (30 %) /B 46/

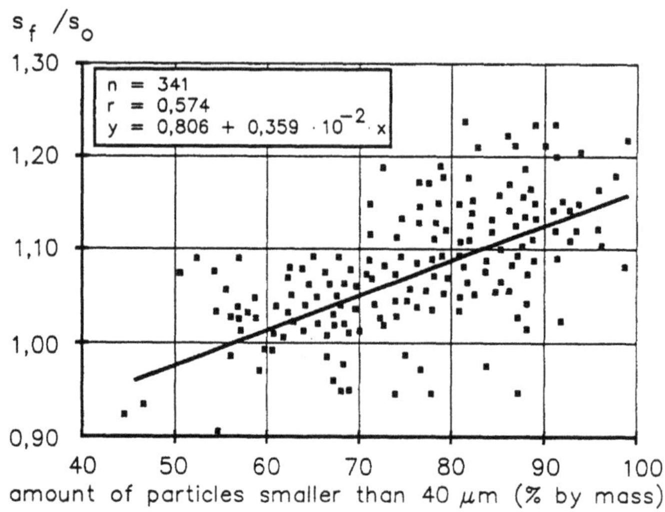

Fig.2.6: Correlation between quantity of particles smaller than 40 μm and the spread at flow table test of fly ash containing mortar s_f related to control mix s_0 /W 11/

Using the method according to Werse to determine the flow time of concrete with a constant spread, Lewandowski /L 33/ demonstrated a means of improving the **flowability** of concrete (*Fig. 2.7*). Different types of concrete exhibiting the same spread showed a decreasing flow time according to Werse with increasing cement/fly-ash exchange rates. This provides some guidance for improving flowability and consequently reducing the work needed for casting and compacting the concrete.

Fly ash has no adverse effects on the **initial setting** of mortar and concrete. Investigations by Lewandowski /L 23/ on different types of mortar with the same $w/(c+f)$ value (*Fig. 2.8*) showed that spread decreased with mortar age in approximately the same manner irrespective of fly-ash content.

According to Bottke /B 55/ and Keller /K 38/, **segregation of water** or "bleeding" is reduced when fly ash is substituted for cement. Venuat and Alexandre /V 5/ examined the relationship between the discharge time of mortar from a vibrating hopper (which denotes a characteristic value for the flowability) and water segregation in the stand cylinder. The authors noted that, given a constant discharge time, water segregation varies inversely with increasing cement/fly-ash exchange (*Fig. 2.9*) and increasing fly-ash fineness (achieved artificially by grinding in the tumbling mill) (*Fig. 2.10*).

Conversely, given a certain bleeding value, discharge time varies inversely with increasing fly-ash content and increasing fineness; the mortar is more workable without intense bleeding. In all cases, workability was improved by using fly ash. The positive effect of the fly ash increased in proportion to fly-ash content.

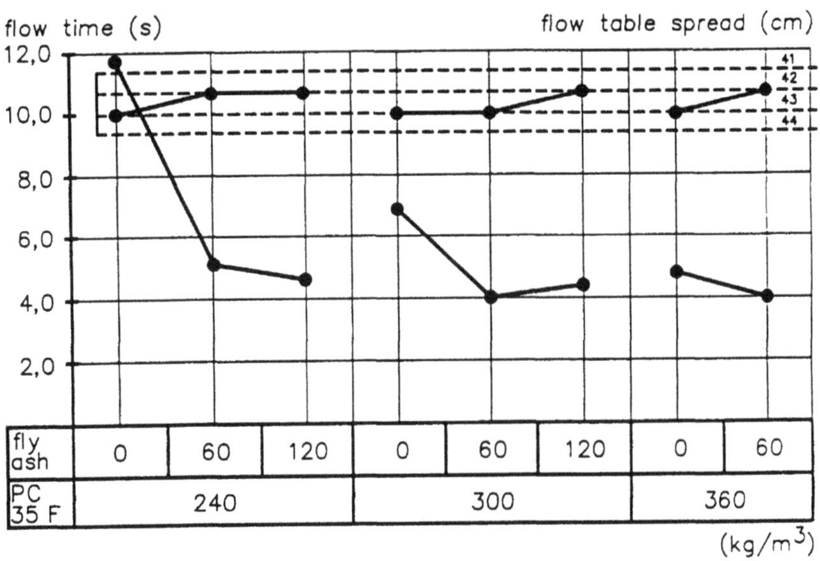

Fig. 2.7: Effect of fly ash added on the flow time and spread of freshly mixed concrete /L 33/

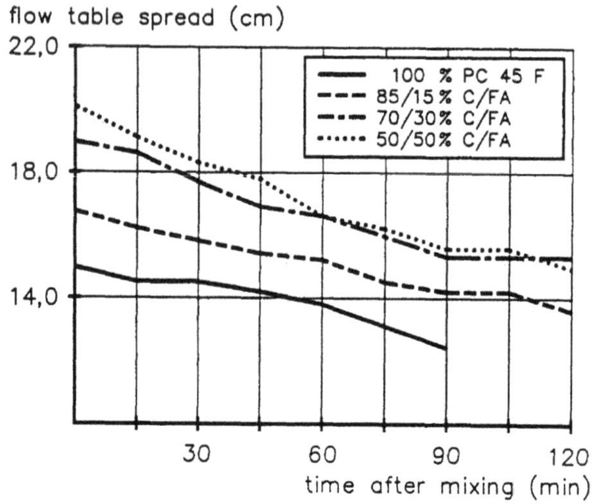

Fig. 2.8: Chronological change of spread according to Haegermann of types of mortars with w/c = w/(c+f) = 0.5 as a measure of water demand and initial setting of the "binding agent mixes" when PC 45 F is replaced by fly ash /L 23/

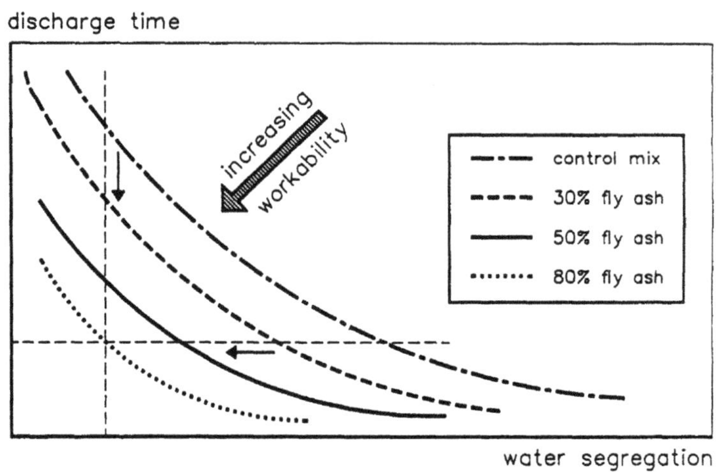

Fig. 2.9: Influence of fly-ash content on the workability of mortar /V 5/

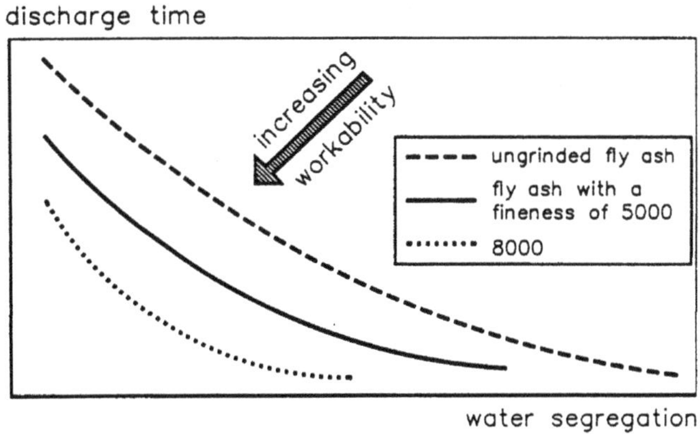

Fig. 2.10: Influence of fly-ash fineness on the workability of mortar /V 5/

2.2 Admixtures and Air Content (Prepared by H. Kukko)

2.2.1 Superplasticizers

In recent years there has been an increase in the use of superplasticizing admixtures, particularly in the production of flowing concrete (concrete with slump values in excess of 250 mm). For correct proportioning of such mixes, it is desirable to use more sand than in conventional concrete. According to Berry and Malhotra /B 12/, it is preferable to provide the fine particles necessary for mix cohesiveness by using fly ash rather than adding excessive amounts of sand.

Eriksen and Nepper-Christensen /E 19/ have studied the water-reducing effects of a sodium naphthalene sulphonate superplasticizer on concretes incorporating two low-calcium fly ashes. They reported a higher dispersive effect for the superplasticizer in fly-ash concrete than in non-fly-ash concrete.

Brooks et al. /B 25/ compared the behaviour of four concrete mixes:

- plain
- plain with superplasticizer
- fly ash
- fly ash with superplasticizer.

The mix proportions used for this study, which were selected to produce a minimum strength at 28 days of 30 MPa with a slump of 49 – 60 mm, are shown in *Table 2.1*. Compressive strength values largely reflected the effects of water reduction for both plain and fly-ash concretes.

Table 2.1: Mix proportions and compressive strength values for superplasticized concretes /B 25/

Mix number	1	2	3	4
Cement (kg/m³)	314	314	219	219
Fly ash (kg/m³)			177.5	177.5
Aggregate/cement ratio	5.98	5.98		
Aggregate/cement + fly ash			5.58	5.58
(% by weight)	33.3	33.3	31.1	31.1
Water/cement ratio	0.57	0.48		
Water/cement + fly ash			0.46	0.35
Admixture (% by weight of cement)		1.60		
Admixture (% by weight of cement + fly ash)				1.60
Compressive strength (MPa)				
1 day	13.0	19.0	11.0	18.5
28 days	48.5	61.0	44.5	53.0

Mix 1 = plain concrete
Mix 2 = plain concrete with admixture
Mix 3 = fly ash concrete
Mix 4 = fly ash concrete with admixture

Swamy et al. /S 61/ reported the data shown in *Table 2.2* for flowing concrete (slump 260 to 280 mm) containing fly ash and proportioned to give compressive strengths at one day comparable to plain concrete. The advantageous effects of moist curing versus air curing on the strength development of fly-ash concretes is clearly apparent from these data.

Mukherjee, Loughborough, and Malhotra /M 55/ have examined the use of superplasticizers to assist incorporation of large percentages of low-calcium fly ash in high-strength concrete. Three types of superplasticizers were examined:

- superplasticizer M, a sulphonated melamine-formaldehyde condensate;
- superplasticizer N, a sulphonated naphthalene-formaldehyde condensate;
- superplasticizer L, a modified naphthalene-formaldehyde condensate.

The following factors were noted:

- Satisfactory high strengths can be achieved with concrete incorporating a high percentage of fly ash and super-plasticizers.
- The mechanical properties of the water-reduced, super-plasticized fly-ash concrete were superior to those of the reference fly-ash concrete.
- The workabability may impose a limitation on use for cast-in-place construction, due to a gluey texture at slumps between 65 and 75 mm.

- Superplasticizers N and L both increased the setting time markedly, but it is not possible to determine from the data whether fly ash also influenced set-time.

It should be borne in mind that the water reductions achieved by superplasticizers and fly ashes when employed individually are not cumulative where both materials are used in the same mix, since only a certain quantity of water can be removed from any given mix. Moreover, the substitution of fly ash for portland cement does not change the optimum superplasticizer dose if dosage is based exclusively on the weight of the portland cement /S 23/.

In their experiment, Uchikava et al. found that a superplasticizer has a strong effect on cement containing fly ash, producing good fluidity /U1/.

Table 2.2: Properties of flowing concrete /S 61/

Curing regime	Slump (mm)	Age (days)	Compressive strength (MPa)	Flexural strength (MPa)
Air	265	1	12.0	1.8
		3	26.4	2.7
		8	36.1	3.3
		28	45.2	3.5
		43	50.8	4.1
3 days water and air	280	1	10.4	1.7
		3	24.6	3.0
		8	34.4	3.4
		28	48.0	4.3
		43	55.0	4.4

Mix proportions (kg/m³)

Cement	287
Fly ash	123
Sand	758
Gravel	881
Water	191

Superplasticizer added at 2.5 % by weight of cement + fly ash

2.2.2 Accelerators

Mailvaganam et al. /M 42/ have studied the effect of chloride and chloride-free accelerators and superplasticizer admixtures on the setting and strengthening properties of fly-ash concrete at normal (22 °C) and low (5 °C) temperatures. At normal temperatures, a reduction of strength values in fly-ash mixes in relation to the control mix (100 % portland cement with no admixture) was noted at all ages, except in the fly-ash mix containing the superplasticizer, which attained the strength of the control mix at 90 days (*Fig. 2.11*). It has also been observed that the effectiveness of chemicals in improving the early strengths of mixtures containing fly ash decreases with increasing ash percentage /S 23/. At a temperature of +5 °C, superplasticizer was again the most effective admixture in the fly-ash mix, but the strength gap was even greater, a superplasticized mix being the only one to achieve strength equivalent to the control at 90 days (*Fig. 2.12*).

Fly-ash concrete requires a longer setting time and has a lower hydration rate than plain portland cement concrete. Superplasticizer and accelerators do not seem to alter setting or hydration to any degree. Fly-ash mixes develop considerably less heat than do mixes without fly ash.

The temperature-time curves are shown in *Fig. 2.13*.

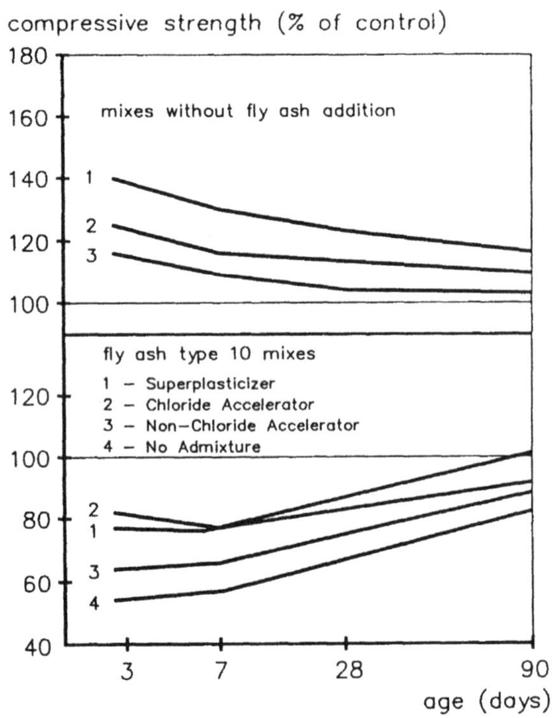

Fig. 2.11: Relative compressive strengths obtained for mix series I (22 °C, 1.5 % admixture dosage).
The slump was maintained at 80 ± 10 mm by varying the w/c ratio
100 %: portland cement control mix, with no admixture

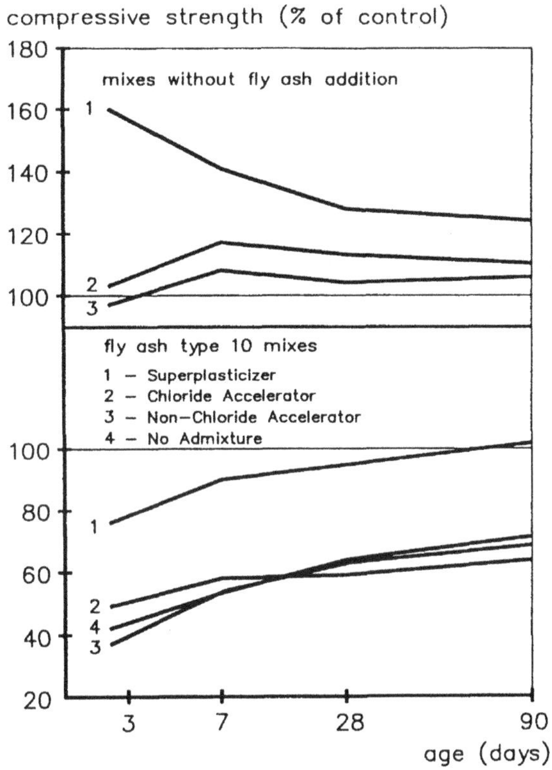

Fig. 2.12: Relative compressive strengths obtained for mix series II (5 °C, 3.0 % admixture dosage) 100 %: portland cement control mix, with no admixture

2.2.3 Air Content

The entrapped air content of fresh portland cement concrete is normally less than 3 %, depending on the fineness of the cement, the grading and shape of the aggregates, and the degree of consolidation. Tests show that concrete containing fly ash has an entrapped air content reduced by 0.5 - 1 % due to the influence of the fines /L 30/.

Concrete containing fly ash must be air-entrained to provide freezing and thawing resistance. According to Sturrup et al. /S 48/ the total air content required for adequate resistance may be less than for non-fly-ash concrete, since fly ash has reduced the entrapped air content. The major factor affecting adequate air-entrainment of fly-ash concrete is the carbon content of the fly ash. Its high surface area absorbs air entraining agents from the concrete mix, resulting in higher dosage requirements to obtain a specified air content. Sturrup et al. /S 48/ state that as long

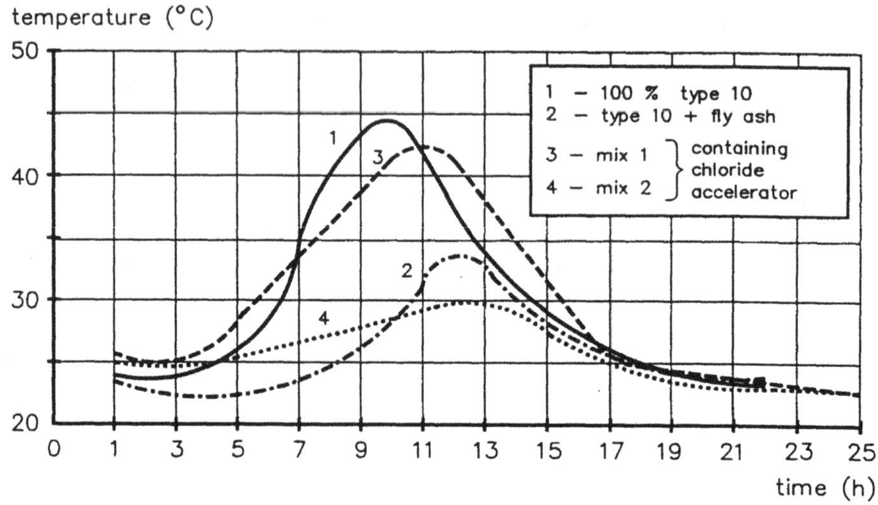

Fig. 2.13: Temperature-time curves at 22 °C for pastes containing admixtures

as the carbon content of the fly ash is known, the required dosages of air-entraining agents can be modified easily.

The terms loss on ignition (LOI) and content of carbon are often used interchangeably. According to Lane and Best /L 38/ the tests show that, beside chemically bound moisture, carbon and sulphur are expelled during ignition. The predominant weight loss is attributable to the carbon. The largest percentage of carbon is usually found in the fraction finer than the 45 μm sieve, but microscopic examination reveals that carbon particles are coarser and usually more porous and amorphous than the finer particles, and may have an adverse effect on the air content in concrete due to adsorption of air-entraining admixture /L 30/. However, a fly ash with a higher LOI usually requires a higher dosage of air-entraining admixture. On the basis of results from test series reported by Lane and Best /L 38/ a linear relationship was found between the LOI of the fly ash and the proportion of air-entraining admixture required to produce a certain air content.

In their air-void stability investigations, Gabler and Klieger /G 21/ found that air contents of concretes containing Class C fly ash appeared to be more stable than those in concretes containing Class F fly ash. Gebler and Klieger also concluded that the higher the organic matter content of a fly ash, the higher will be the air-entraining admixture requirement for concrete in which the admixture is used. In addition, the higher the air-entraining admixture requirement, the greater is the air loss on extended mixing.

Gebler and Klieger /G 21/ performed regression analyses in order to establish whether there is a correlation between the chemical and physical properties of fly ash and the air-entraining admixture requirement for concrete containing fly ash. Analyses indicated that the most significant components of fly ash affecting air-entraining admixture requirements were the organic matter content, the carbon content, the loss on ignition and the alkali content. The air-entraining admixture requirement generally decreases as total alkalis in the fly ash increase. Both the air-entraining admixture requirement (*Fig. 2.14*) and the loss of air in plastic concrete (*Figures 2.15 and 2.16*) increase in proportion to the organic matter content, the carbon content and the loss on ignition of fly ash.

According to Gebler and Klieger /G 21/, the total carbon content and loss on ignition of the fly ash are less clearly correlated with plastic air content retention than organic matter content. Gebler and Klieger also found that the retention of air in concrete increases with the specific gravity of the fly ash. The retained air in the concrete also increases with the SO_3 content of fly ash.

For each concrete mixture, the freshly mixed (plastic) air content was measured and a 76 x 76 x 286 mm^3 prism was cast immediately upon completion of initial mixing. A 29 minute rest period followed. Thereafter, the remaining concrete was mixed for 1 minute and water was added to retemper the mixture and to maintain slump within ± 1/2 in of the initial slump measurement. An additional prism was then cast for linear traverse measurement, and the air content of the plastic concrete determined. The rest period and mixing cycle were continued at 30 minute intervals for a total of 90 minutes. Prisms were rodded and moist cured and 76 · 19 · 254 mm^3 slabs were subsequently cut along the major axis for linear traverse measurements. Results are presented in *Fig. 2.17*.

Air contents of hardened concretes indicate that concretes containing Class C fly ash retained air content better than concretes with Class F fly ash. Concretes with Class F fly ash were also subject to significantly higher variability in air content retention than concretes with Class C fly ash /G 21/.

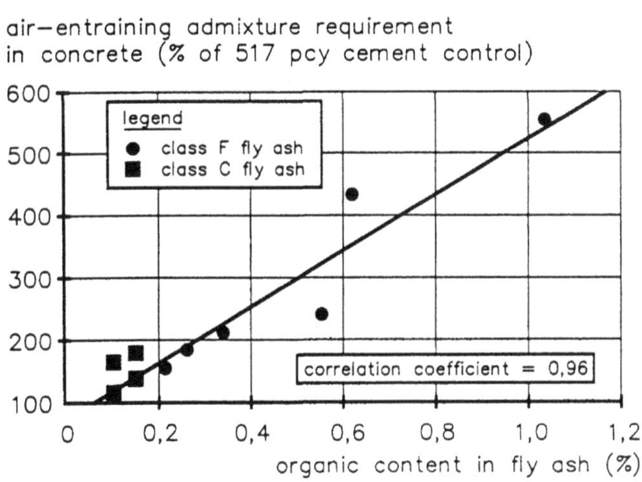

Fig. 2.14: Effect of organics in fly ash on air-entraining admixture dosage /G 21/

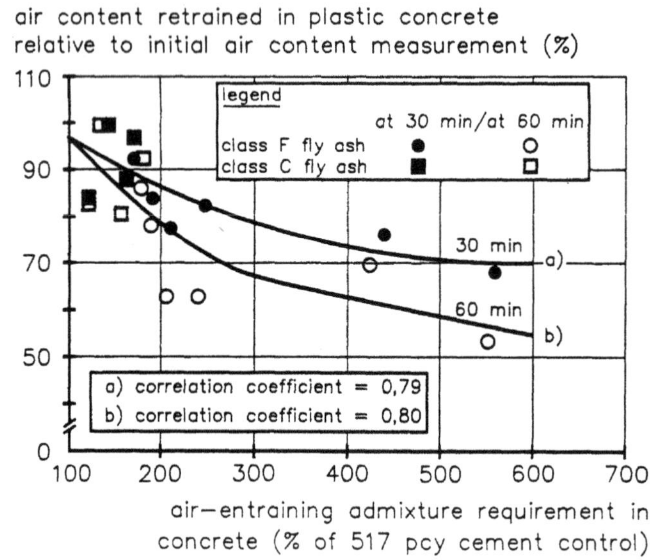

Fig. 2.15: Relationship between air-entraining admixture dosage and retention of air content in plastic concrete at 30 and 60 minutes /G 21/

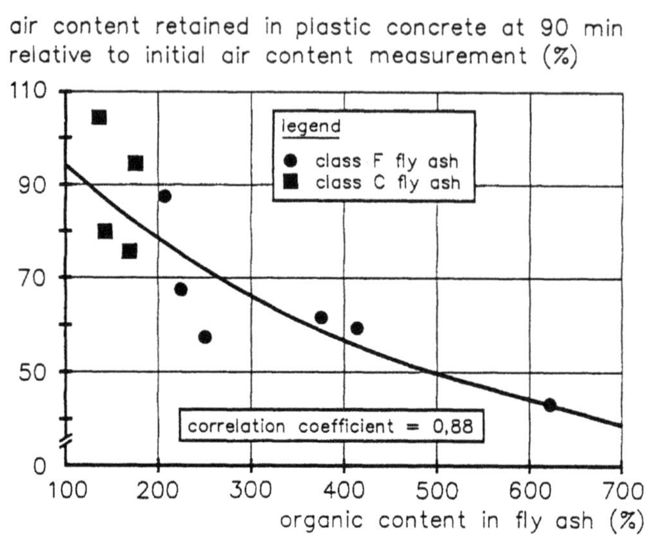

Fig. 2.16: Relationship between organics in fly ash and retention of air content in plastic concrete at 90 minutes /G 21/

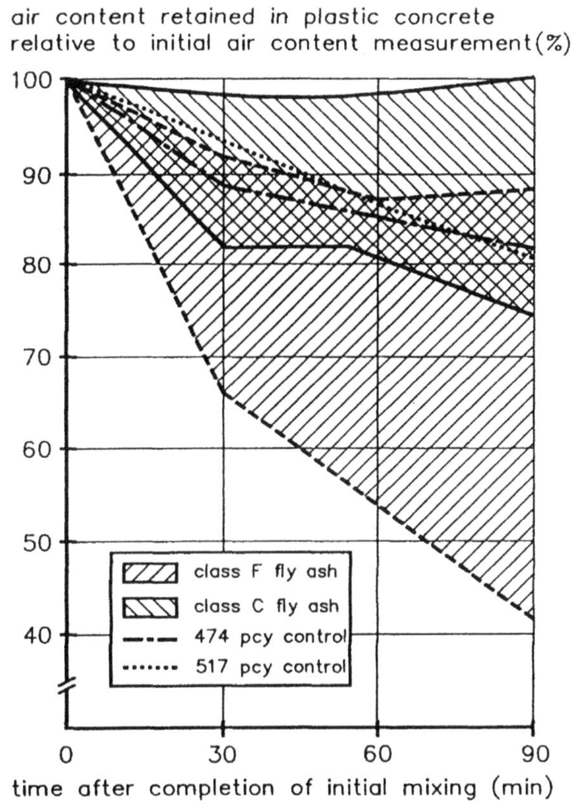

Fig. 2.17: Air content in plastic concrete with fly ash versus time /G 21/

2.3. Setting (Prepared by H. Kukko)

Class F fly ash generally prolongs concrete setting, although both initial and final setting times remain within the limits specified in cement standards. Retardation of setting due to fly ash may be affected by the proportion, fineness and chemical composition of the ash, although the cement fineness, the water content of the paste and the ambient temperature usually have a much greater effect on setting times than does the addition of fly ash /L 30, L 37, V 14/.

The chemical composition of fly ash (Class F) has been observed to influence the setting time of mortars, particularly where ashes with high carbon contents are concerned. A test series shown in *Table 2.3.* included ashes with varying carbon contents in order to determine the effect on setting time. In the blend with high carbon ash, substitution of fly ash for roughly one third by weight of the cement resulted in an increase of 100 % in the time of final setting as compared to the control mixture,

Table 2.3: Effect of fly-ash carbon content on setting time /L 30/

Fly ash source	Cement (g)	Fly ash (g)	Water (ml)	LOI	Time of setting (h:min)	
					Initial	Final
Control	650	0	165	-	2:10	3:55
Gallatin	406	244	169	2.35	3:25	4:40
Kingston	406	244	169	3.16	4:00	5:05
Colbert	406	244	186	10.45	4:20	6:35
Watts bar	406	244	243	15.68	6:25	7:10

Mixture data: Ideal portland cement Type I.
All tests were conducted to ASTM C 191 at normal consistency

without exceeding the ASTM C 150 maximum limit of 8 hours. Increases in water content may have contributed significantly to the increase in setting time for carbon levels above 10 % /L 30, L 38/.

Properties of pozzolanic cements made by mixing 30 wt.% of low calcium fly ash with 70 wt.% portland cement were studied as a function of fineness by Costa and Massazza /C 22/. The addition of fly ash prolonged the initial and final setting times of cements. This effect varied inversely with increasing cement fineness. Even very high fly-ash fineness did not modify the setting times.

On the basis of laboratory and field test data, Samarin et al. /S 49/ concluded that concrete temperature plays a dominant role in determining setting time of all concretes and that fly ash does not appear to have any primary effect in prolonging the setting time. It is also notable that, according to Montgomery et al. /M 23/, there would appear to be some evidence that fly-ash particles act as nuclei for the formation of hydration products in portland cement, thus actually accelerating the cement setting process. The fly-ash cement mortar contains less water in consequence of the presence of fly ash, particularly when its carbon content is low, and this will likewise influence the rate of stiffening.

Smith /S 60/ has studied the influence of Class C fly ash on the setting of concrete at different temperatures. The setting times were increased at all temperatures in such a way that the maximum retardation occurred when the fly-ash replacement value was 50 %. For practical replacement values of up to 40%, the effect of fly ash on the setting time was minimal (*Table 2.4*). According to Smith /S 60/, the fast initial setting for fly ash alone indicated that the fly ash under examination was cementitious as well as pozzolanic. Although the fly ash alone produced a rapid initial setting action, ultimate strength development was poor.

Diamond and Lopez-Flores /D 21/ have studied both high-calcium and low-calcium fly ashes, all of which retarded the initial set by 2 hours or more, and the final set by least 5 hours.

Table 2.4: Effect of high calcium fly ash on the initial setting time of mortar /S 60/

The setting time determinations were conducted on mortars consisting of 2.5 parts of sand and 1 part of total cementitious material, with water sufficient to produce a certain flow.

Cement content (%)	Fly ash content (%)	Water requirement (% of control)	Initial setting time (hours) at a temperature of 11,7 °C	22,8 °C	30 °C
100	0	100	7.9	4.5	3.8
80	20	91	9.8	5.6	4.2
60	40	86	11.7	7.1	5.4
40	60	81	20.9	10.2	6.6
20	80	77	30.9	15.3	9.1
0	100	75	1.5	<0.8	<0.5

The results of different studies on different types of fly ash seem to differ markedly. Prior to use of an unknown fly ash, its influence should therefore be investigated and if necessary suitably controlled by application of set-modifiers.

2.4 Plastic Shrinkage (Prepared by H. Kukko)

The effect of fly-ash content on shrinkage deformation is not as significant as the effect of the water/cement ratio /Y 3/. On the other hand, the amount of water needed to obtain a required workability is lower for high quality fly-ash concrete than for concrete without fly ash, and the reduction in the amount of water affects concrete shrinkage.

Practical experience tends to show that the use of fly ash reduces plastic shrinkage to a greater or lesser degree. This is probably due to the combination of decreased bleeding with the filler effect /M 34/. However, there would appear to be a lack of research results concerning plastic shrinkage.

3 HARDENED MORTAR AND CONCRETE WITH FLY ASH

I. JAWED, J. SKALNY, Th. BACH, P. SCHUBERT, J. BIJEN, H. GRUBE, S. NAGATAKI, H. OHGA and M.A. WARD

3.1 Hydration and Strength

3.1.1 Effect of Fly Ash on the Hydration of Cement and Concrete (Prepared by I. Jawed and J. Skalny)

Although a great deal of information on fly ash and its use in concrete is available, very little of it contributes to an understanding of the interactions between portland cement and fly ash during the hydration and the hardening process which could provide a basis for predicting and improving the performance of fly ash in concrete. It is generally accepted that, in the pozzolanic reaction of fly ash, the $Ca(OH)_2$ produced during cement hydration reacts with the silicate and aluminate phases of fly ash to produce calcium silicate and aluminate hydrates /L 39/. However, the cement hydration and the pozzolanic reactions do not proceed independently. Water-soluble alkalis, sulphates, lime and organics from the fly ash may affect the surface reactions and the nucleation and crystallization processes, especially in the early stages of cement hydration. Similarly, the pozzolanic reactions will depend on the amount of calcium, alkalis, sulphates, silicate and aluminate ions released into the liquid phase from cement and fly ash. In view of these complexities, studies of the interactions between individual clinker minerals and fly ash, as well as those between cement and fly ash, are especially useful in developing an understanding of the phenomena resulting from several independent and inter-dependent processes.

3.1.1.1 Effect of Fly Ash on C_3S Hydration

Recent work reports some conflicting data on the effect of fly ash on C_3S hydration. Takemoto and Uchikawa /T 1/ found that C_3S hydrated more quickly in the presence of fly ash both in the very early stages and after the induction period.

About 55 % C_3S hydrated in 24 hours in the presence of fly ash as compared to about 38 % in pure C_3S pastes. They attributed the accelerated C_3S hydration to its increased dissolution by adsorption of Ca^{+2} in the fly-ash particles (thus decreasing Ca^{+2} concentration in the liquid phase) and to the additional surface availability of fly-ash particles on which C-S-H can precipitate. Mohan and Taylor /M 26/ also found that more than 45 % of C_3S hydrated in one day in the presence of fly ash as compared to about 35 % in its absence. The amount of $Ca(OH)_2$ as determined by thermogravimetry and expressed as g/g C_3S was higher in C_3S fly-ash paste than in pure C_3S paste, confirming that more of the C_3S had hydrated in the presence of fly ash. Fly ash was also found to accelerate the polymerization of hydrated silicates; about 60 % of Si in the C-S-H of C_3S fly-ash paste was present as polymers compared with 40 % for pure C_3S paste. Costa and Massazza /C 13/ also noted an accelerating effect of fly ash on the hydration of C_3S.

Huang /H 37/ reported further acceleration of C_3S hydration by fly ash in the presence of gypsum (*Fig. 3.1.1*). Some differences in the hydration rate of C_3S in C_3S-fly-ash mix and cement-fly-ash mix were noted. Whereas hydration of C_3S in cement-fly-ash mix was roughly equivalent to that in pure cement paste after 3 days, it accelerated throughout the hydration period in C_3S-fly-ash paste. Addition of 4 % gypsum decreased the degree of hydration at early ages but increased it at later ages. However, an excessive amount of gypsum (8 %) inhibited the hydration at all ages. Lukas /L 18/ also noted a difference in the hydration rates of C_3S and C_2S in the presence of fly ash and in cement paste. After 18 months, there was still unhydrated C_3S in pure cement paste whereas C_3S was fully hydrated in C_3S-fly-ash mix. Interestingly, the reverse was true for C_2S.

Jawed and Skalny /J 16/ on the other hand, observed a pronounced delay of the main heat evolution peak of C_3S in the presence of fly ash in both water and NaOH solution (*Fig. 3.1.2*). Two fly ashes of similar chemical and mineralogical composition retarded the heat evolution peak to the same extent in water but to a different extent in NaOH solution, the one with a higher surface area showing higher retardation. The authors suggested that retardation was perhaps due to the delayed nucleation and crystallization of $Ca(OH)_2$ and C-S-H by the soluble aluminate species released from

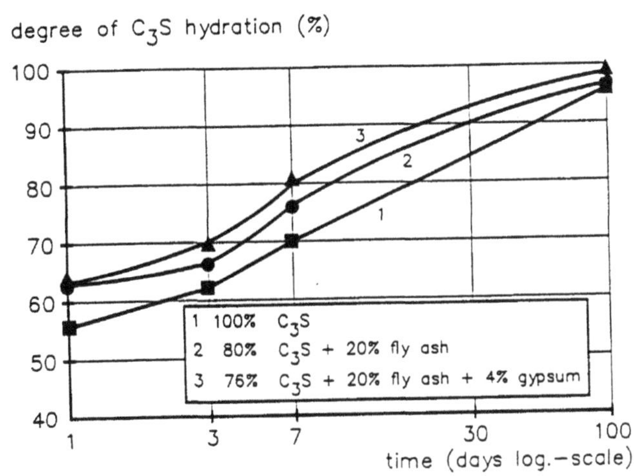

Fig. 3.1.1: Degree of C_3S hydration in the presence of fly ash and gypsum /H 37/.

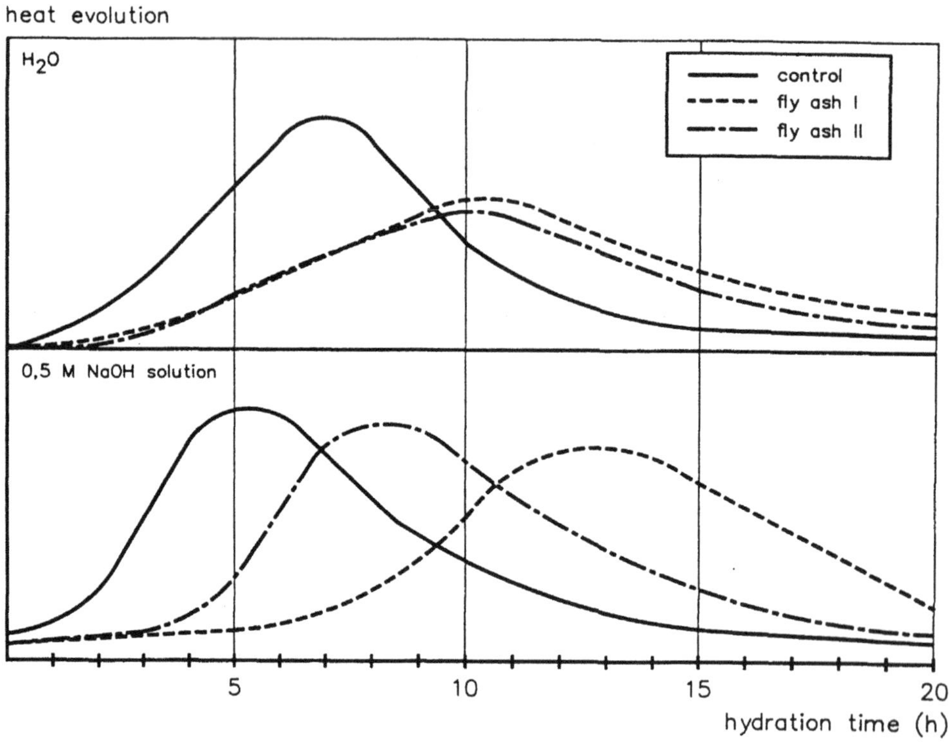

Fig. 3.1.2: Rate of heat evolution of C_3S hydrated in the presence of fly ash in water and NaOH solution /J 16/.

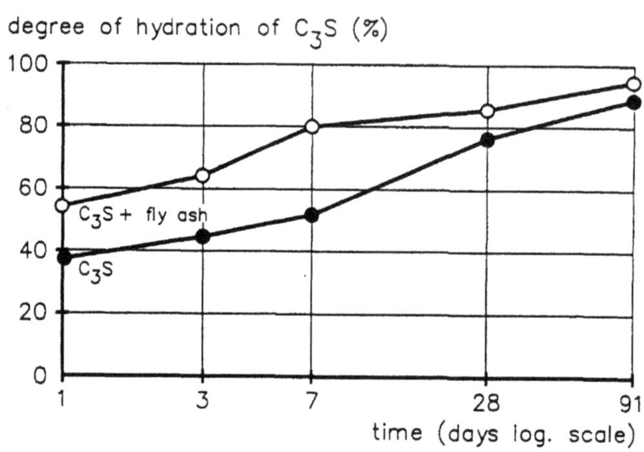

Fig. 3.1.3: Degree of C_3S hydration in the presence of fly ash /O 18/.

the fly ash. The fly ash which yielded more aluminate species in NaOH solution retarded the C_3S hydration more effectively. Ogawa et al. /O 18/ also reported a significant delay (more than 12 hours) in the maximum heat evolution peak in the presence of fly ash. However, their XRD results indicated that more C_3S (about 55 %) hydrated in 24 hours in the presence of fly ash than in pure C_3S paste (about 40 %) (*Fig. 3.1.3*). Fly ash was found to decrease the Ca^{+2} and SiO_2 concentrations in the liquid phase of a hydrating C_3S-fly-ash system significantly (by about 20 and 50 % respectively) in the early stages /C 18/.

The presence of fly ash is expected to affect the composition of C-S-H produced in the hydrating system. Analytical electron microscopic measurements on C_3S-fly-ash pastes by Mohan and Taylor /M 26/ showed an apparently non-time-dependent reduction in the mean Ca/Si ratio in the C-S-H particles from 1.51 to 1.43 from one day onwards in the presence of fly ash. The compositional variation was, however, the same as in pure C_3S pastes. Small amounts of ettringite (AFt) and monosulphoaluminate (AFm) phases were found at 7-28 days but, within 3 months, all the Fe, Al and SO_4^{-2} supplied by the fly ash was incorporated in the C-S-H. Ogawa et al. /O 18/ reported a constant value of 2.0 for the Ca/Si ratio from their scanning electron microscopic/energy dispersive spectroscopic measurements. Presumably their measured value was high, because the solid material between the particles contained both the C-S-H and $Ca(OH)_2$, whereas Mohan and Taylor were able to separate the particles more readily after grinding their samples to a fine powder. Close to the fly-ash particles, Ogawa et al. /O 18/ found a lower value of about 1.5 for the Ca/Si ratio, in better agreement with Mohan and Taylor.

3.1.1.2 Effect of Fly Ash on C3A and C4AF Hydration

Plowman and Cabrera /P 16, C 27, C 8/ found fly ash a more effective retarder of C_3A and C_4AF hydration than an equivalent quantity of gypsum. XRD and SEM data indicated that fly ash retarded the conversion of hexagonal to cubic hydrates. The hexagonal hydrates were found to incorporate sulphate ions which were believed to stabilize them and delay their transformation to cubic hydrates. Analysis of the liquid phase of the hydrating system showed that saturation in terms of gypsum occurred within a few seconds of water being added to fly ash. The SO_4^{-2} and Ca^{+2} dissolved from fly ash may also partly explain the retarding effect of fly ash.

According to Cabrera and Plowman /C 27/, the interaction of C_3A with fly ash probably involves the following processes:

- initial adsorption of sulphate ions, which reduces its active dissolution sites,
- formation of ettringite at an early age, which reduces available water migration of sulphate,
- migration of sulphate ions through the foils which are the first hydration products, and stabilization of hexagonal structures, of which the foils are the precursors,
- eventual transformation of hexagonal hydrates to cubic hydrates, much delayed in comparison with a system not containing fly ash.

It was duly pointed out that the rate and amount of sulphate released from fly ash are a consequence of its history and type and that the extent of fly-ash interaction with C_3A would consequently vary with the type. Uchikawa and Uchida /U 2/ reported that pozzolanic materials including fly ash accelerated not only the formation of ettringite and its conversion to monosulphoaluminate, but also the hydration of C_3A in the presence of gypsum. Higher amounts of alkalis in the pozzolanic materials promoted the formation of cubic hydrates. The formation of ettringite and its conversion to monosulphoaluminate were retarded by $Ca(OH)_2$. They concluded that fly ash increased the dissolution and hence the hydration of C_3A by providing surface for ettringite precipitation and Ca^{2+} absorption. However, the authors' heat evaluation curves showed a retardation of the second peak in the presence of gypsum, which was further retarded by $Ca(OH)_2$. Huang /H 37/ and Lukas /L 18/ also reported that fly ash accelerated the hydration of C_3A in cement. The former author noted that the interaction between C_3A fly ash and gypsum in cement produced ettringite but that monosulphoaluminate could not be detected before 28 days.

3.1.1.3 Effect of Fly Ash on Cement Hydration

Hydration Rate

The hydration of portland cement may be affected by fly ash in much the same way as the individual clinker minerals. However, the individual reactions and their kinetics may change in different ways when they occur simultaneously. The literature contains many conflicting reports on the effect of fly ash on cement hydration. Lukas /L 18/ observed increased formation of $Ca(OH)_2$ in pastes of fly ash and cement as compared to pure cement pastes up to 3 days, and attributed it to the accelerated formation of C_3S in cement. The $Ca(OH)_2$ content decreased with time, indicating that it had been used for the pozzolanic reaction of fly ash. Abdul-Maula and Odler /A 4/ also observed a distinct acceleration of C_3S hydration in cement in the presence of fly ash. The hydration was affected in a similar way by fly ashes of different composition. Within the first 28 days, only a moderate reaction of fly ash with $Ca(OH)_2$ was observed. Huang /H 37/ found no difference between the hydration of C_3S in pure cement paste and in pastes of cement and fly ash, whereas the hydration of C_3A was accelerated slightly in the latter case. The amount of $Ca(OH)_2$ in the cement-fly-ash paste reached a maximum value which was roughly the same as in pure cement paste, and remained constant up to 28 days before decreasing significantly. Adding gypsum decreased the early hydration of C_3S cement-fly-ash paste but increased it at later ages. However, excessive amounts of gypsum retarded the hydration at all ages. Adding gypsum also appeared to accelerate the pozzolanic reaction of fly ash.

On the other hand, Takemoto and Uchikawa /T 1/ found that the main C_3S evolution peak for cement + fly-ash pastes was retarded (*Fig. 3.1.4.*). However, the degree of hydration of C_3S in cement-fly-ash paste was higher than in pure cement pastes from one day onwards. Ghose and Pratt /G 14/ also reported a retardation of both C_3A and C_3S heat evolution peak maxima for cement-fly-ash pastes (*Fig. 3.1.5.*):

the rate of heat evolution was also decreased by fly ash. The initial retardation of C_3S was attributed to the aluminate ions in solution released from fly ash, and the subsequent acceleration at the end of the induction period to the increased surface available for precipitation of hydration products. The C_3A retardation was attributed to Ca^{+2} and SO_4^{-2} in solution produced by dissolution of fly ash.

The shoulder observed on the main peak (generally associated with ettringite to monosulphoaluminate transformation) was accentuated in the presence of fly ash. The appearance of a broad peak after 1 to 2 days indicated some exothermic reactions involving fly ash and cement hydration products. Diamond and Lopez-Flores /D 21/ noted a strong tendency for fly ash to retard setting, extend the induction period and develop a less intense main heat evolution peak. Long-term pore solution analysis showed that high-calcium fly ashes contributed substantial alkalis to the pore solution, whereas low-calcium fly ashes did not release alkalis despite their substantial K_2O contents. The low-calcium fly ash appeared to act as an inert diluent as regards early heat generation, but the high-calcium fly ash contributed significantly to early heat evolution.

The heat of hydration of cement + low-calcium fly ash is generally lower than that of portland cement alone /C 14, B 20, T 8, S 49, L 31, S 48, V 2/. The pozzolanic reactions of the aluminosilicates in the fly ash with the $Ca(OH)_2$ liberated by the hydration of C_3S and C_2S phases of portland cement are considered to take place more slowly than C_3S hydration and approximate to the reaction rate of C_2S /L 39, M 56/. Costa and Massazza /C 22/ however, reported higher than expected heat of

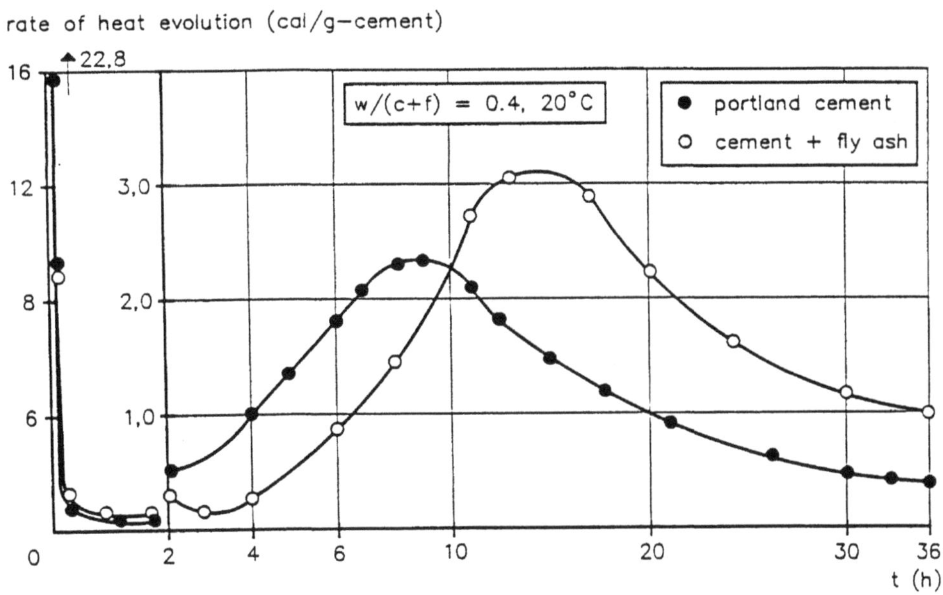

Fig. 3.1.4: Effect of fly ash on the rate of heat evolution of cement /T 1/.

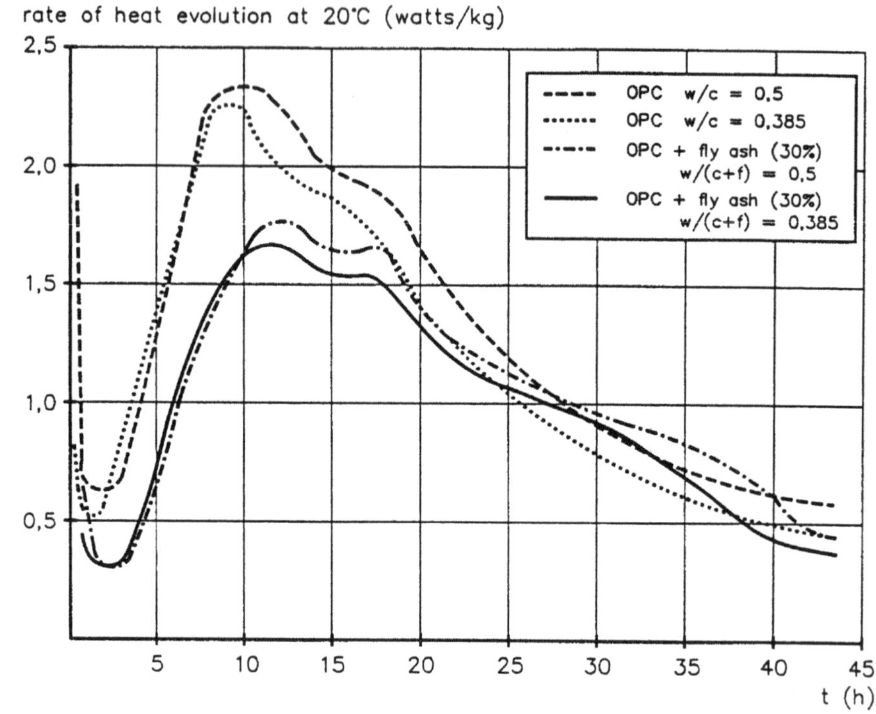

Fig. 3.1.5: Rate of heat evolution of portland cement in the presence of fly ash /G 14/.

hydration for cement + fly-ash pastes, based on the dilution ratio. Sorensen /S 45/ found no significant effect of fly ash on the heat of hydration of cement at 28 days.

Hydration Products

The hydration products of fly-ash-cement mix are essentially the same as those of portland cement under normal conditions of curing. The rate of development of hydration products is slow for low-calcium fly ash but about the same as with portland cement for high-calcium fly ashes /D 21/. The pozzolanic reactions of fly ash are very slow at normal conditions of curing.

Takemoto and Uchikawa /T 1/ detected Type I C-S-H and ettringite at one day and massive $Ca(OH)_2$ and monosulpholuminate after 7 days, with a clear space separating the hydration products from the fly-ash particles. Abdul-Maula and Odler /A 5/ reported continued ettringite formation up to 28 days; the amount of ettringite was found to be especially high for fly ashes containing anhydrite. Conversion of ettringite to monosulphoaluminate occurred only in cases of low-SO_3 fly ashes. However, other authors have detected ettringite at early stages and monosulphoaluminate at later ages in hardened cement-fly-ash pastes /T 1, H 37, G 14, S 62/.

The cement-fly-ash paste contains more C-S-H gel and less $Ca(OH)_2$ than portland cement. Somewhat more $CaCO_3$ has been reported in cement-fly-ash pastes, which may actually be the result of carbonation /K 45/. The C-S-H gel content is found to increase, especially if the water content of the system is lowered and as the C_3S/C_2S molar ratio of the system decreases /B 34/. At higher curing temperatures, Nagataki et al /N 10/ identified hydrogarnet and C-S-H at various cement/fly-ash ratios. At 25 % fly ash, α-C_2SH was identified and at 45 % replacement tobermorite was observed. The Ca/Si ratio for C-S-H in fly-ash-cement paste is expected to differ from that in cement paste, since C-S-H is also formed by the reaction of $Ca(OH)_2$ with the aluminosilicate phase of the fly ash.

Rayment /R 25/ reported that fly ash decreased the Ca/Si ratio of the inner hydrate around C_3S grains in cement from 1.71 to 1.55 and attributed this to an increase in the Si content. More potassium was found in the hydrates with lower Ca/Si ratio. Hydrated rims around the C_3S grains were slightly more developed in cement-fly-ash paste than in portland cement paste. The greater proportion of C-S-H gel in the hydrated fly-ash cement results in decreased permeability, which, together with the reduction in the $Ca(OH)_2$ content, offers an explanation for the improved resistance to chemical attack, particularly by sulphates, observed for fly ash as opposed to plain concrete.

Hydration Mechanism

The effect of fly ash on the hydration of cement and clinker minerals appears to be complex, and may depend greatly on the chemical and physical nature of the fly ash. The observed changes may also depend on the water/cement ratio of the system. There appears to be retardation of the very early hydration of both C_3S and C_3A, as shown by heat evolution profiles over time. After the induction period, however, this is followed by increased formation of $Ca(OH)_2$ and C-S-H and also by increased formation of ettringite and its subsequent transformation to monosulphoaluminates. The prolongation of the induction period in C_3S hydration is probably due to the species dissolved from the fly ash into the aqueous phase of the hydrating system such as aluminate ions and organics which could delay the nucleation and crystallization of $Ca(OH)_2$ and/or C-S-H. There may also be a physical effect in which the fine fly-ash particles adhere to the surface of cement grain and thus hinder its interaction with water.

Once the nucleation and crystallization of hydration products end the induction period, hydration is accelerated by the presence of fly ash. The fly-ash particles provide additional surfaces for the precipitation of the hydration products which would otherwise be formed on the surface of the C_3S, and hinder its interaction with water. Similar arguments may apply to the hydration of C_3A, where the initial retardation is probably due to calcium sulphate and alkalis (in addition to organics) dissolved from the fly ash. The precipitation of the hydration products on the fly-ash spheres may hinder the pozzolanic reaction. However, the alkaline solution may attack the glassy phase of the spheres beneath the coating of hydration products, leaving a clear space between the coating and the sphere.

3.1.2 Pore Size Distribution (Prepared by Th. Bach)

Pore size analysis on cement paste, mortars and concretes is performed in two different ways. Some kind of optical method can be employed, either directly using a light-microscope and reflected light (plane polished section) or transmitted light (thin polished section), or indirectly using an electron microscope and scanning electron microscopy (SEM – on lumps of material) or transmission electron microscopy (THM – on ultra thin sections). Alternatively, pore analysis can be carried out by some type of indirect method, using a physical relation between pore size, the medium in the pores, and a variable factor controlled during the operation, such as mercury-porosimetry and low-temperature calorimetry.

The optical methods do give real impressions of the pores and their size distribution, but tend to be applied qualitatively, since quantification is very tedious, whereas the indirect methods of analysis are used for quantification through an idealized relation between the externally controlled variable factor, the medium and the pore size. The pore sizes and their distributions also change continually, e.g. due to progressing reactions (hydration of cement) and variations in temperature and humidity. Thus, although the results obtained by characterizing pore size distributions may be very informative to the investigator, they will certainly be more susceptible to variations than will other properties, e.g. strengths and heats of hydration, and care should be exercised when comparing results obtained by two different investigators. The most valuable use of the methods is to compare the effect on pore size distribution from different variations in material composition and/or treatment.

3.1.2.1 Microscopy

Light microscopy can reveal information about the macroinhomogeneities and their distribution, such as size and distribution of entrained and entrapped air (down to a few micrometers in size), as well as the existence and density of cracks. Furthermore the technique may provide information on the capillary porosity of the paste itself (by using transmitted light on thin polished sections). Equally, these techniques may reveal information on the existence and possible distribution of fly ash in the concrete as well as information about the influence of fly ash on the density of cracks. The former information is primarily influenced by the method of mixing and pouring, whereas the latter is primarily related to the method of hardening and sample preparation.

Thus, as the fly-ash particles are easily visible under light microscopes, it is well-established that fly ash can be homogeneously distributed between the cement particles by proper proportioning and mixing, usually benefiting from the use of dispersing agents (plasticizers). The use of microscopes to characterize microstructure has hitherto been possible neither with respect to evaluation of the degree of hydration of cement and/or fly ash nor with respect to evaluation of the ratio between water and the amount of cementitious material in a sample.

Electron microscopes can reveal information about the microinhomogeneities and their distribution, including the formation of reaction products. A number of workers /G 14, C 8, J 3, K 14, P 16, D 7, M 23/ have studied the microstructure of fly-ash-cement pastes. There is general agreement that fly ash reacts with lime in the alkaline environment of hydrating cement. However, different reports quote highly divergent time scales at which the pozzolanic reaction products are detected, varying from 3 days to 28 days. It is difficult to draw general conclusions. Generally, the microstructural development takes more than a year. However, the most significant morphology is developed during the first 6 months. The work described in /G 14/ is outstanding and in good agreement with the findings of most other researchers.

Studies of pastes made from 70 % Ordinary Portland Cement and an (ASTM) class F fly ash with water to solid ratios of 0.5 and 0.385 show that noticeable bonding may start to develop 12 to 18 hours after mixing, while from 18 hours onwards specimens may normally be fractured without crumbling.

After as little as 1 hour, there may be evidence of reaction on the cement grains (*Fig. 3.1.6*) in the form of AFt needles and small granular products of early hydration together with larger crystals of secondary gypsum. Some of the fly-ash spheres may also have hydration products on them; it is difficult to say whether these are due to surface hydration or precipitation from the drying pore solution. After 4 hours, some of the fly ashes may show definite signs of surface pitting with granular hydration products growing at the pits (*Fig. 3.1.7*), in addition to the CH and AFt seen lying on the surface; this is even more pronounced at 8 hours (*Fig. 3.1.8*). By this time, the cement grains are covered with reaction products including C-S-H and AFt with some large CH crystals (*Fig. 3.1.9*) and after 12 hours this type of coating may envelop even fly-ash spheres as shown in *Fig. 3.1.10*. This resembles the duplex coating found by Diamond et al. on fly ash.

After 18 hours, the paste may acquire some cohesion, the AFt rods being better crystallized, and Hadley grains being more numerous (*Fig. 3.1.11*). Large CH crystals may also start to grow (*Fig. 3.1.12*), while some of the fly-ash grains show very different morphologies of the coating on different parts of their surface (*Fig. 3.1.13*). These coatings are generally found to be intact over the entire grain surface.

In 1-day old specimens, AFt phases may be found to grow longer and interlock in some of the void spaces. In general, the fly-ash particles are found to be coated but some appeared to have reacted. Flocculation of the fly ash may be evident in places. Cleavage through CH crystals may first be seen at this age and few spheres may be embedded in the CH (*Fig. 3.1.14*). These fly ashes do not have any coating and appear to have been etched in the pore solution. In some regions, large well-formed crystals of CH may be found in intimate contact with big fly-ash spheres, and reactions may take place at the interface (*Fig. 3.1.15*).

The growth of massive CH in the form of stacked platelets which cleave during fracture will be much more pronounced at 3 days. Most of the fly-ash spheres embedded in this CH growth again look etched rather than reacted on the surface. Heavily coated fly ashes may also be observed, but not in the vicinity of these massive CH deposits. Some of these may show reaction coating or product over part of the surface (*Fig. 3.1.16*), and this can also be evident at longer ages. It is difficult to imagine part of the coating tearing off during fracture in this particular case. This sphere may be partly glassy and partly crystalline, the glass in the fly ash being the reactive component according to Kokubu /K 14/; the demarcation of surface products may reveal the glass/crystal interfaces. In resin-impregnated, cut, polished and etched sections of 2-month old specimens, this may sometimes be evident after differential etching of the fly ash.

Up to 14 days, the structure becomes denser with continuing infilling from CH and C-S-H; Hadley grains disappear, probably being engulfed in the developing structure, and evidence of AFt-AFm conversion may be seen in open places. Sometimes fly-ash spheres are seen to be encapsulated by the reaction products. Between 3 days and 2 months, fly-ash spheres with a variety of reaction and etch patterns are found (*Figures 3.1.16* and *3.1.17*), the extent of the attack increasing with the time. At longer times of up to 5 months, some of the fly-ash spheres have reacted significantly (*Fig. 3.1.18*) and some have been eaten away (*Fig. 3.1.19*), while at the same time, some spheres which have not reacted very much are found. Microanalysis in fracture faces is not always very convincing, but there appears to be some indication from microanalytical studies on these cement fly-ash pastes that the reacted fly ashes are generally richer in alumina and poorer in silica than non-reacted ashes.

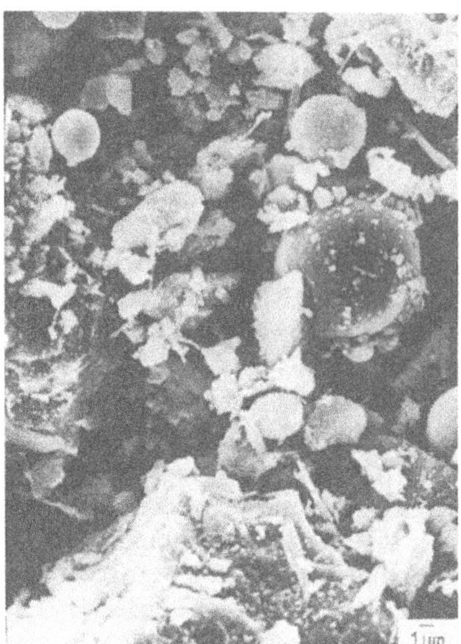

Fig. 3.1.6: Cement + fly-ash paste after 1 hour

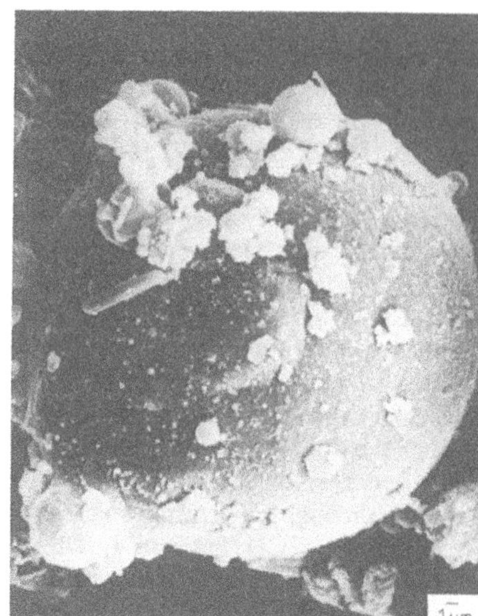

Fig. 3.1.7: Cement + fly-ash paste after 4 hours

The SEM images (Figures 3.1.6 to 3.1.19) were kindly supplied by Prof. P. L. Pratt, Imperial College of Science and Technology, Dept. of Materials, London, United Kingdom /G 14/.

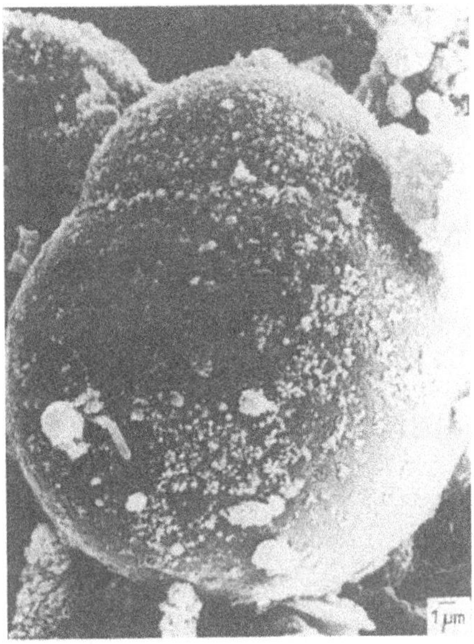

Fig. 3.1.8: Cement + fly-ash paste after 8 hours

Fig. 3.1.9: Cement grain in fly-ash paste after 8 hours

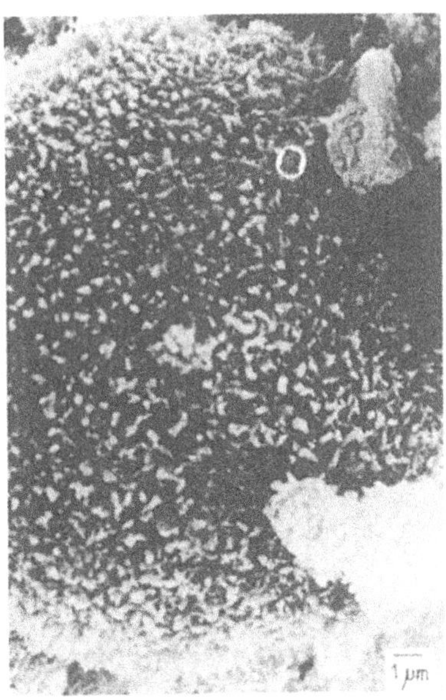

Fig. 3.1.10: Fly-ash sphere with duplex film after 12 hours

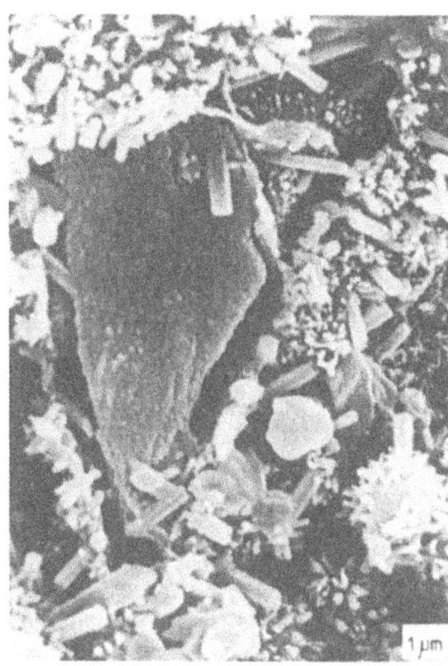

Fig. 3.1.11: Hadley grain and AFt rods in cement + fly-ash paste after 18 hours

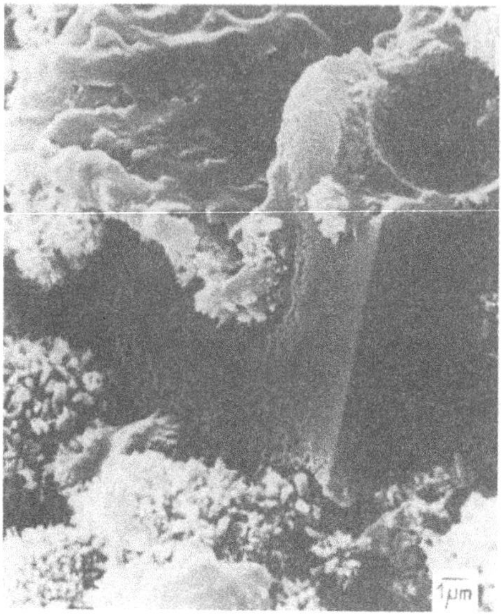

Fig. 3.1.12: Large CH crystal in cement + fly-ash paste after 18 hours

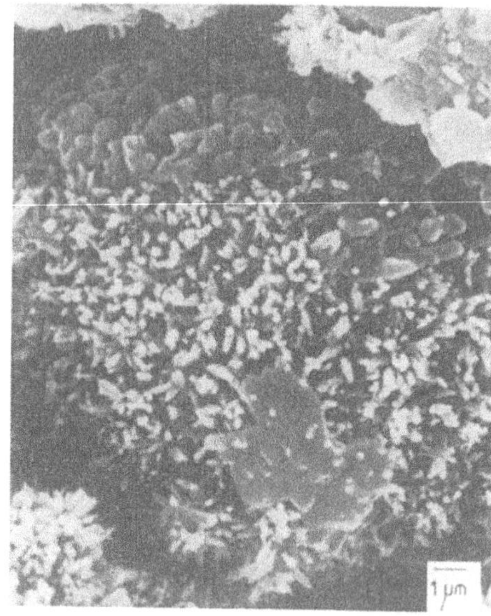

Fig. 3.1.13: Fly-ash sphere in the paste after 18 hours

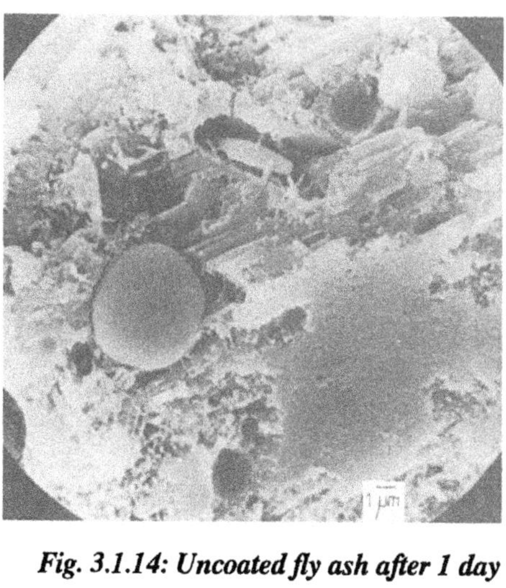

Fig. 3.1.14: Uncoated fly ash after 1 day

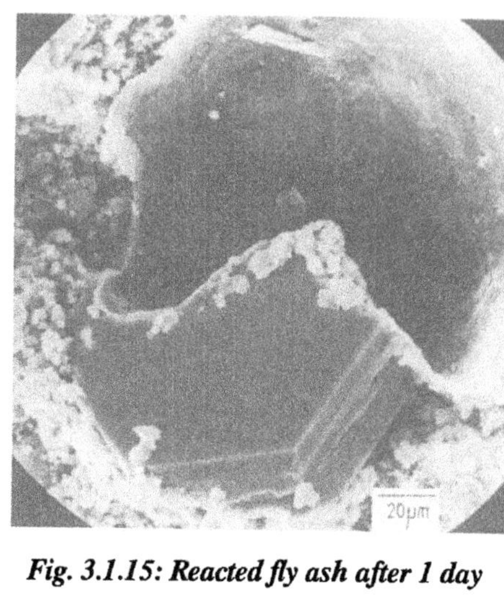

Fig. 3.1.15: Reacted fly ash after 1 day

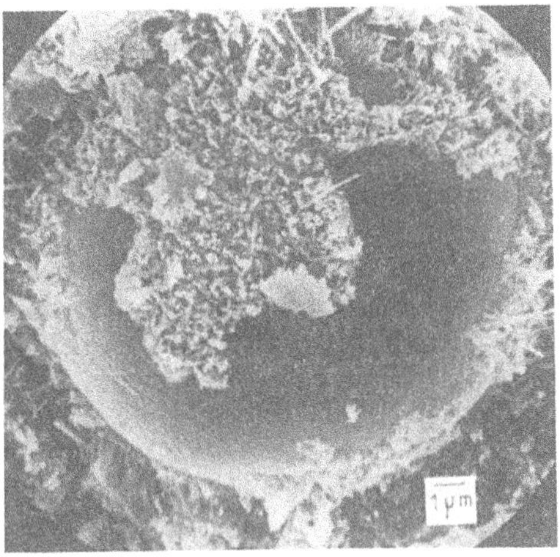

Fig. 3.1.16: Partly coated fly-ash sphere after 3 days

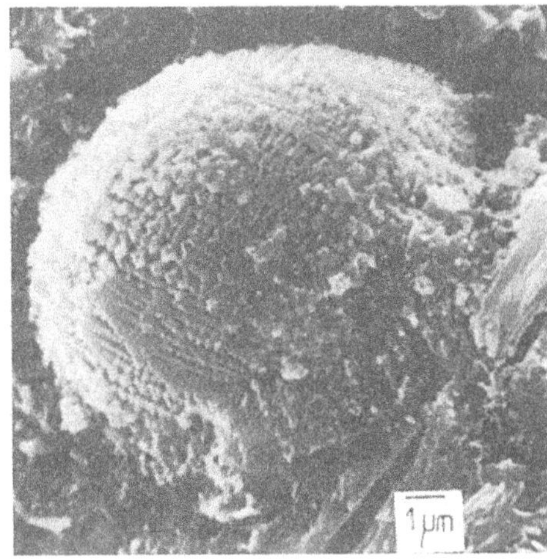

Fig. 3.1.17: Etched fly ash after 2 months

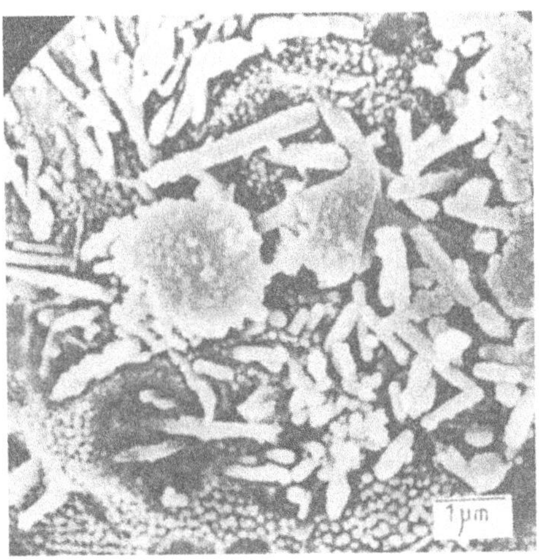

Fig. 3.1.18: Reacted fly ash after 5 months

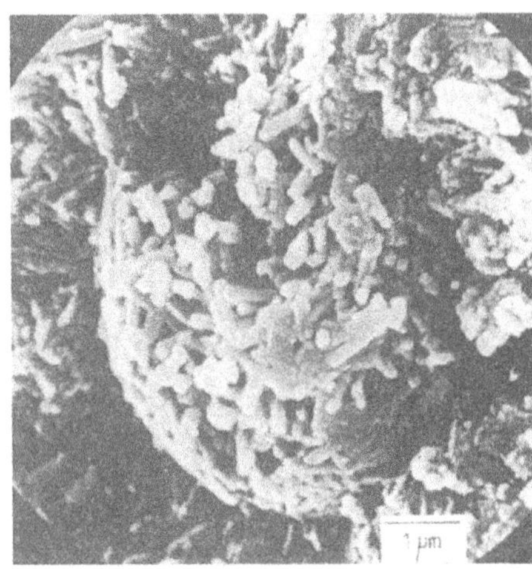

Fig. 3.1.19: Heavily reacted fly-ash sphere after 5 months

3.1.2.2 Other Methods

The use of low temperature calorimetry for characterizing cement paste is under development, and so far it has not been used to characterize the influence of fly ash on the microstructure of cement-bound materials.

The mercury-calorimeter has, however, long been used in characterizing hydrated cement pastes. This technique has therefore also been adapted to characterize the influence of fly ash on the microstructure of materials bound by mixtures of fly-ash cement /F 14, F 26, M 47, M 36/.

Fig. 3.1.20 shows the effect of adding different amounts of fly ash (ASTM class F) at ages of 28 days and one year on the pore size distribution of cement pastes, while *Fig. 3.1.21* shows the effect of curing time on pore size distribution of cement and fly-ash-cement pastes. It is evident from the figures that the pore sizes become finer and finer as hydration proceeds. It is also apparent that the initial pore sizes of cements containing fly ash are coarser than those of pure cements, whereas the ultimate pore sizes of cements containing fly ash are finer than those of pure cements. The shift from fly-ash-cements giving coarser pores than pure cements seems to occur at a time of between 1 and 3 months. This observation is in good agreement with /M 36/.

Structural investigations thus seem to indicate that the effect of fly ash on structure formation at early ages is a simple question of thinning the cement particles with an almost inert material; the reaction then begins very slowly after several days, is clear at about one month, yields a cement-fly-ash reaction product which after 1 – 6 months is of almost the same character as that with a pure cement of the same age, and eventually produces a reaction product between cement and fly ash which is denser than that with a pure cement.

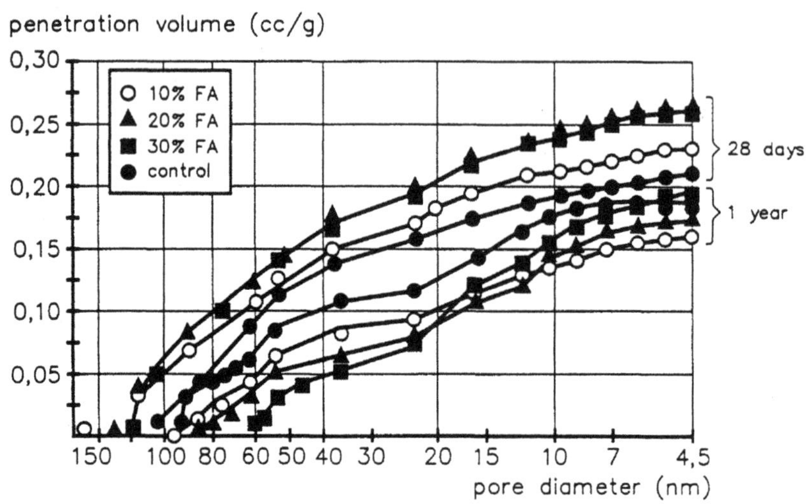

Fig. 3.1.20: Effect of various proportions of fly ash on pore size distribution of cement pastes /M 47/

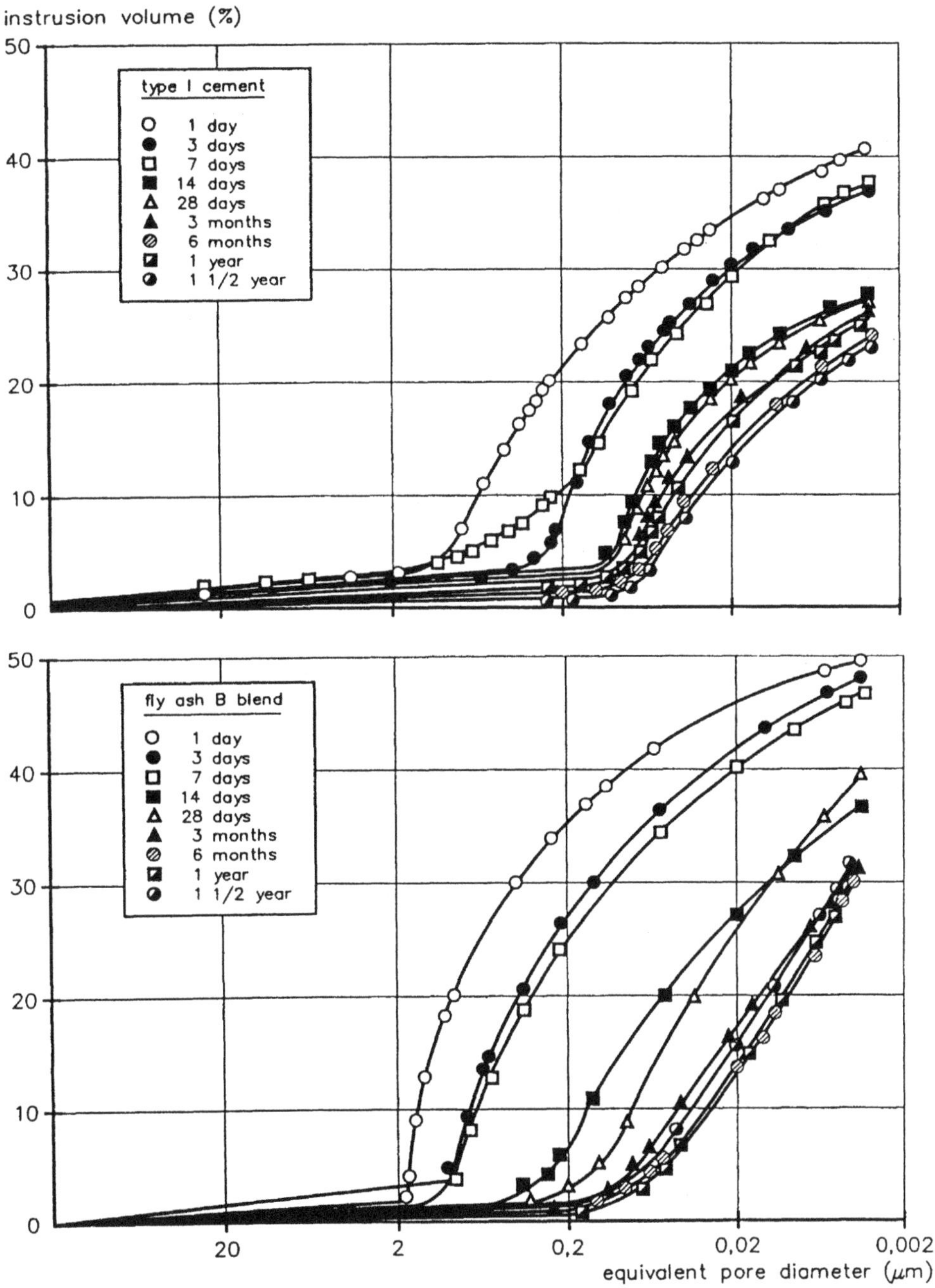

Fig. 3.1.21: Change in pore size distribution of pure cement paste and cement + fly-ash paste with hydration time at 21 °C /F 14/

As reactions take place through the aqueous phase, lack of water at a certain time will lead to an interruption in the process of structure formation at that time, so that a variety of final structures may be observed under real conditions. When lack of water occurs at early ages, a very porous structure may be expected, whereas a dense structure may be expected when lack of water occurs at later ages.

3.1.3 Reactions of Fly Ash in Mortars and Concrete (Prepared by I. Jawed and J. Skalny)

The progress of chemical reactions involving fly ash may vary considerably with the type of fly ash. Low-calcium fly ash apparently does not react appreciably at an early age under normal curing conditions, but the reaction is accelerated at higher temperatures /D 19, C 22, M 23, B 32, T 10,/. After a few weeks, pitting and eroding of individual spheres is seen and the results of various chemical studies suggest that the $Ca(OH)_2$ content of the cement paste or concrete begins to decline. Only the glassy portion seems to be attacked, and only on some particles. The amount of fly ash reacted depends on its glass content and on the amount of $Ca(OH)_2$ present in the system. It appears that the glass is dissolved by the alkaline pore solution that builds up with time and reacts with $Ca(OH)_2$ to produce a gel of calcium alkali silicate hydrate (with Al, Fe, SO_4^{-2} and possibly other species) not readily distinguishable from the ordinary C-S-H gel /D 19/.

Deposition of a thin duplex film of $Ca(OH)_2$ and a single layer of C-S-H gel around the fly-ash sphere has been observed /D 7/. The film is presumably not really a result of fly-ash reaction but represents precipitation from the supersaturated pore solution onto the fly-ash surface. In addition to $Ca(OH)_2$ and C-S-H, ettringite is also detected on the fly-ash surface /G 14, D 31, J 3/. It appears that the reaction usually occurs on a more or less particle-by-particle basis with some particles never reacting while others react at different rates. Several authors noticed the separation of ash particles from the hydrated mass by a gap of about 1 to 2 μm thickness during the early stages of hydration /T 1, E 14, K 14, E 5,/. This has been attributed to the increased osmotic pressure due to migration of alkali metal ions to the surface of fly-ash particles. These ions, penetrating through the hydrated film, entrain water and form a thin layer of liquid with high ionic concentration between the particle and the film of new reaction products /T 1/.

High-calcium fly ash is capable of independent setting when mixed with water. The hydraulic minerals usually present in high-calcium fly ash react in a manner entirely analogous to their reaction in portland cement /D 19, D 21, E 14, D 32, H 38, C 28, G 15/. The soluble components of high-calcium fly ash such as alkali salts, anhydrite and free CaO affect the course of cement hydration from the very beginning. The alkalis may tend to promote the early hydration of cement. CaO and anhydrite produce an equivalent amount of $Ca(OH)_2$ and gypsum when mixed with water, and much additional ettringite would precipitate in the early stages of hydration. It also appears that glass in high-calcium fly ash may be more reactive than glass in low-calcium fly ash leading to an overall acceleration of the glass reaction process /D 19/.

3.1.4 Autogenous Shrinkage (Prepared by Th. Bach)

Very little work has been done in order actually to measure the effect of fly ash on autogenous shrinkage of mortar and concrete. However, what has been written indicates that the autogenous shrinkage of mortars and concretes with fly ash included in the cementitious material will be less than that of mortars and concretes of the same composition but without fly ash /G 13, M 34/. It is not possible to quantify the reduction on the basis of the literature available.

3.1.5 Effect of Fly Ash on Strength Development of Mortars and Concretes (Prepared by I. Jawed and J. Skalny)

The compressive strength and other mechanical properties of mortars and concretes containing fly ash will depend on the pozzolanic reactivity of the fly ash, the richness of the mix, the character and grading of the aggregate, the water content of the mix and the curing conditions. Fly ash may be added separately to the mix or it may be added to cement clinker during or after grinding. However, the latter procedure restricts the designer to a fixed cement to fly-ash ratio.

3.1.5.1 Effect of Chemical Composition

The main components of fly ash contributing to its pozzolanic reactivity are its reactive silicates and aluminates (contained in the glass phase), whereas mullite and quartz are ineffective. It has been suggested that the reactivity of fly ash depends on the temperature at which the coal is burned rather on than the quality of coal /K 39/. A number of authors found no clear correlation between the inorganic constituents of low-calcium fly ash and the strength development of mortars and concretes /D 17, A 4, C 14, B 19, G 6, W 1–3/. Joshi et al. /J 25/ noted poor correlations between the reactivity and the CaO, Al_2O_3, SiO_2 and Fe_2O_3 contents of fly ash, although the form and not the amount of CaO in fly ash was considered important. The water-soluble fraction did not appear to be related to the fly-ash reactivity. Beretka and Brown /B 8/, on the other hand, found poor strength for fly ashes with a high content of water-soluble materials and low SiO_2 and Al_2O_3 contents.

The sulphate content of a fly ash has been reported to contribute significantly to the early hardening of fly-ash mortars /O 3, C 4/. Increased carbon content (indicated by high loss on ignition) is found to affect the strength of fly-ash concrete adversely. This is attributed to an increased water requirement and an increased amount of air-entraining agent (needed to maintain the same workability and air content as in control concretes) as the carbon content of fly ash increases /S 45, M 57, C 15, L 4/. Since high-calcium fly ash contains hydraulic cement minerals, its reactivity may be better correlated with its composition. However, it is the mineralogical rather than the chemical composition which is important in determining its reactivity and its effect on the strength development of concrete /J 26/.

3.1.5.2 Effect of Fineness

The fineness of fly ash, particularly that of its glassy phase, is considered to be more important than its chemical composition in determining its reactivity and improving the strength characteristics of mortars and concretes. Finer material will dissolve and react faster in the liquid phase of the hydrating system. Many authors have reported a direct correlation between the fineness of fly ash and its reactivity and effect on the strength development of mortar and concretes, although the effect of fineness may not be evident at the early ages /C 14, C 22, B 19, G 6, W 1–3, J 25, B 8, M 57, C 15, D 12, K 46, R 6, W 11/. The lower the residue above 45 μm sieve, the greater is the reactivity. Krueger et al. /K 46/ reported that the fly-ash reactivity correlated better with the 45 μm residue than with the specific surface of fly ash whereas Ravina /R 6/ found the reverse to be true. Some authors, however, noted only a limited effect of the specific surface or particle size distribution of fly ash on the strength development of concrete /M 27, I 12/. The coarse fraction of fly ash contributes significantly to early concrete strength only under thermal curing /R 24/.

Grinding of fly ash increases its fineness and reduces its porosity, both of which have a positive effect on fly-ash reactivity and strength development of concrete /S 26, Y 4, O 15/. However, it is also found to increase the water requirement in the concrete mix, possible entailing an adverse effect on strength development /C 22, E 14, Y 4/. The increased water requirement is attributed to the destruction of the spherically-shaped particles of fly ash, which can no longer contribute favorably to the workability.

3.1.5.3 Effect of Mix Design

Apart from the quality of fly ash and cement, the method of mix design is the single most important factor influencing the properties of fly-ash concrete. In general, these methods can be classified /B 12, M 41/ as

- partial replacement of cement,
- partial replacement of cement and fine aggregate,
- partial replacement of fine aggregate,
- partial replacement of fine and coarse aggregate.

In addition, many authors have suggested numerical relationships for the proportioning of fly-ash concrete mixes /W 14, H 25, P 10/ or developed a series of empirical tables and graphs to define the required mix proportions to give any specific strength /O 8, C 1, G 3, S 16, R 44/. However, these numerical relationships and graphs do not eliminate the need for trial mixes in proportioning, although they may reduce the number of trials required.

There is an enormous amount of work reported in the literature on the effect of fly-ash incorporation on the development of the compressive strength and other mechanical properties of mortars and concrete. It would be neither possible nor valuable to review all this work, much of which is specific to certain fly ashes or certain construction projects. However, certain generalizations can safely be made on the basis of this work.

A substantial amount of work in the literature suggests that the partial replacement of cement (either by weight or by volume) in mortar or concrete by fly ash results in lower compressive strength at early ages (about 3 to 6 months), with development of greater strength as compared to control concrete at and beyond 6 months (*Fig. 3.1.22*). The higher later strength is the result of increased pozzolanic reaction at later ages, producing an increasing amount of C-S-H at the expense of $Ca(OH)_2$. The time at which the strength of fly-ash concrete will catch up with that of plain concrete will generally depend on the amount, reactivity and fineness of the fly ash, the water to solid ratio, and curing conditions such as humidity and temperature.

An increased amount of fly ash in the mix delays setting and may slow down strength development at the early ages /C 14, T 8, S 49, L 31, C 22, B 23, W 12, R 29, D 22, R 28, S 11/. High-calcium fly ashes develop better early strength than low-calcium fly ashes, although they may cause false setting /C 15, Y 3/. Faster strength development is achieved by reducing water content /C 14, T 8, S 49, L 31, C 22, B 19, W 14, H 25, B 23, K 3, O 9/ and raising curing temperature /C 15, R 24, B 23, G 5, D 5, F 14, O 3, D 33, M 58, O 17, R 27/. However, Butler et al. /B 32/ have reported lower ultimate strengths for mortars cured at higher temperatures. Moist or wet curing is found to be beneficial for the strength development of fly-ash concretes /M 57, R 27, K 40/.

Partial replacement of cement and fine aggregate by fly ash in the concrete mix results in a concrete with early strengths usually comparable to those of control mixes, but with higher strengths at later ages. Partial replacement of aggregate, whether fine or both fine and coarse, by fly ash generally produces concrete with increased strength at all ages as compared to control mixes (*Fig. 3.1.23*). These observations obviously

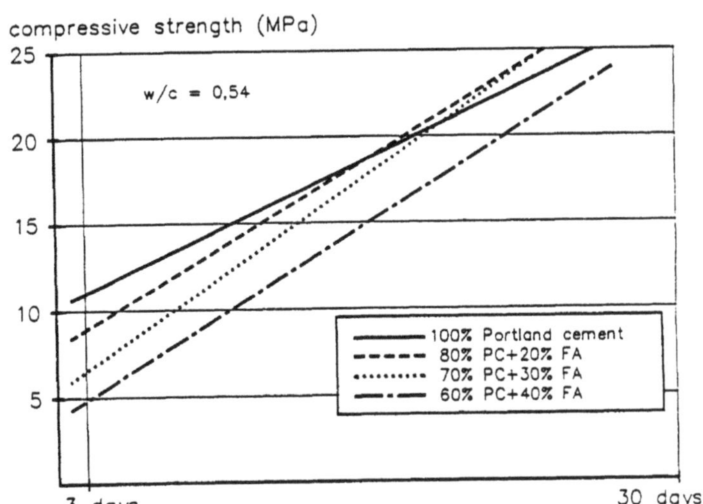

Fig. 3.1.22: Strength development in fly-ash concretes /L 31/.

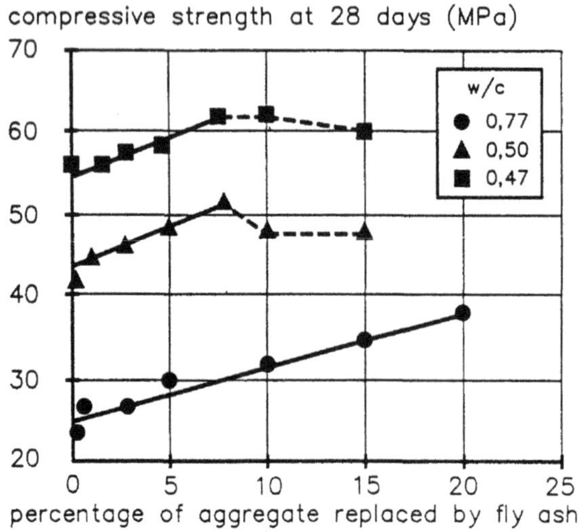

Fig. 3.1.23: Variation of compressive strength at 28 days with amount of aggregate replaced by fly ash /H 25/

result from the fact that there is no reduction in cement content in the mix, and that there is increased cement hydration after one day.

It has been reported that the use of water reducing admixtures and superplasticizers in fly-ash concrete generally results in early strength comparable to (or higher than) that of plain concretes /S 63, L 40, B 25, M 59, M 55, M 42, E 19/. The workability of fly-ash concrete may not be very significantly improved by water-reducing admixtures /L 40, N 23/. This is attributed to the observation that the fly-ash particle surface already carries a negative charge which is not affected appreciably by the negative charge of the water-reducing admixtures /N 23/.

3.1.6 Flexural and Tensile Strength (Prepared by Th. Bach)

When using fly ash as a partial or complete substitute for portland cement, data obtained from both laboratory experiments and field experiments generally indicate that flexural strength (or the modulus of rupture) and indirect tensile strength (or splitting strength) can be predicted from compressive strength results /F 23, B 25, C 23, S 11, S 49, J 22, L 31, R 37, T 6, G 19, V 3, V 6/. The ratio of flexural to compressive strength is generally found to be slightly higher than the ratio of indirect tensile to compressive strength /S11, S54/.

The relationship between flexural and compressive strength is found to be independent both of fly-ash class, whether ASTM class F or C /K 40/, and of the method of proportioning, whether compared on the basis of a constant ratio between water and cementitious material or on the basis of constant workability /V 3, K 39/. The relationship generally found is shown in *Fig. 3.1.24*.

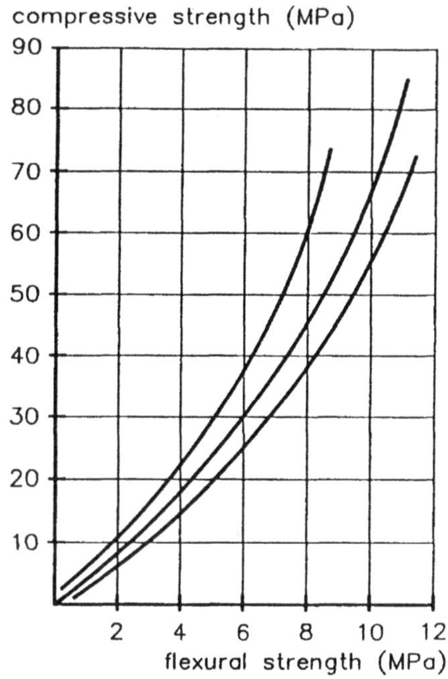

Fig. 3.1.24: Relationship between flexural and compressive strength as generally obtained. The vast majority of results fall in between the bounding lines.

In some cases it has been found that the ratio of flexural to compressive strength of pozzolanic cements containing fly ashes is higher than the corresponding one obtained with portland cements /C 22, K 14/. In other cases this observation seems to be related to situations in which the strength development is very slow for some reason, i.e. due to using rather coarse fly ash /V 3, V 6/, due to using excessive amounts of fly ash /V 3, V 5, K 39/, or due to using a fly ash with a rather high amount of unburned matter /G 19/.

3.1.7 Conclusions (Prepared by I. Jawed and J. Skalny)

The interaction between fly ash and cement is a fairly complex phenomenon which involves several independent and interdependent processes. The inhomogeneity and variability of fly ash as regards its chemical and mineralogical composition and physical characteristics further complicates the picture. Certain conclusions can, however, be drawn on the basis of the recent literature reviewed above.

The hydration of C_3A, C_3S and portland cement appears to be retarded in the early stages but accelerated from one day onwards. Water-soluble constituents of fly ash such as alkalis, sulphates, lime, organics, soluble aluminate and silicate species play a decisive part in determining the course of reaction and the nature and quantity of reaction products. The pozzolanic reaction of fly ash reduces the amount of $Ca(OH)_2$ produced and lowers the Ca/Si ratio of the C-S-H in the cement/fly-ash mix.

Under normal conditions of curing, the strength development of mortar or concrete in which cement is partially replaced by fly ash is lower than that of plain mortar or concrete in the early period (presumably a consequence of decreased cement content rather than slower hydration and pozzolanic reaction) but higher at later ages. The reduction in early strength may be avoided by partially replacing aggregate with fly ash while keeping cement content unchanged. Concrete containing high-calcium fly ash may not show any reduction in early strength. Given a certain fly ash and cement, it should be possible to design a suitable concrete mix to yield any specified strength.

There is a direct correlation between the fineness of fly ash and the strength development of mortars and concrete. However, there seems to be no clear correlation between the chemical composition of fly ash and its reactivity and effect on strength development of concrete. Presumably the mineralogical rather than the chemical composition is more pertinent. An increase in the carbon content of fly ash adversely affects the air entrainment, workability and strength of fly-ash concrete.

3.2 Deformations

3.2.1 Deformation Behaviour under Compressive Strength (Prepared by P. Schubert)

3.2.1.1 Stress-Strain-Curve ($\sigma-\varepsilon$)

Apart from the modulus of elasticity E and Poisson's ratio μ, $\sigma-\varepsilon$ graphs are widely used for assessing the deformation behaviour of building materials.

Although the shape of the $\sigma-\varepsilon$ graph for concrete depends upon several factors such as concrete mix, age, compressive strength, temperature, test and storage conditions, there are few results for concretes and mortars containing fly ash.

Effect of Concrete Age

In /L 5/ $\sigma-\varepsilon$ graphs were determined at 3, 7 and 21 days for concretes of about the same workability with and without fly ash (f/c = 0.30) (*Fig. 3.2.1*). At 28 days, fly-ash concrete had a compressive strength of 18.7 MPa and non-fly-ash concrete a compressive strength of 26.8 MPa. At 3 and 7 days, the $\sigma-\varepsilon$ graphs were virtually identical with those at 21 days. The fly-ash concrete was more deformable when young than the non-fly-ash concrete, which had a higher 28-day strength. In /L 5/, priority was given to determining the modulus of elasticity, and the $\sigma-\varepsilon$ curves were consequently not determined up to failure.

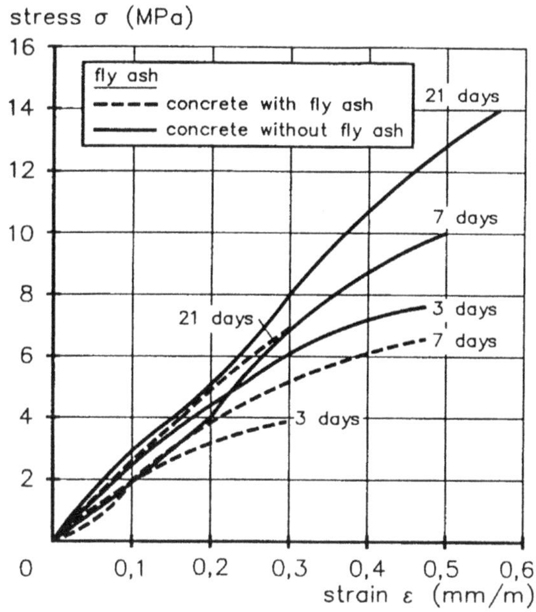

Fig. 3.2.1: ***$\sigma - \varepsilon$ curves of concrete without fly ash and with fly ash (f/c - 0.30) at an age of 3, 7 and 21 days /L 5/***

Effect of Temperature

To ascertain in detail the deformation behaviour of fly-ash concrete used under high thermal loads in applications such as nuclear reactors, portland cement concretes containing brown coal ash (f/c = 0.25) were investigated in /N 1/ at temperatures of 21.4, 71, 121, 149, 177, and 232 °C (*Fig. 3.2.2*).

As compared to concrete without fly ash, there was a greater increase in strength from 121 to 149 °C. This is attributable to the transformation of the tobermorite gel – formed at atmospheric pressure and room temperature by the reaction of $Ca(OH)_2$ with fly ash – to a tobermorite with between two and three times the strength of the gel. At 177°C (not represented here) and 232°C, the $\sigma - \varepsilon$ graphs for fly-ash and non-fly-ash concretes are similar, i.e. flatter and more elongated. According to /N 1/, this is because of the transformation of tobermorite gel and tobermorite into compounds of weaker alpha dicalcium silicates (α-Ca_2SiO_3).

The shape of the $\sigma - \varepsilon$ curves was determined mathematically from a polynomial equation which also describes the sloping leg of the curve:

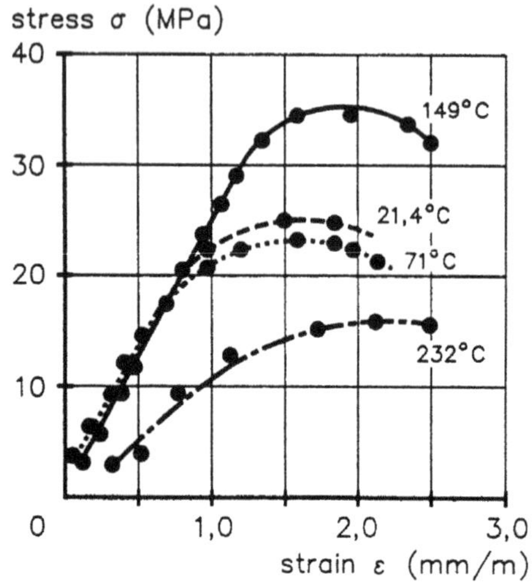

Fig. 3.2.2: ***σ–ε curves of concrete with brown coal fly ash (F/c = 0.25) at an age of 28 days at temperatures of 21.4, 71, 149 and 232 °C /N 1/***

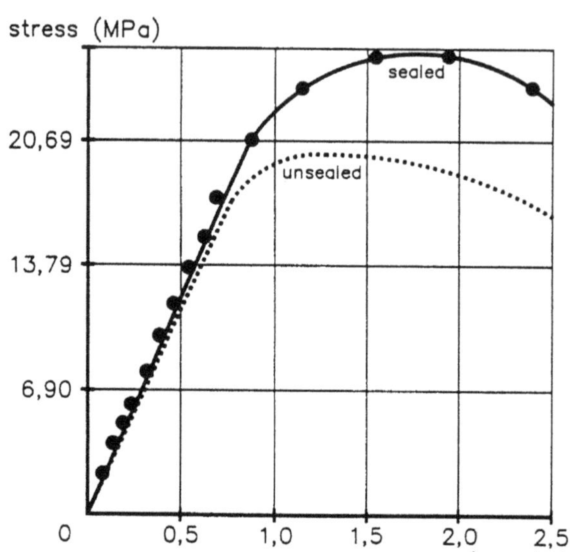

Fig. 3.2.3: ***σ–ε curves of concrete containing fly ash (f/c = 0.25) with blast furnace slag cement at an age of 200 days at a storage temperature of 21.4 °C /N 2/***

$$\sigma/\sigma_0 = a(\varepsilon/\varepsilon_0) + b(\varepsilon/\varepsilon_0)^2 + c(\varepsilon/\varepsilon_0)^3 \quad (3.2.1)$$

where
- σ_0 is the maximum stress
- σ is the stress
- ε_0 is the strain at maximum stress
- ε is the strain and
- a, b, c are constants depending on the temperature range and the corresponding storage time.

Both authors in /N 1/ report in /N 19/ on similar investigations made on concrete with brown coal fly ash and blast furnace slag cement /Type V/. The specimens were tested under conditions similar to those in mass concrete, i.e. at high steam pressure (sealed) as well as at atmospheric pressure (unsealed). The compressive strength of concrete with fly ash at 200 days (storage temperature 21.4 °C) was 22.3 MPa unsealed and 27.8 MPa sealed. According to /N 2/, the higher temperature does not affect the typical parabolic shape of the $\sigma-\varepsilon$ graph *(Fig. 3.2.3)*.

Effect of Admixtures

/S 23/ reports on the effects of superplasticizers on the early strength of concrete containing fly ash and on the $\sigma-\varepsilon$ curves for various fly-ash contents (f/c = 0, 0.20, 0.40). Two different types of plasticizer were used, proportioned for maximum strength. The $\sigma-\varepsilon$ curves for the concretes with and without fly ash were similar both for the melamin-based plasticizer and for the naphthalene-based type.

Other Influences

It is impossible to make general statements concerning the shape of the $\sigma-\varepsilon$ curve for fly-ash concrete as a function of compressive strength, f/c ratio, pore volume of the hardened cement paste, w/(c + f) ratio and fly-ash composition in the basis of current literature.

3.2.1.2 Modulus of Elasticity

The modulus of elasticity is an essential characteristic of a building material, indicating the relation between stress and the resultant elastic strain. In addition, the modulus of elasticity is also relevant to time-dependent deformation (creep), since creep is generally related to elastic deformation.

3.2.1.2.1 Effect of Compressive Strength

The moduli of fly-ash and non-fly-ash concretes which have the same compressive strength at the same time would seem to be of interest. However, much research has centred on fly-ash and non-fly-ash concretes of identical workability, making it impossible to assess this aspect.

Fig. 3.2.4 from /G 4/, shows the modulus of elasticity as a function of compressive strength for concretes with and without fly ash. As will be apparent, there is little difference between the two concretes for a given compressive strength.

According to /S 49/, there are slight differences between the elastic properties of fly-ash and non-fly-ash concretes at 28 days and beyond. The equation as a function of compressive strength and the bulk density (Australian Standard 1470)

$$E_c = 0.043 \cdot \rho^{1.5} \cdot f'_c{}^{0.5} \text{ in MPa} \tag{3.2.2}$$

fails to show a definitive difference between calculated moduli of elasticity and those determined by tests. The same report states that the fall in bulk density when part of the portland cement is replaced by fly ash is counterbalanced by the lower water demand of the fly-ash concrete.

In /H 30/, the dynamic moduli of concretes containing high proportions of fly ash were measured. (No tests were carried out on reference concretes without fly ash). *Fig. 3.2.5* shows the results together with a regression curve. Since the bulk density varied only slightly ($\bar{x}$ = 2260 kg/m^3, v = 1.7%), only the modulus of elasticity was correlated to the compressive strength.

In /F 27/, very lean concrete (cement content from 38 to 103 kg/m^3) with a high fly-ash content (f/c = 1.5 to 4.0) was tested for its suitability for use in road base construction. *Fig. 3.2.6* shows the relation between the dynamic modulus and compressive strength for coarse sand and crushed limestone aggregate.

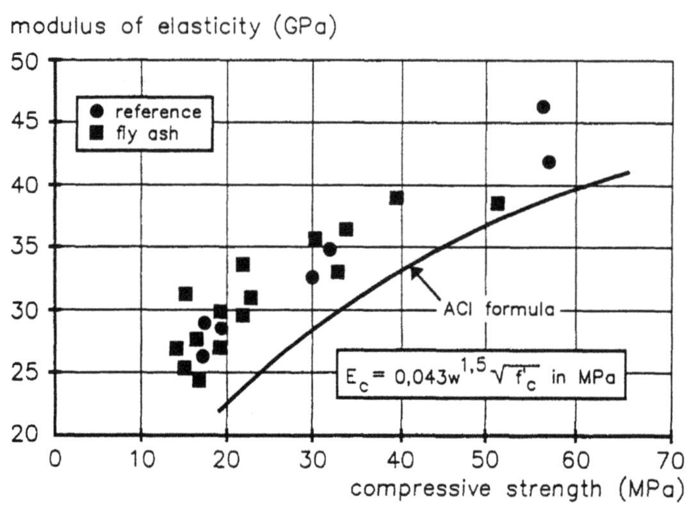

Fig. 3.2.4: Relationship between modulus of elasticity and compressive strength of concretes without and with fly ash /G 4/

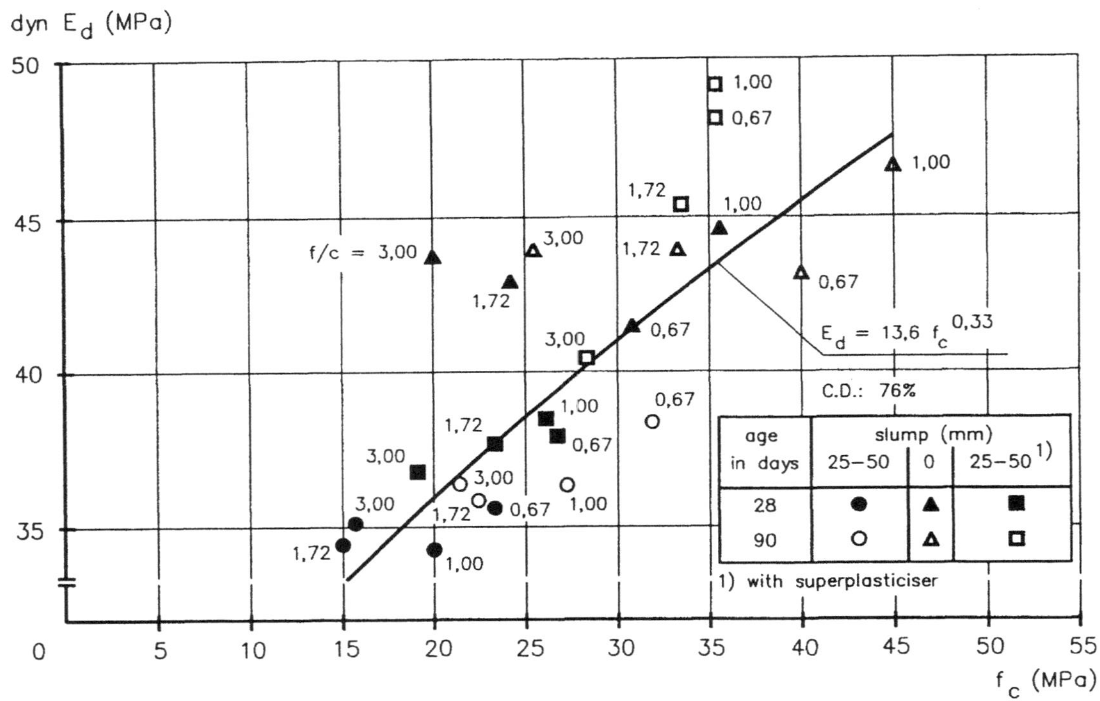

Fig. 3.2.5: Dynamic modulus of elasticity dyn E as a function of the compressive strength f_c of concrete containing fly ash at various ages /H 30/

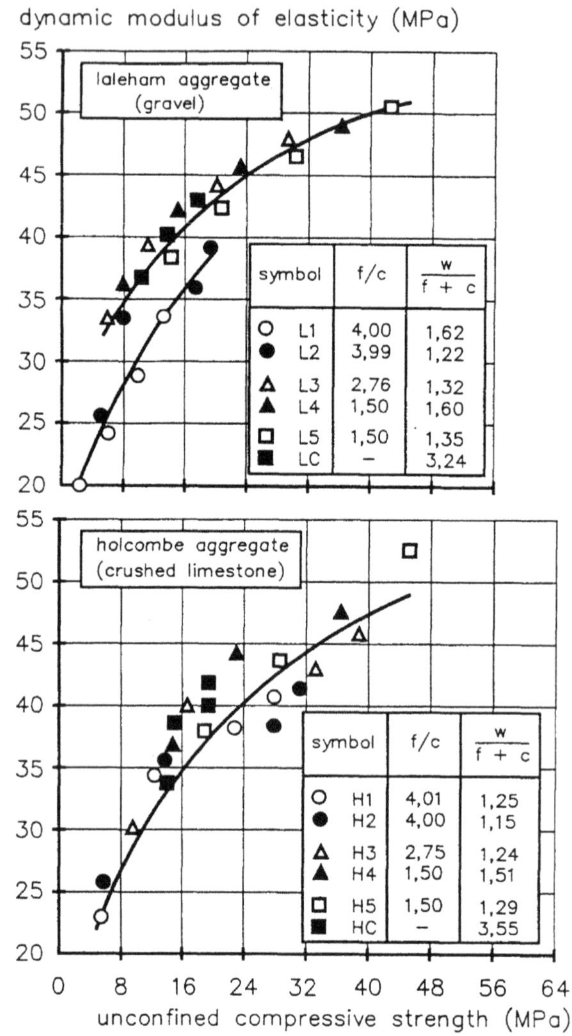

Fig. 3.2.6: Relation between the dynamic modulus of elasticity and compressive strength of lean concrete with high fly-ash content /F 27/

In /W 12/, it is concluded from an examination of the tests in /L 8/ that the moduli of elasticity of concrete with fly ash (f/c ≤ 0.75) are usually about 5 to 10 % higher at 28 and 90 days than the moduli of concretes of equivalent strength which do not contain fly ash.

Results from several publications /A 3, B 4, B 12, G 4, L 8/ are indicated in *Fig. 3.2.14*, which shows that the moduli of elasticity of fly-ash concretes are about the same or up to 10 % higher than those for ordinary concrete. There is no evidence of any effect on the modulus of elasticity at relative compressive strengths in the 0.8 to

1.2 range. Consideration of these results together with the other values in this figure does, however, indicate a slightly increasing relation between rel E and rel f_c.

3.2.1.2.2 Influence of Age

The effect of age on the modulus of elasticity of concretes containing plasticizer was investigated in /B 25/. It is apparent from *Fig. 3.2.7* that the modulus of elasticity of concrete with fly ash (f/c = 0.30) is reduced at early ages but overtakes that of non-fly-ash concrete at greater ages.

The relation between the modulus of elasticity and age can be expressed /W 20/ by the formula:

$$E_t/E_{28} = a \cdot \exp(b/t^c) \qquad (3.2.3)$$

where
a is the ratio of final elasticity (E_∞/E_{28})
b is a parameter controlling the function
c can according to /W 20/, be taken as 0.60 for normal concrete. Since no exact values are available, c was here assumed to be 0.60 for concrete containing fly ash.

The values for the parameter (*Table 3.2.1*)show that the rate of increase in the modulus of elasticity beyond 28 days is greatest for concrete containing fly ash.

According to /L 27/, the moduli of elasticity of concretes containing fly ash increase more with advancing age than those of concretes with no fly ash, echoing the development of compressive strength. This may be because of the pozzolanicity /W 1, W 2/ of the fly ash, its gradual incorporation into the binder matrix often resulting in a large increase in the modulus of elasticity between 28 and 90 days.

Table 3.2.1: Parameters a and b from Equation (3.2.3) for results in /B 25/ (see Fig. 3.2.7)

Parameter	OPC	OPC/Ad	OPC/fly ash	OPC/fly ash/Ad
a	1.16	1.20	1.42	1.35
b	-1.07	-1.32	-2.06	-2.23
CD*	88.7	99.0	95.7	91.0

* Coefficient of determination (%)

3.2.1.2.3 Effect of Fly-Ash Content

Fig. 3.2.8, based on results from several sources, shows that the modulus of elasticity is little affected by the fly-ash content. According to /H 33/, the modulus of elasticity is slightly reduced by increasing fly-ash content.

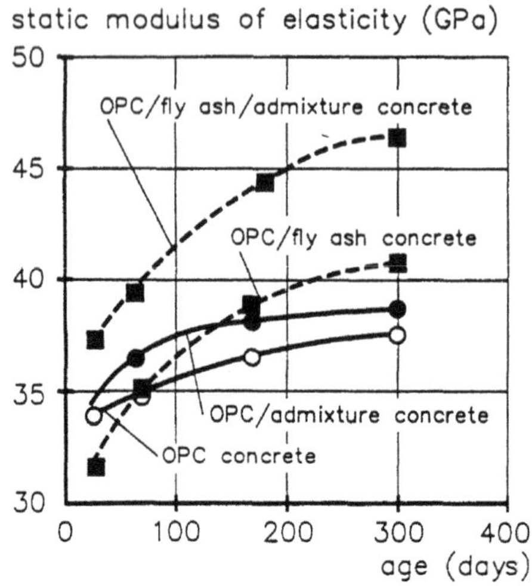

Fig. 3.2.7: Modulus of elasticity of concrete with and without fly ash as a function of age /B 25/

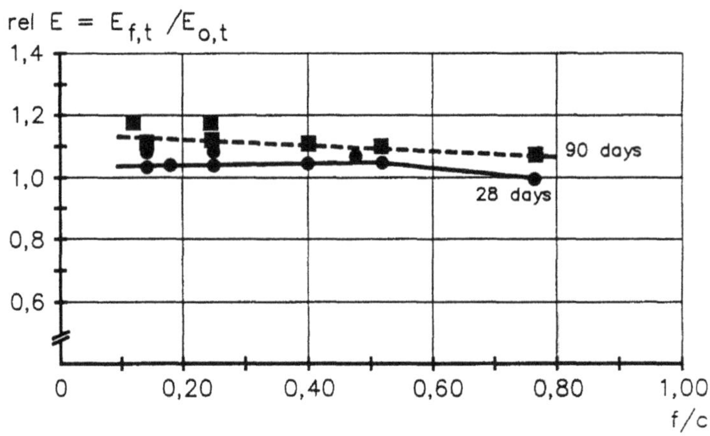

Fig. 3.2.8: Relative modulus of elasticity rel E as a function of fly-ash content f/c at an age of 28 and 90 days /B 25, G 4, L 8, Y 1/

3.2.1.2.4 Effect of Storage Conditions

Tests in /B 25/ were carried out on concretes with and without fly ash, the specimens being cured after demoulding (one day after casting) to an age of 28 days in an environment of 18 °C/95 % relative humidity and then either in water at 20 °C or in a 20/70 environment. The results reveal no relationship between the storage conditions and the modulus of elasticity for concretes not containing fly ash (*Fig. 3.2.9*).

However, for concretes containing fly ash, the modulus of elasticity increased little in the 20/70 environment but quite considerably when the specimens were stored in water. /B 25/ attributes the difference in behaviour to the slow pozzolanic reaction of the fly ash and hence to possibly greater water adsorption when stored dry.

Early heat treatment of concrete /B 41/ containing fly ash often leads to a higher modulus of elasticity for the same compressive strength (*Fig. 3.2.10*) and, from 7 days onwards, to a higher modulus of elasticity at the same age. The heat treatment simulated the conditions obtaining at the centre of a 2.5 m thick concrete section.

3.2.1.2.5 Effect of Temperature

In /F 14/, the modulus of elasticity of hardened cement paste made from pure portland cement and from a mixture of portland cement and fly ash (f/c = 0.35) was determined as a function of age at temperatures of 21, 35 and 55 °C. The specimens were stored at one of the three temperatures immediately after fabrication. /F 14/ does not, however, indicate whether the storage was in water or air. It is evident from *Fig. 3.2.11* that the modulus of elasticity of cement paste with fly ash increases with temperature up to 28 days and that the moduli of elasticity at 35 °C exceeded those at 55 °C. At all times, the moduli of elasticity at 35 °C were higher than those at 21 °C.

By contrast, the modulus of elasticity of hardened cement paste with portland cement not containing fly ash decreases with rising temperature. In /F 1/, this is attributed to delayed hydration resulting from the accumulation of hydration products already developed at higher temperatures around as yet unhydrated cement particles. Such coatings might be reduced by reaction of $Ca(OH)_2$ with fly ash,

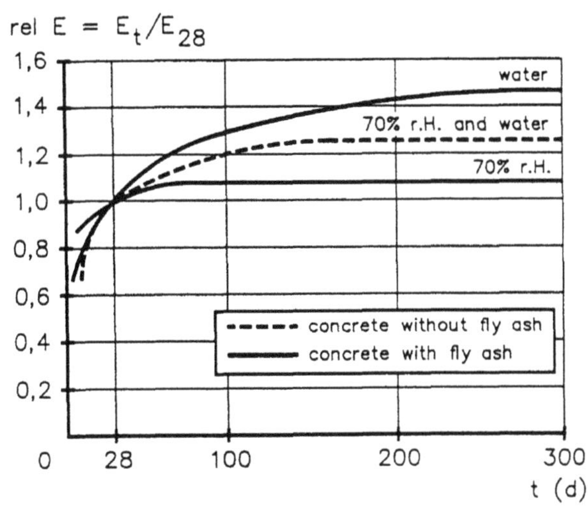

Fig. 3.2.9: Relative modulus of elasticity rel E of concrete without and with fly ash (f/c = 0.30) as a function of the storage method and the age t /B 25/

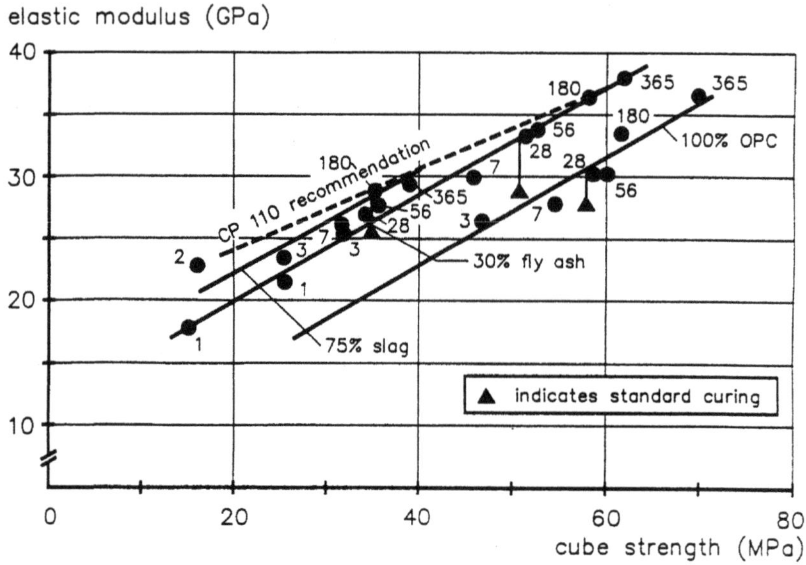

Fig. 3.2.10: Influence of heat treatment on the modulus of elasticity and compressive strength of concrete without and with fly ash /B 41/

allowing the hydration to proceed more rapidly.

It is essential to be able to forecast the deformation behaviour of mass concrete containing fly ash at higher temperatures. In /N 1/, concretes containing brown coal fly ash (f/c = 0.25) were exposed to various temperatures (21 to 232 °C) for between 7 and 180 days at an age of 28 days. Fly-ash-free concretes were not tested. *Fig. 3.2.12* shows that the modulus of elasticity fell steadily in relation to that at 21 °C after 28 days storage as the temperature rose. However, the compressive strength increased – by up to 48 % at 121 °C (*Fig. 3.2.13*).

The modulus of elasticity of concrete exposed to temperatures of 177 °C and 232 °C was much lower, falling by up to 75 % as compared to that at 21.4 °C; at 21.4 and 71 °C, the modulus of elasticity generally increased with storage time. According to /S 1/, the ACI Building Code relation between the modulus of elasticity and compressive strength is not affected by temperatures up to 232 °C for concretes without fly ash.

Results from /N 1/ show, however, that the relation between compressive strength and the modulus of elasticity for mass concrete containing fly ash is affected by temperature, particularly in the 121 to 149 °C range. The inference is that, as the temperature rises, the compressive strength initially increases, while the modulus of elasticity falls as a result of two overlapping chemo-physical processes. Firstly, under the conditions prevailing in mass concrete (high steam pressure and high temperature), high-strength tobermorite phases are formed in addition to the similar phases resulting from the reaction of fly ash and $Ca(OH)_2$, contributing to the increased

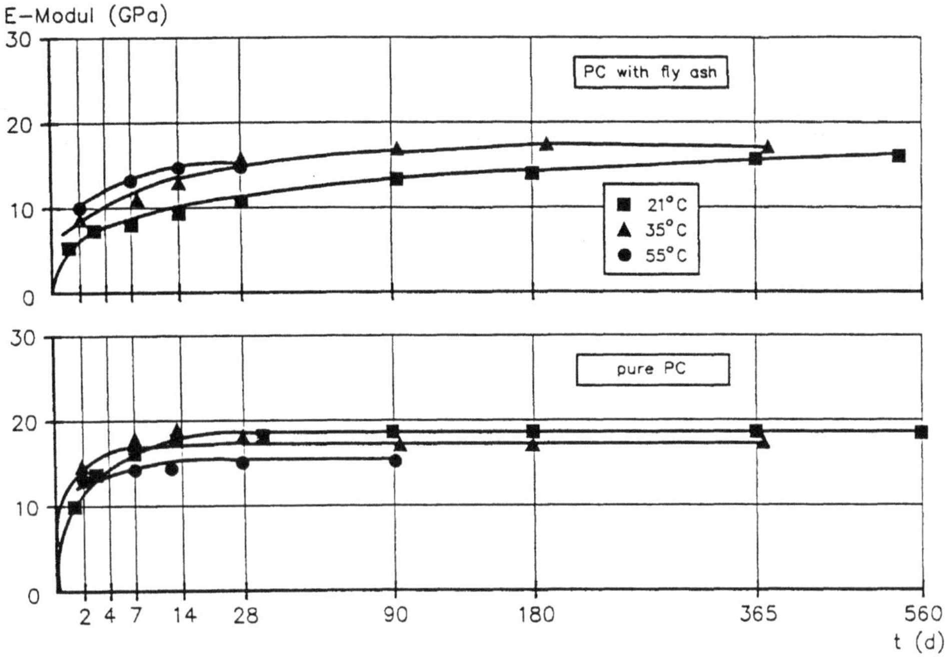

Fig. 3.2.11: Influence of temperature and hydration time on the modulus of elasticity of hardened cement paste with portland cement with and without fly ash

strength of concrete containing fly ash (cf. Section 3.2.1.1). Secondly, similar tobermorite phases are transformed into weaker α-Ca_2SiO_3 from about 100 °C upwards in concretes with and without fly ash, causing the modulus of elasticity to fall with increasing temperature. At even higher temperature (177 and 232 °C), the tobermorite is changed into crystalline α-Ca_2SiO_3, lowering the modulus of elasticity even further.

In another report /N 19/, the elastic properties of concrete containing a sulphate-resisting cement and brown coal fly ash (f/c = 0.25) were investigated. Two series of tests were carried out at temperatures from 21.4 to 232 °C: one under atmospheric pressure and the other under high steam pressure (simulating mass concrete). Both showed a fall in the modulus of elasticity with increasing temperature. Concrete without fly ash was not tested. The results in /N 19/ thus differ from those in /L 1/ for concretes made from portland cement (Types I and II) and fly ash, which showed that temperature had no effect on elastic properties at atmospheric pressure and temperatures up to 260 °C.

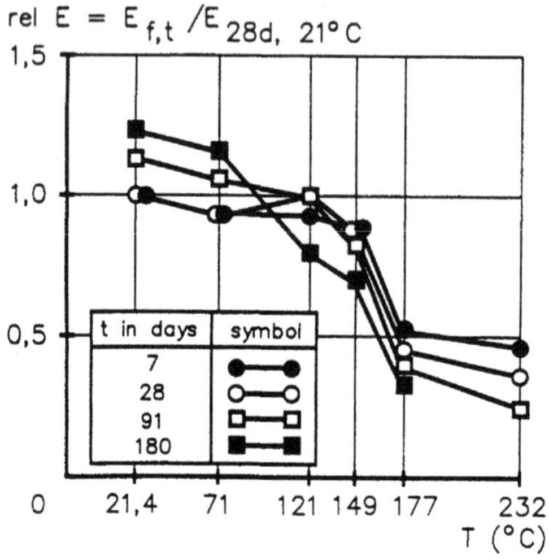

Fig: 3.2.12: Relative modulus of elasticity rel E of concrete containing fly ash (f/c = 0.25) as a function of the temperature T and the storage time t /N 1/

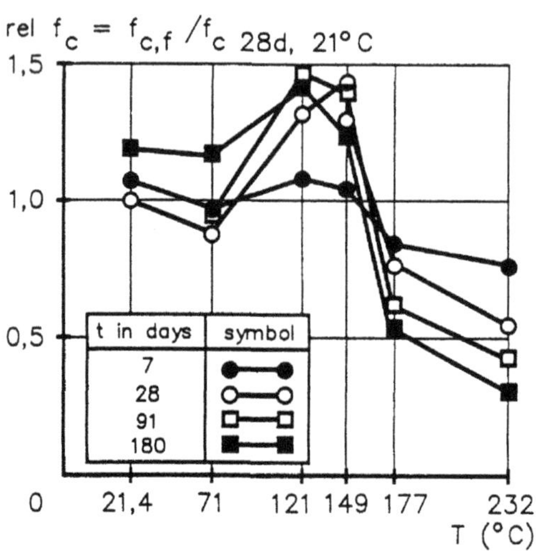

Fig: 3.2.13: Relative compressive strength rel f_c of concrete comtaining fly ash (f/c = 0.25) as a function of temperature T and storage time t /N 1/

3.2.1.2.6 Effect of Loss on Ignition of Fly Ash

Tests in /L 27/ on concretes containing fly ash with a loss on ignition (LOI) between 3.6 and 9.34 % by weight showed that the higher LOI led to a lower modulus of elasticity (*Fig. 3.2.14*). This was attributed to the higher water demand and consequently reduced compressive strength.

3.2.1.3 Ultimate Strain

The literature so far surveyed gives no information on the ultimate strain of concrete containing fly ash.

3.2.1.4 Summary

Various conclusions on the behaviour of concrete with and without fly ash can be drawn from the available literature.

Stress-Strain Curve

The $\sigma-\varepsilon$ curve of concrete containing fly ash is flatter at early ages. At later ages, the difference is less.

Storage temperature does not affect the typical shape of the $\sigma-\varepsilon$ curve. From 121 to 149 °C, the $\sigma-\varepsilon$ curve is more elongated, perhaps because of the increasing strength brought about by crystalline transformation.

Modulus of Elasticity

Beyond 28 days, the modulus of elasticity of concrete containing fly ash is not less than, and possibly up to 10 % higher than that of normal concrete of the same compressive strength.

The modulus of elasticity of concrete containing fly ash increases slowly with time, but more quickly than that of concrete without fly ash if no benefit has been obtained through the reduced water demand resulting from the use of fly ash.

The fly-ash content has little effect on the modulus of elasticity.

Storage conditions have a marked effect on the modulus of elasticity of concrete containing fly ash. For concrete stored in water, the modulus of elasticity is higher than that of concrete without fly ash. For concrete stored in air, the opposite is true – again provided that there is no benefit from the water-reducing properties of fly ash.

The influence of higher temperatures on the modulus of elasticity of fly-ash concrete seems uncertain. Some researchers found that the modulus of elasticity was unaffected by temperature (up to 260 °C) while others found it was reduced considerably (up to 230 °C).

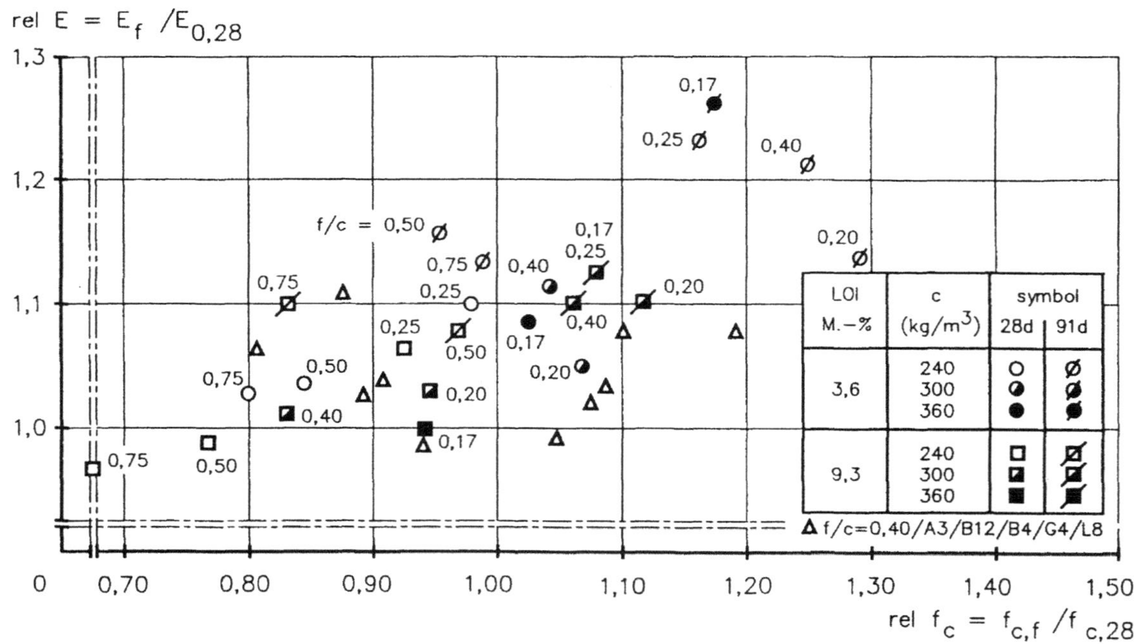

Fig. 3.2.14: Relationship between the relative compressive strength rel f_c and relative modulus of elasticity rel E of concrete without and with fly ash as a function of loss on ignition LOI /L 27/

Fly ash with high LOI (about 10 % by weight) yielded a lower modulus of elasticity.

3.2.2 Deformation Behaviour in Tension (Prepared by P. Schubert)

3.2.2.1 Modulus of Elasticity

The deformation behaviour in tension is usually determined by the so-called modulus of rupture, which is the ratio of maximum tensile stress to strain.

Tests in /G 13/ on the thermal behaviour of mass concrete with a high content of fly ash (f/c = 0,67) investigated the tensile strength of concrete with and without fly ash.

At 28 days, the compressive strengths of concretes with and without fly ash were 7.5 and 13.8 N/mm^2 respectively. Adding fly ash produced lower values for the modulus of elasticity, particularly with young concretes (see *Fig. 3.2.15*). In /G 13/, this is attributed to slower hydration of the concrete containing fly ash. At greater ages, the modulus of rupture approaches that of comparable concretes without fly ash.

There are no reports of other factors affecting the stress-strain behaviour of concrete with fly ash.

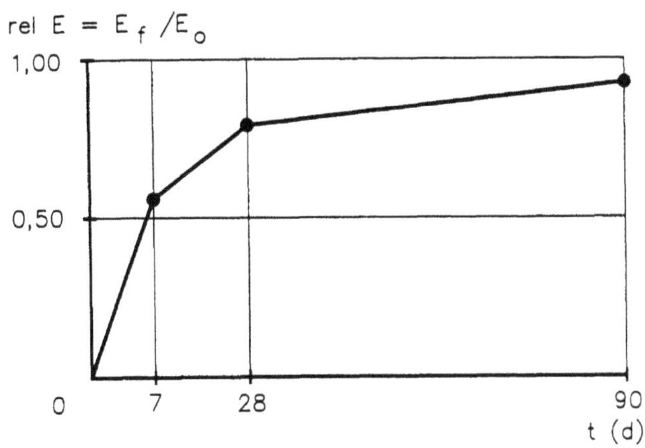

Fig.3.2.15: Relative modulus of rupture of concrete without and with fly ash (f/c = 0.67) as a function of age t /G 13/

3.2.2.2 Ultimate Strain

Effect of Age

Ultimate strain almost invariably increases with age (*Fig. 3.2.17*). Ultimate strain at early ages is less for fly-ash than for non-fly-ash concretes, but the difference decreases with time (*Fig. 3.2.16*). Eventually, the ultimate strain of concrete containing fly ash is about the same as or even greater than that of concrete without fly ash (*Fig. 3.2.17*).

Ultimate strain was calculated as a function of the strength in bending and the dynamic modulus of elasticity on the basis of tests in /F 23/ (*Fig. 3.2.17*). Some of the cement was replaced by fly ash (to reduce water demand and contribute to strength by pozzolanic reaction) and fly ash was also used as an admixture to obtain a predetermined 28-day strength. Whenever fly ash was added, the sand content was reduced to allow for the volume of the fly ash (lower apparent density).

Effect of Fly-Ash Content

Since the ultimate strain increases with higher cement contents /H 1/, it is suggested in /G 13/ that the same will be true for higher contents of cement + fly ash. The results from /F 23/ in *Fig. 3.2.17* support this hypothesis only after an age of about one year.

Other investigations reported in /H 30/ show that the ultimate strain is lowered by increasing contents of fly ash (*Fig. 3.2.18*). The ultimate strain was hence determined as a function of the ratio of bending strength to the dynamic modulus of elasticity.

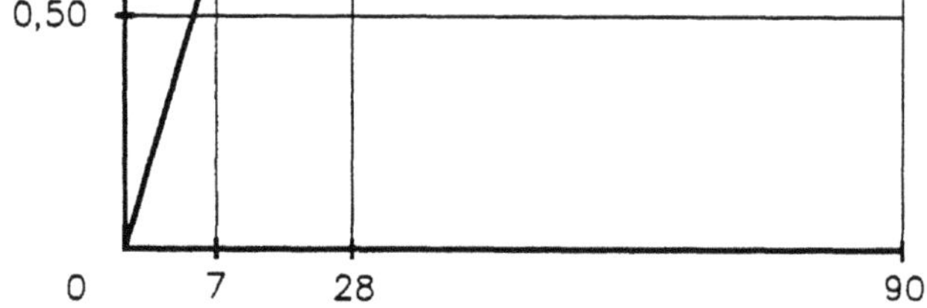

Fig. 3.2.16: Relative ultimate tensile strain rel ε_u of concrete with and without fly ash (f/c = 0.67) as a function of age t /G 13/

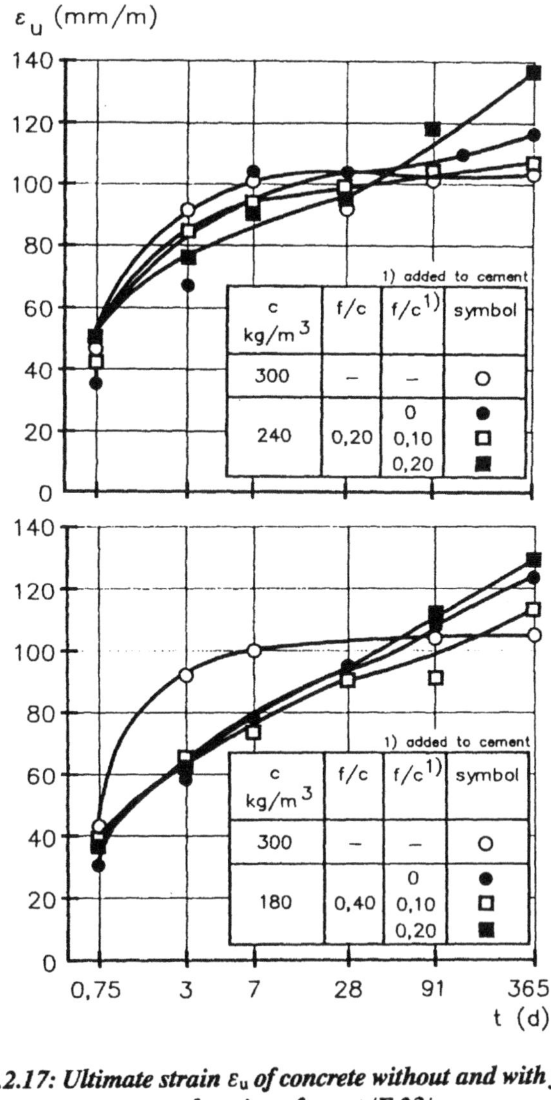

Fig.3.2.17: Ultimate strain ε_u of concrete without and with fly ash as a function of age t /F 23/

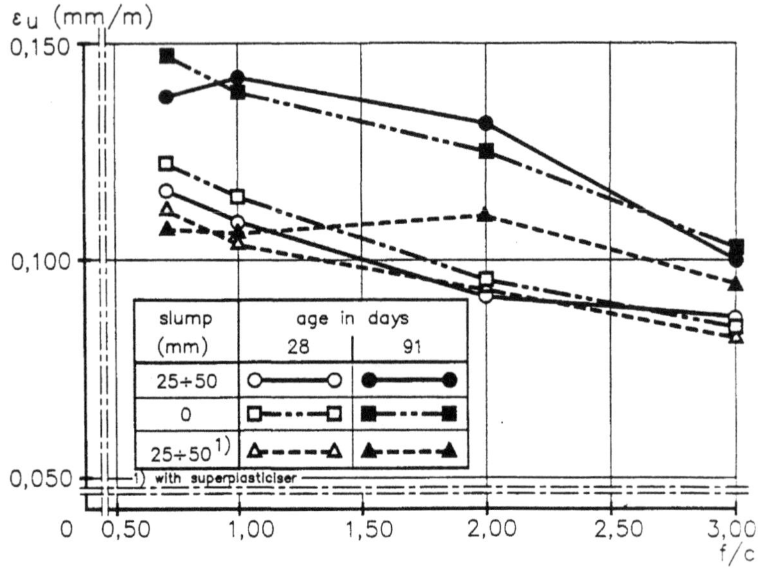

Fig. 3.2.18: Ultimate strain ε_u *as a function of the proportion of fly ash f/c and age /H 30/*

3.2.2.3 Summary

Specific results from the publications reviewed here are:

Modulus of elasticity

The modulus of elasticity (modulus of rupture) of concretes containing fly ash is lower at early ages. However, the difference decreases with advancing age.

Ultimate tensile strain

The ultimate tensile strain of concretes containing fly ash is lower at early ages. The difference reduces with time; the ultimate tensile strain of concretes with fly ash is eventually greater than for those without.

The ultimate tensile strain is reduced by high fly-ash contents.

3.2.3 Creep (Prepared by P. Schubert)

3.2.3.1 Terminology

Creep is defined as the increase in deformation with time as a result of a constantly applied stress. Within the range of permissible stress (up to about 40% compressive strength), the rate of creep diminishes steadily with time and after several years becomes almost zero. Whether the decrease is asymptotic to a final value is still disputed, and the causes of creep have as yet not been completely explained.

In tests, the creep strain ε_c is derived from the total strain ε_{tot} by subtracting the elastic deformation ε_{tot} under stress and the deformation ε_h due to moisture uptake without any stress being applied:

$$\varepsilon_c = \varepsilon_{tot} - \varepsilon_{el} - \varepsilon_h \qquad (3.2.4)$$

Depending upon the drying characteristics and concrete composition, the creep strain can be anything up to five times the elastic strain. This ratio of creep strain to elastic strain is known as the creep factor (coefficient of creep) φ:

$$\varphi = \varepsilon_c/\varepsilon_{el} \qquad (3.2.5)$$

Another characteristic quantity is the ratio of creep strain to creep stress:

$$\alpha_c = \varepsilon_c/\sigma_c \text{ (specific creep)} \qquad (3.2.6)$$

It is important to differentiate between basic creep (without simultaneous drying) and drying creep (with simultaneous drying).

Among the factors affecting creep strain are;

- climate (temperature, relative humidity)
- effective thickness of test specimen/structural member 2 x A/u (A = area of section, u = circumference of section)
- age of concrete on loading
- type and strength grade of cement
- loading.

It is assumed that the creep increases in proportion to stress within the serviceable stress range.

3.2.3.2 Time Dependence

In the literature surveyed, creep was investigated over periods up to one year. A semi-logarithmic graph is often used to characterize time-dependence or creep /B 41,

L 9, Y 2/. The plot is almost linear. To allow for the discontinuity at t = 0, time-dependence is expressed in terms of (t + 1):

$$\varepsilon_c = a + b \cdot \ln(t + 1) \qquad (3.2.7)$$

In /Y 2/ an expression containing three parameters is used:

$$\varepsilon_c + \varepsilon_{el} = a \cdot [\ln(t + 1)]^2 + b \cdot \ln(t + 1) + c \qquad (3.2.8)$$

The expression from the ASTM standard

$$\varepsilon_c = (t^{0,6}/(10 + t^{0,6})) \cdot K \qquad (3.2.9)$$

is investigated in /Y 1/. For $f/c \leq 0.25$, K was 2.4 and, for $f/c \leq 0.43$, K was 3.6.

To permit comparison of results from several sources, the data were extrapolated to one year, using the factor $\ln(365)/\ln(t_n)$.

3.2.3.3 Influences Investigated

Comparison between basic and drying creep

Tests in /B 25/ on concrete with and without fly ash (f/c = 0.54) yielded a ratio of 0.5 between the creep factors for basic and drying creep. The fly ash had no practical effect.

Workability

In most of the work reported, workability (slump) was kept constant /B 25, B 41, G 4, L 9, N 19/. In /G 4/, a fall in creep was observed for concrete containing fly ash, given equal slump. This was attributed to the reduction in water/cement ratio, leading to increased gel formation and hence to a rapid reduction in the free water content responsible for creep. As the proportion of gel increases, the stress resulting from constant load on the gel decreases and the creep falls. On the other hand, the cement and fly ash produce more of the gel, which is susceptible to creep. The net result of these two opposing phenomena seems to be that the reduction of stress on the gel is dominant.

Fig. 3.2.19 shows the creep factor as a function of the w/(c + f) ratio. At constant slump, the ratio – and hence the creep factor – fall due to the fly ash. The increase in creep factor in the basic creep research reported in /B 25/ is attributed to extraordinary swelling of the specimen, which was stored unsealed in water.

Fig. 3.2.20 is based on results from /Y 1/ and /Y 3/. Since no data are given on the calculation of the creep factor, only the creep strain can be shown. In the tests, w/(c + f) was kept constant at 0.38, so that the slump increased with rising fly-ash content. The creep factor increased with the f/c ratio.

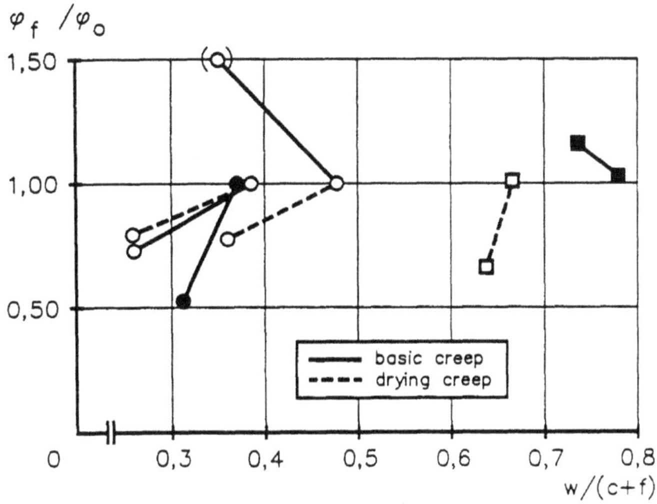

source	symbol	slump	f/c
B25	○	40–60	0,54
G13	■	25–75	0,67
B41	●	80–85	0,43
G4	□	83–121	0,40

Fig. 3.2.19: Effect of the w/(f + c) ratio on the creep factors of concrete without (φ_0) and with (φ_f) fly ash at equal consistency (slump)

Plasticizer and air-entraining agents were used in concretes with and without fly ash and with fly-ash cement (f/c = 0.25) having the same w/(c + f) ratio and the same slump /Y 2/. Three different aggregates were employed. In all cases, the creep factor of the concrete containing fly ash was 14 to 38 % lower than that of the non-fly-ash concrete. Little difference was found between concretes made with fly-ash cement and with fly-ash additive respectively.

Type of fly ash

Concretes with and without fly ash were tested in /G 4/ and /G 13/. A range of fly ashes (LOI 6.3 to 18.2% by weight) were used, with f/c = 0.40; other mixes had very different properties (f/c = 0.67) but equal slump (85 to 95 mm). One mix had a slump of 120 mm. In /G 4/, the drying creep factor for concrete with fly ash was 12 to 37 % lower than for concretes without. The concretes made from the two fly ashes with the highest LOI (13.1 and 18.2 %) had larger creep factors than those containing the fly ashes with lower LOI (6.3 and 7.2 %) – see *Fig. 3.2.21.*

In /G 13/ no real difference could be found in the basic creep of mass concretes having low cement contents and made from two different fly ashes. In all three test series, the creep factors were very similar.

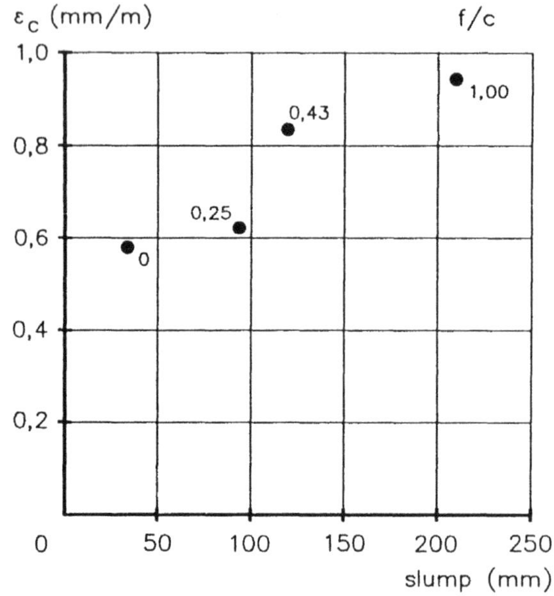

Fig. 3.2.20: Effect of the addition of fly ash on the creep strain ε_c at constant w/(c + f) ratio /Y 1/

Fly-ash content

Reference has already been made to the work reported in /Y 1/ (*Fig. 3.2.20*). At constant w/(c + f) ratio the creep strain increased with fly-ash content. Unfortunately, the creep factor could not be determined. In the investigations in /L 9/, the sum of the elastic and creep strains was determined on concretes with and without fly ash (f/c ≤ 0.33, with constant cement proportion) after 150 days under load in a 27/90 environment. The slump was 40 mm in each test concrete. Compared to concrete without fly ash, the total strain increased only slightly – 9% at the most.

In /G 4/ (cf. above), where the slump was also constant but the environment (23/50) drier than in /L 9/, the creep factor at f/c = 0.4 and 1.0 was up to 40 % lower than for concrete without fly ash. /G 4/ explains this as being due largely to the greater strength development of the concrete containing fly ash and to the attendant fall in the ratio of creep stress to compressive strength.

Type of aggregate

The effect on creep of various different sands as constituents of concretes without and with fly ash (generally as a replacement for cement) was examined in /Y 2/. For all three types of sand, creep was reduced when fly ash was included either as fly-ash cement or as an admixture.

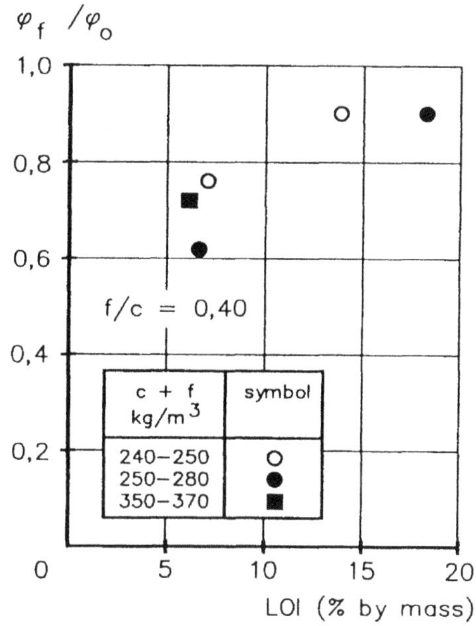

Fig. 3.2.21: Effect of fly ash of varying amount of loss on ignition LOI on the creep factors of concretes without (φ_o) and with (φ_f) fly ash at fairly equal consistency (slump) /G 4, G 13/

Admixtures

The effect of plasticizers on the basic and drying creep of concrete with and without fly ash as a cement substitute was one of the aspects examined in /B 25/. Unfortunately, the specimens stored in water swelled considerably, making it hard to evaluate and compare basic creep. As regards drying creep, however, the creep factors of the concretes with and without plasticizer were the same for the reference concrete as for the concrete containing fly ash.

Age at loading

In /B 41/, heat-treated concretes with and without fly ash (f/c = 0.43 as a substitute for cement) and of approximately the same slump were subjected to creep stress at various ages. The test specimens were sealed and basic creep was examined. For concrete with fly ash, the effect of age at loading was much higher than for concrete without fly ash (*Fig. 3.2.22*). The creep strain and the rate of creep were higher at early loading and lower at later loading. This was attributed to the lower early strength and the higher later strength of the fly-ash concrete.

In /G 13/, which examined the basic creep of concretes with a low cement content, with and without fly ash (f/c = 0.25 and 0.67 as cement substitute), and of equal slump, the creep factors for concrete with fly ash were up to 40 % higher at 7 days.

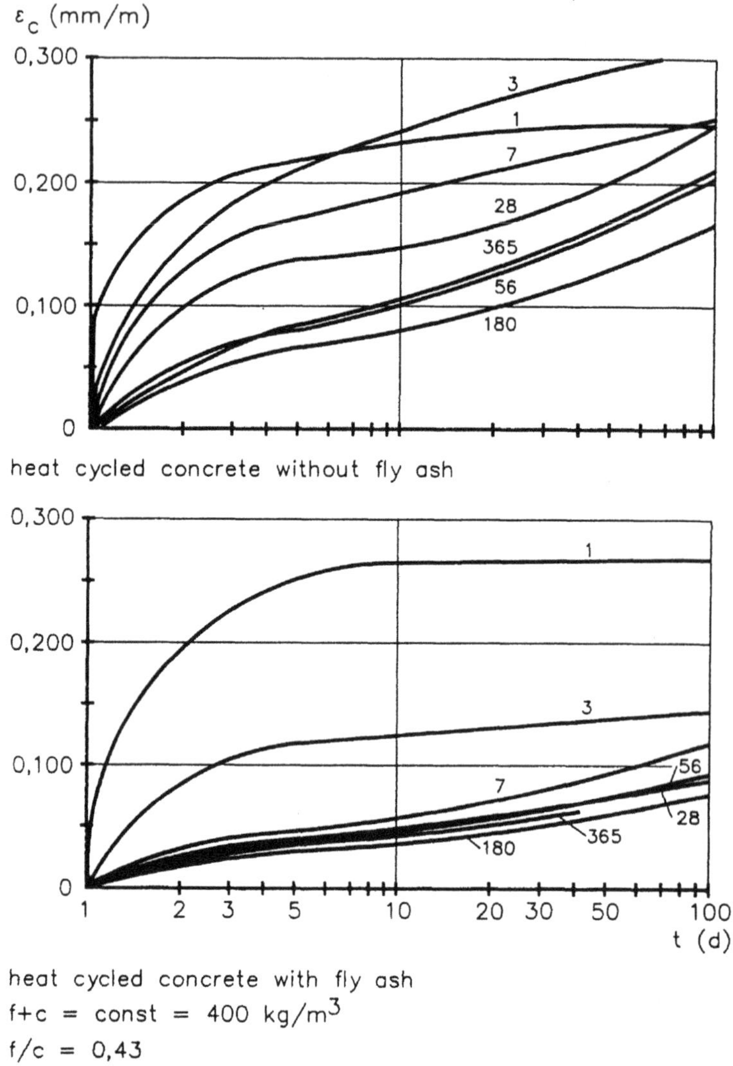

Fig. 3.2.22: Effect of the age of loading on the creep strain ε_c in relation to period of loading t /B 41/ (σ/f_c = 0.25, numbers on the curves indicate the age of loading in days)

At 90 days, the creep factors for fly-ash concrete were usually lower than for non-fly-ash concrete. The elastic and creep strains for fly-ash concretes were lower than those for non-fly-ash concretes, particularly with early loading. According to /G 13/, this is because the fly ash behaves as aggregate at early ages but later reduces the cement lime content, causing the creep.

Effect of the degree of stress

In /L 9/, concretes with and without fly ash ($f/c \leq 0.33$, fly ash as admixture) and of equal slump showed a relative creep stress σ_c/f_c of between 0.20 and 0.35. The results show that, in this range, the creep stress and the total strain (elastic plus creep strain) are roughly proportional for fly-ash and non-fly-ash concretes.

Temperature

The effect of temperature on the basic and drying creep of concrete made from sulphate-resisting cement and containing fly ash was investigated in /N 19/. *Fig. 3.2.23* shows the results.

For basic creep, the creep strain fell steadily with increasing temperature apart from an increase at 177 °C. According to the authors, this behaviour differs from that of concrete made from cement Types I and III and containing fly ash. The behaviour was, however, the same in the case of drying creep. In the latter case, with one exception, creep also fell with rising temperature. With drying creep, the maximum occurred at 71 °C, a temperature much lower than for basic creep.

In /N 19/, the effect of temperature at atmospheric pressure is explained as follows:

Temperature range	*Cause of creep*
$T \leq 70\ ^\circ C$	Diffusion of adsorbed moisture, physical change in gel
$T > 70\ ^\circ C$	Some of the adsorbed water evaporates and the gel remains the only deformable phase
$T > 120\ ^\circ C$	Mobility of the solid part of the gel, loss of bond between the hardened cement paste and aggregates.

3.2.3.4 Recovery from Creep

Because of the increase in strength during the creep tests and the change of texture resulting from creep, the modulus of elasticity is greater at relaxation than at loading. /G 4/ (in which f/c was generally 0.40 as a cement substitute and the slump was constant) reported that the elastic recovery ε_{rel} was 74 to 93 % of the elastic deformation at loading ε_{el}. The type and quantity of fly ash had no appreciable effect.

In the tests reported in /Y 3/ ($f/c \leq 1.0$ as a cement substitute, $w/(c + f)$ = constant), $\varepsilon_{rel}/\varepsilon_{el}$ was greater for concretes with fly ash than for those without.

Creep recovery stops more quickly than creep. Tests in /L 9/ indicated less recovery for $f/c > 0.18$ than for smaller amounts of fly ash.

The effect of temperature (21 to 232 °C) on fly-ash concretes was studied in /N 19/. Following heat treatment, the scatter in the deformation due to creep recovery was greater than in the creep strains. This was explained as being due to the development of microcracks between the aggregates and the hardened cement paste. In tests on

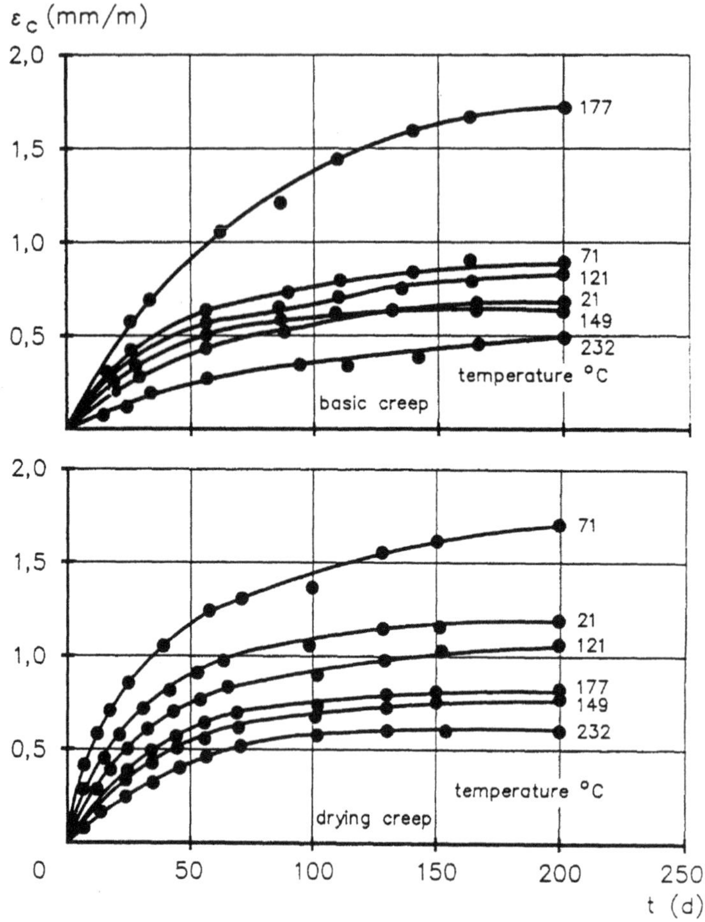

Fig 3.2.23: Basic and drying creep ε_c of fly-ash concretes as a function of time t at various temperatures /N 19/

creep at temperatures of 20 to 70 °C, the creep recovery was attributed to the combined effect of the movement of adsorbed water into the gel and the delayed elastic deformations of the solid phase. In creep tests at temperatures above 120 °C, the effect of adsorbed water movement was negligible. Creep recovery is essentially due to the delayed elastic effect of the crystalline gel phase.

3.2.3.5 Summary

No systematic fundamental research has been carried out into the effect of fly ash on the creep behaviour of concrete. Most tests have been limited to special applications

tions (mass concrete, reactor concrete, economies in cement, etc.). In most cases, therefore, results have to be evaluated individually.

Essentially, the creep behaviour of concrete with and without fly ash is the same.

Because of the fluidizing effect of fly ash, at constant slump the w/(c + f) ratio falls as the fly-ash content increases, leading to a lower creep factor than for concretes without fly ash. This also occurs if the w/(c + f) ratio **and** the slump are kept constant by using air-entraining agents and plasticizers. As compared to those for ordinary concretes, however, the creep factors of fly-ash concretes increase with increasing fly-ash content for a given w/(c + f) ratio.

Because of the generally lower early and higher later strength development of concretes made with fly ash, their creep factors at early loading are greater than for ordinary concrete (and lower at later loading).

The type of fly ash in terms of its LOI was not found to affect creep.

To summarize, the following are not influenced by the incorporation of fly ash in concrete:

(a) the ratio of the creep factors of basic and drying creep
(b) the behaviour of creep recovery (the relation between elastic creep recovery and elastic strain on loading)
(c) the proportional relationship between creep strain ε_c and relative creep stress σ_c/f_c.

3.2.4 Moisture Deformation (Prepared by P. Schubert)

3.2.4.1 Definitions and Process

The deformation due to moisture movement (ε_h) consists of shrinkage ε_s due to moisture evaporation and swelling ε_{sw} due to moisture adsorption. Because of the higher moisture content of mortar and concrete at the time of placing and the subsequent moisture evaporation, shrinkage is more common than swelling. Moreover, since tensile stresses and cracks can arise from contraction, shrinkage is more important than swelling.

The parameters influencing moisture deformation (shrinkage) are:

- environment (temperature, relative humidity)
- size of the element
- age at which shrinkage begins
- cement and binder content
- w/c ratio and w/(c + f) ratio
- cement characteristics
- presence and amount of ultrafine aggregate particles

where c, w and f denote the cement content, water content and fly-ash content.

Shrinkage and swelling are time-dependent. For a constant environment, the rates of shrinkage and swelling decrease with time; the decrease is usually asymptotic toward a final value.

3.2.4.2 Evaluation of Available Results

In the literature studied, shrinkage was investigated up to a maximum age of one year. Swelling tests covered periods of up to 270 days. The development in time of the moisture deformation has not yet been described mathematically in the published literature. In general, the deformations determined at the end of the tests are compared for various mixtures and storage conditions. The test results are evaluated in the following sections.

In publications with more than three shrinkage values at different ages, the final shrinkage value was here determined according to the hyperbola formulation by Ross:

$$\varepsilon_s(t) = t/(a + b \cdot t) \qquad (3.2.10)$$

where $\varepsilon_{s\infty} = 1/b$.

The final shrinkage values determined in this way were invariably higher than the values at the end of the tests.

The following observations refer solely to moisture deformation. It was impossible to estimate superimposed influences such as shrinkage due to carbonation or chemical shrinkage on the basis of existing publications.

3.2.4.3 Influences on Shrinkage

3.2.4.3.1 Workability

In most cases, tests were carried out at constant flow and slump values /B 25, B 41, G 13, N 11, Y 2/.

In /B 41/ (*Fig. 3.2.24a*), there was slower shrinkage of fly-ash than for non-fly-ash concrete at a storage period between approximately 2 weeks and one year. However, at an age of approximately one year the shrinkage values for both concretes were equal. An extrapolation according to the CEB-FIP recommendations for an element 800 mm thick indicated less effect of fly ash up to an age of 30 years (*Fig. 3.3.24 b*).

In /B 25/, shrinkage and swelling of concretes (stored in water) with and without fly ash as well as with and without plasticizer were studied. Cement was replaced by fly ash (f/c = 0.54) and an extra, but smaller, amount was included as an admixture. All the mixes were of roughly equal slump and were sealed for a preliminary storage period of 28 days. After three months, the shrinkage of concretes containing fly ash was always lower than for those without (*Fig. 3.2.25*) This is attributed to the lower water content of the concretes containing fly ash and to their slower hydration. They

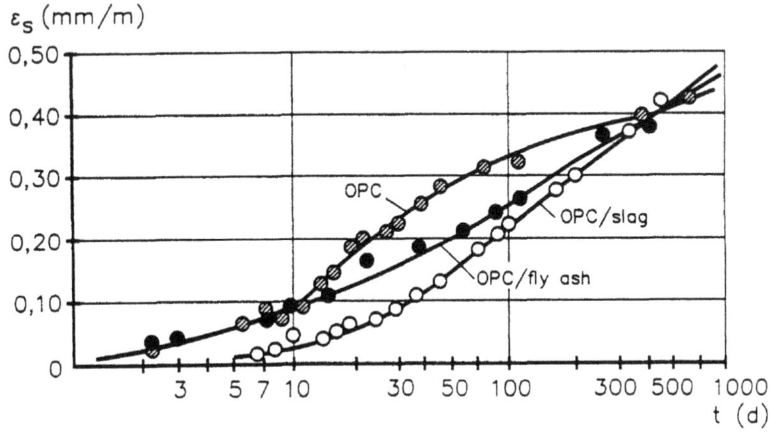

Fig.3.2.24a: Shrinkage ε_s of concretes with fly ash and blast furnace slag compared with control concrete as a function of storage time t /B 41/
Prisms:100 mm x 100 mm x 300 mm
Preliminary storage: 28 d at 20 °C, sealed
Shrinkage environment: 20/65

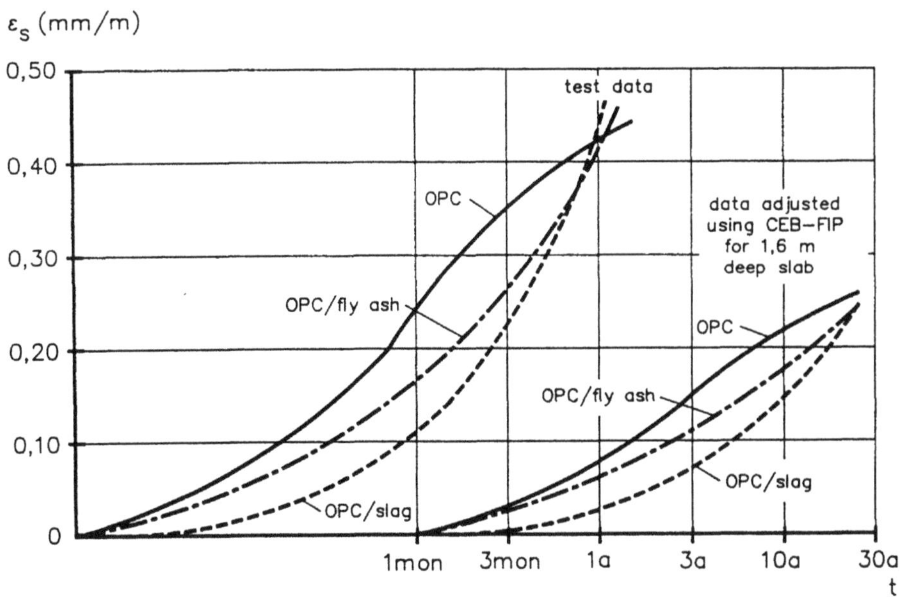

Fig. 3.2.24b: Shrinkage ε_s of an 800 mm slab as a function of storage time t, derived according to CEB-FIP-Recommendations /B 41/

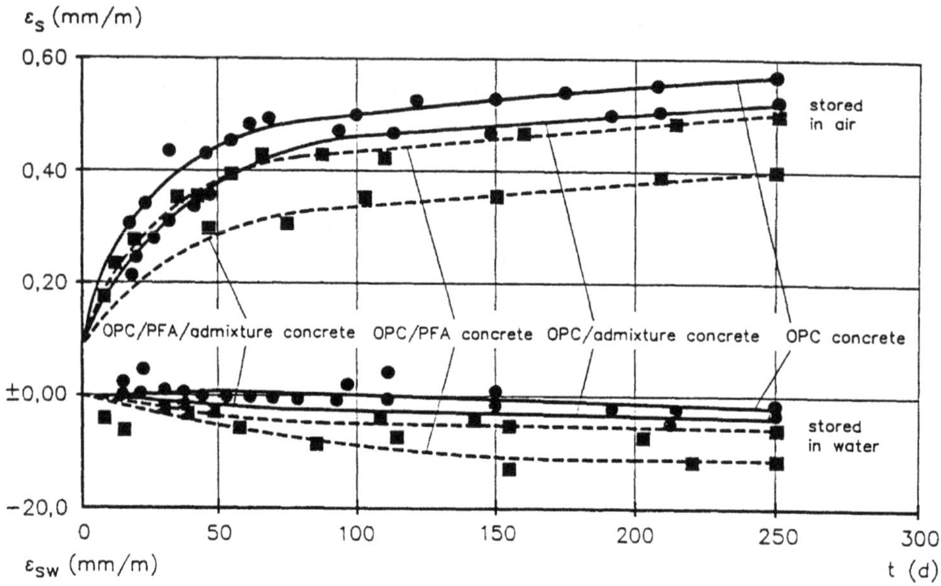

Fig. 3.2.25: Shrinkage and swelling ε_s, ε_{sw} of concrete without and with fly ash and without and with plasticizer as a function of storage time t /B 25/ Cylinder: 76 mm x 255 mm

also contained more of the components tending to reduce creep, i.e. fly ash, dehydrated cement particles, etc.

In /N 11/ and other publications the lower shrinkage of concretes containing fly ash is ascribed to the plasticizing effect of the fly ash and hence to the reduced w/(c + f) ratio.

In /G 13/, concretes containing fly ash were found to undergo greater shrinkage in spite of the water-reducing effect of the fly ash at equivalent workability. This could be explained by the fact that the initial measurement was made two days before 26 days of damp storage.

In /G 4, L 27, M 24/, tests were carried out on concretes with equal slump and compressive strength. In /G 4/ this was achieved by increasing the w/(c + f) ratio. With one exception, shrinkage of the concretes containing fly ash (f/c = 0.20, 0.40 and 1; LOI ≤ 7.3% by weight) was lower – by up to 30% – than that of concretes without fly ash.

The test series in /L 27/ (*Fig. 3.2.26*) revealed a maximum difference of 10 % in the shrinkage behaviour of concretes of equal consistency with and without fly ash. Test series with high fly-ash contents (in excess of 200 kg/m^3) showed up to 40 % more shrinkage than the corresponding reference concretes with no fly ash.

When the w/(c + f) ratio is kept constant and the fly-ash content is increased (f/c = 0.25, 0.43, 1.00), there is no practical difference in shrinkage for concretes with and without fly ash, according to the tests reported in /Y 1/ (*Fig. 3.2.27*).

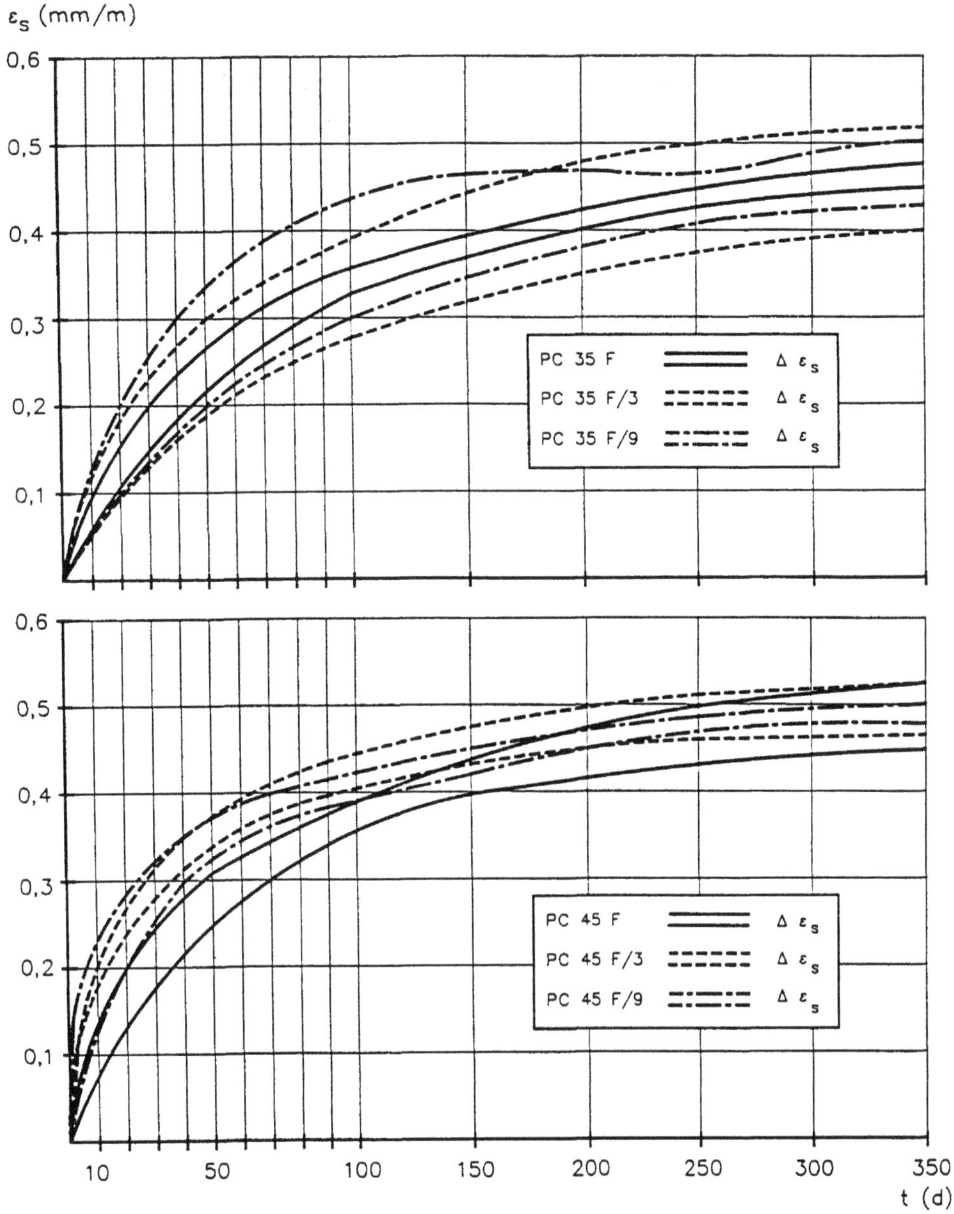

Fig. 3.2.26: Shrinkage ε_s of concrete (strength classes B15 up to B45 according to DIN 1045) without and with fly ash (LOI = 3.6 % (/3) and 9.3% (/9)) and PZ 35 F or PZ 45 F as a function of storage time t /L 27/. In the tests, great care was taken to maintain equivalent values for workability and strength. $\Delta\varepsilon_s$: Minimum — maximum range of ε_s values

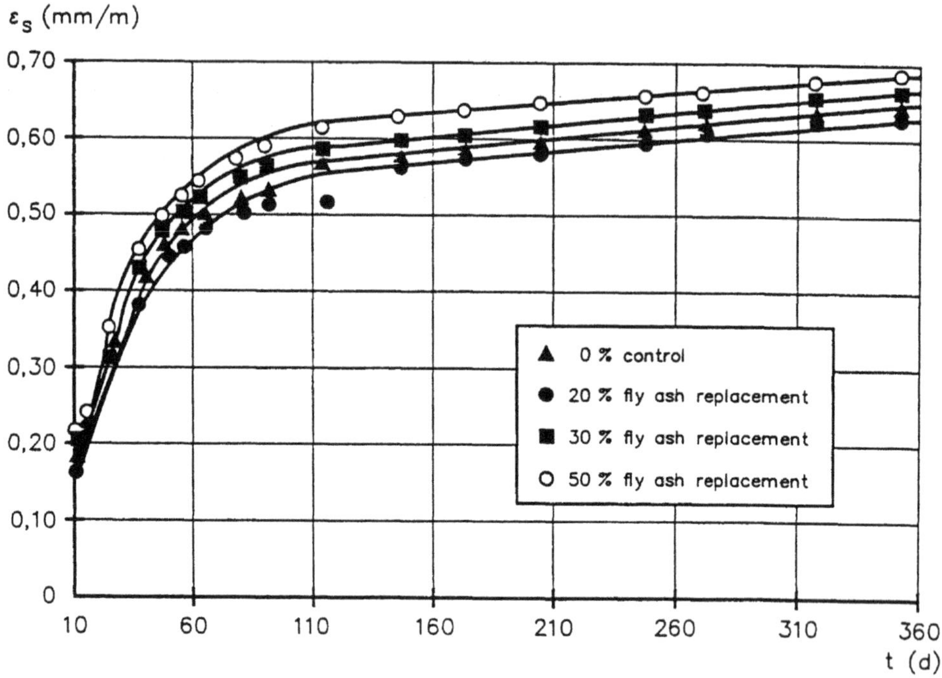

Fig. 3.2.27: Influence of fly-ash content on the shrinkage ε_s of concrete (f + c = const.) as a function of the storage time t /Y 1, Y 3/

In /V 2/, shrinkage tests were carried out on mortars of portland cement and fly-ash cements (two different portland cement clinkers, three different fly ashes, f/c = 0.25 to 9.00) at water/cement ratios of 0.50 and 0.60. In no case was the shrinkage of fly-ash cement mortar greater than that of portland cement mortar, and it was generally up to 35 % less.

3.2.4.3.2 Type of Fly Ash

A range of fly ashes was studied in tests reported in /F 2, F 6, G 4, L 27, M 24, V 2). The LOI and carbon content were assumed to be typical.

In the tests in /F 2/, fly ashes with lower carbon contents entailed lower shrinkage in the mortars made from them. In /L 27/ (see *Fig. 3.2.26*), a change in LOI from 3% to 9 % had scarcely any effect on shrinkage. Even concretes containing fly ash with 15 % LOI had virtually the same shrinkage.

In /F 6, G 4, M 24/, shrinkage was greater when fly ashes with higher LOI and carbon content were used. It is, however, important to note that the increased shrinkage is partly due to the higher w/(c + f) ratio.

In /V 2, shrinkage strains in mortars made with fly-ash cement were sometimes higher and sometimes lower, depending on the clinker type and fly-ash content, but were always less than those of the control mixes. The maximum LOI was in fact quite low, at 6.8 %.

3.2.4.3.3 Fly-Ash Content

In /A 3/ (*Fig. 3.2.28*), tests on lightweight concrete containing fly ash showed that shrinkage at high fly-ash contents (f/c = 1.08) was much lower than at f/c = 0.43. It should, however, be remembered that an increase in the f/c ratio also entails an increase in the cement and fly-ash contents, and that the w/(c + f) ratio was reduced.

In the tests in /B 16, N 13, P 5/ (*Figures 3.2.29 to 3.2.31*), adding fly ash reduced the shrinkage. *Fig. 3.2.30* from /N 13/ shows clearly that the shrinkage of concrete made from portland cement and fly-ash cement was reduced at higher f/c ratios.

In tests /F 6/ on concretes containing various different fly ashes (carbon content 1 to 17 % by weight) with f/c = 0.1 and 0.2 and at roughly constant w/(c + f) ratio, no clear effect of f/c on shrinkage could be deduced. This agrees with the results in /V 2/ on fly-ash cement mortars and in /F 2/ on mortars containing fly ash. There may be a minimum shrinkage for a given f/c ratio, but it should be pointed out that no details of either the test conditions or *Figures 3.2.29 and 3.2.31* were available, so that the minimum shrinkage rate could be due to a reduction in the w(c + f) ratio.

Tests in /H 60/ on concretes with high fly-ash contents (f/c up to 0.75) and roughly constant w/(c + f) ratio showed that the higher contents led to both higher and lower shrinkage strains than in the equivalent fly-ash-free concretes. The fly ash used had a high lime content (15 % CaO by weight, 8.1 % CaO_{free} by weight). Tests in /H 33/ on fly-ash and non-fly-ash concretes at three different w/(c + f) ratios (0.45, 0.55, 0.65) and with f/c = 0.2 and 0.6 showed both increased and reduced shrinkage. The differences were small (*Fig. 3.2.32*).

In /M 24/, shrinkages for concretes were mainly lower with increasing f/c ratio. This may, however, have been due to a higher aggregate content associated with the increasing f/c ratio. The concretes had equivalent strength and slump and a constant w(c + f) ratio, though with different amounts of c + f.

Conversely, tests in /G 13/ on fly-ash concretes with a low cement content indicated a generally higher shrinkage for concretes with a high (f/c = 0.67) than with a low (f/c = 0.25) fly-ash content.

/Y 1/ and /Y 3/, which tested concretes with f/c = 0.25, 0.43 and 1.00 and with constant c + f content and w/(c + f) ratio, found that fly-ash content did not affect shrinkage.

3.2.4.3.4 Type of Cement

The tests reported in /G 4, M 24, N 11, V 2, Y 2/ employed a range of different cements. The effect of fly ash can therefore be regarded as being independent of the cement type.

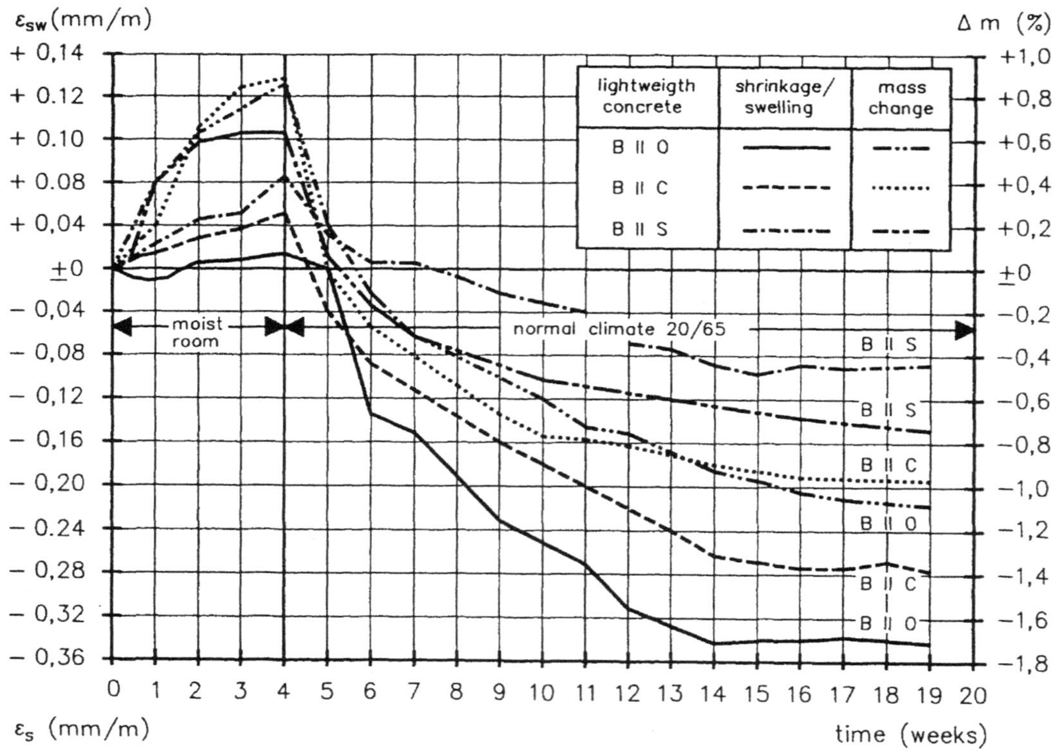

Fig. 3.2.28: Shrinkage ε_s, swelling ε_{sw} and mass change Δm as a function of storage time t /A 3/
Lightweight concrete with and without fly ash; Prisms: 100 mm x 100 mm x 400 mm;
Storage: 28 d in moist chamber, environment 20/99, covered with plastic sheetings, subsequently at an environment of 20/65

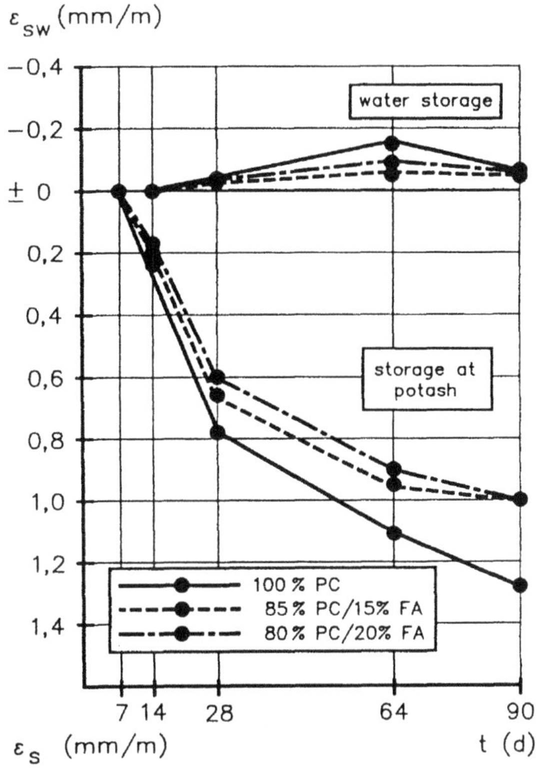

Fig. 3.2.29: Shrinkage and swelling ε_s, ε_{sw} of mortar prisms without and with fly ash as a function of storage time t (SiO_2 = 48.9 %, CaO = 7.92 %, LOI = 1.02 % by weight) /B 16/

3.2.4.3.5 Type of Aggregate

It can be assumed from /N 11, Y 2/ that the effect of the fly ash on shrinkage will be essentially the same regardless of the type of aggregate concerned.

3.2.4.3.6 Admixtures

/B 25/ tested concrete of the same workability with and without fly ash and with and without plasticizer. Results show much lower shrinkage for fly-ash concretes with plasticizer than for similar concretes without fly ash. The shrinkage reducing effect of the fly ash was greater for concretes containing plasticizer than for those without.

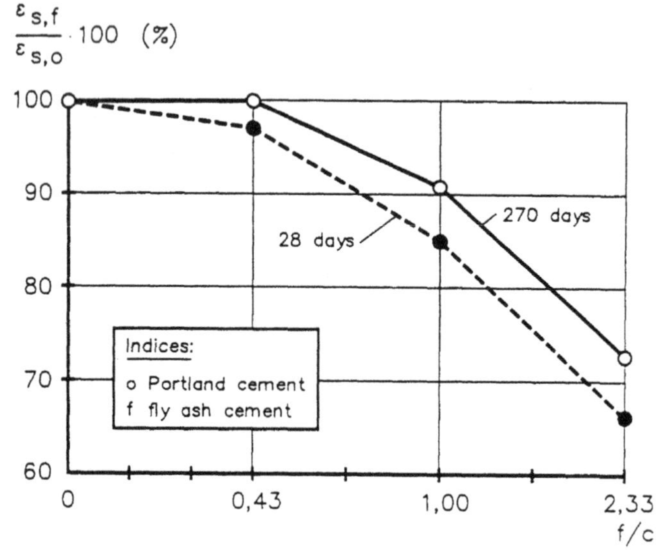

Fig. 3.2.30: Relationship between fly-ash content f/c and relative shrinkage of concrete when using fly-ash cement (no further conclusions available) /N 13/

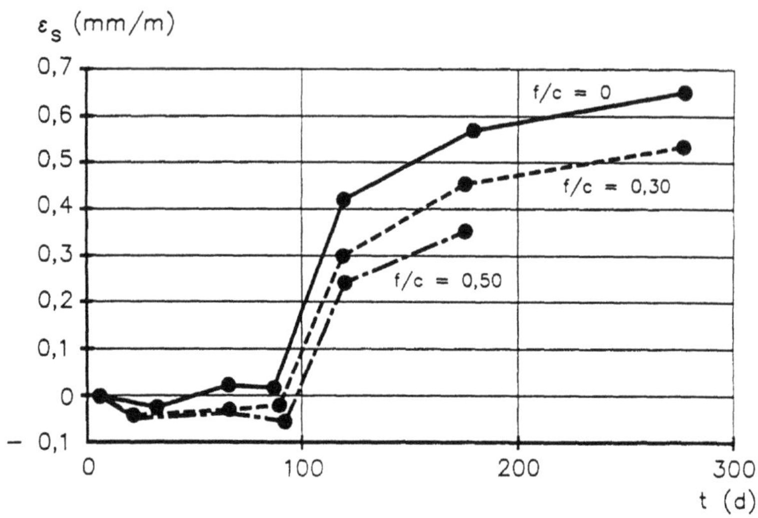

Fig. 3.2.31: Shrinkage ε_s of concrete without and with fly ash ($f + c = 170\ kg/m^3$) /P 5/
Beams: 100 mm x 100 mm x 760 mm
Preliminary storage: 90 d in moisture
Main storage environment: 21/50

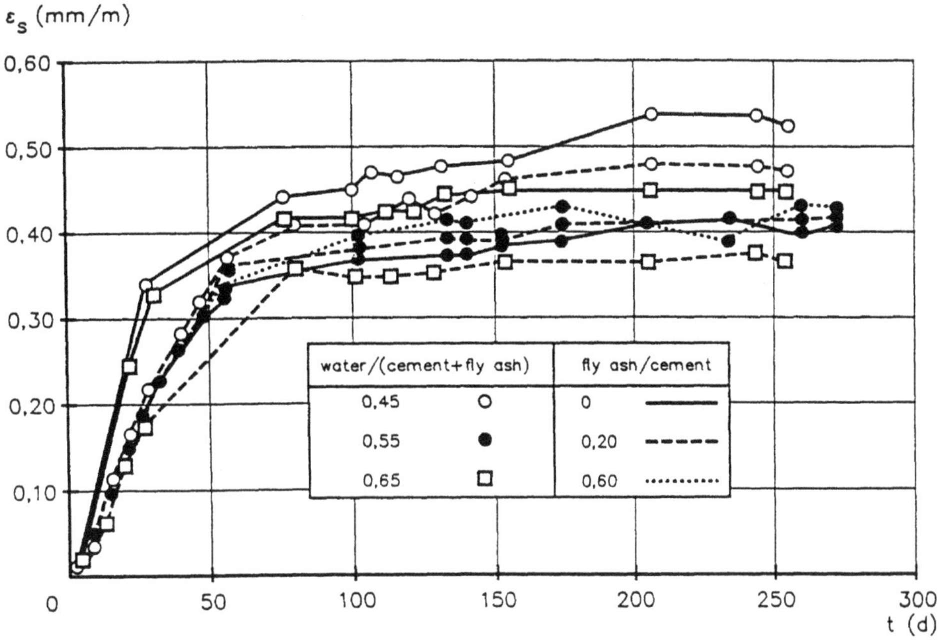

Fig. 3.2.32: ***Shrinkage ε_s of concrete without and with fly ash (LOI = 3.8 % by weight, Blaine value: 3770 cm^2/g) as a function of storage time t /H 33/ Cylinder: 150 mm x 300 mm, storage environment: 20/65***

3.2.4.3.7 Size of the Specimen

The shrinkage tests in /B 25/ on prisms and cylinders revealed no substantial influence of the fly ash.

3.2.4.4 Influences on Swelling

Workability

Tests on concretes with constant slump /B 25/ showed greater swelling after 270 days' water storage (*Fig. 3.2.25*) for fly-ash than for non-fly-ash concretes. The swelling is attributed chiefly to the adsorption of water by the cement gel. As the pozzolanic reaction is slow, hydration – and hence gel formation – takes longer in fly-ash than in non-fly-ash concrete, and final swelling is consequently greater. Swelling was in fact less in prisms tested at the same time as the cylinders.

In the lightweight concrete tests carried out in /A 3/ *(Fig. 3.2.28)* – presumably at constant workability – there was greater swelling in fly-ash concrete after 28 days' storage in a 20/99 environment.

Mortar prisms complying with DIN 1164, with and without fly ash, /B 16/ (see *Fig. 3.2.29*) swelled to approximately the same extent after 90 days' water storage.

Type of fly ash

Both LOI and carbon content have been regarded as factors affecting shrinkage. Tests with different fly ashes failed to demonstrate any appreciable effect of carbon content on swelling. Swelling did, however, tend to be greater with higher carbon contents.

Following one year's shrinkage storage, specimens in /M 24/ were subjected to alternating storage conditions (water at 20 °C, drying oven at 50 °C and 20 % relative humidity). Moisture deformation after 36 cycles was determined for the initial wet and dry conditions. In general, the fly ashes with the greatest ignition losses produced the highest moisture deformations. The report also stated that moisture deformation under alternating storage conditions was far more dependent on the type of cement (two types were used) than on the type and content of fly ash.

Fly-ash content

In the tests reported in /A 3/ (*Fig. 3.2.28*) on lightweight concrete with the highest fly-ash content, the greatest swelling was obtained after 28 days' storage in a 20/99 environment. In fact, this concrete had the smallest increase in weight, despite its large amount of expansion. This was attributed to the increased density of the hardened cement paste, even though the high fly-ash content of the hardened cement paste presumably prevented higher water adsorption.

Results from 28 day moist storage tests in /F 6/ were erratic. Since higher fly-ash content of f/c = 0.25 resulted in both increased and reduced swelling as compared to a lower fly-ash content (f/c = 0.11), no conclusions can be drawn.

In /P 5/ (Fig. 3.2.31), greater swelling was also observed with increasing fly-ash content after moist storage for 90 days. The same trend was found in the tests given in /B 16/ (Fig. 3.2.29), although the difference between the fly-ash contents (f/c = 0.18 and 0.25) was less.

Admixtures

/B 25/ considered the swelling of fly-ash and non-fly-ash concrete with and without plasticizer up to 270 days (*Fig. 3.2.25*). The fly ash reduced swelling in cylindrical specimens; no data were reported for prism specimens.

3.2.4.5 Summary

The remarkable results indicated by the currently available data are as follows:

- The addition of fly ash generally reduces shrinkage in mortars and concretes of constant workability. This is widely attributed to the water-reducing effect of the fly ash.

- There is little difference in the shrinkage behaviour of fly-ash and non-fly-ash mortars and concretes with a constant w/(c + f) ratio.
- Given constant workability and equal compressive strength, the shrinkages of fly-ash and non-fly-ash concretes are roughly the same.
- Fly ashes with high LOI (> 10 % by weight) increase the water absorbency and hence the w/(c + f) ratio at constant workability, explaining the greater shrinkage of fly-ash concretes as opposed to comparable non-fly-ash concretes.
- No precise effect of fly-ash content on shrinkage can be inferred.
- Plasticizing admixtures generally seem to enhance the shrinkage-reducing effect of fly ash.
- Fly-ash concrete which is moist-stored or stored in water exhibits greater swelling than concretes and mortars without fly ash.

3.2.5 Cracking (Prepared by J. Bijen)

Blended cements containing fly ash are recommended whenever there is a risk of an expansive alkali-aggregate which could lead to cracking. A UK working party on alkali-aggregate reaction recently recommended replacing at least 25 % of the portland cement by fly ash and limiting the alkali content of the concrete to 3 kg/m^3 or less where aggregates were suspect.

Two explanations have been offered for the beneficial effect of fly ash. The first hypothesis is that the pozzolana ties up the alkalis by preferential reaction, making them unavailable for alkali-aggregate reactions. The second hypothesis is that the blended cements have a denser cement gel than portland cements, retarding the movement of the ions involved and preventing destructive osmotic pressures from building up.

Investigators disagree on the vulnerability of fly-ash concretes to other problems such as sulphate attack, some stating that resistance is improved and others maintaining precisely the opposite. The effect seems to depend on the silica/alumina ratio of the fly ash /F 15, H 10, H 17, K 25, S 21, B 40, B 41, R 33/.

3.2.6 Coefficient of Thermal Expansion (Prepared by J. Bijen)

Data on the effect of fly ash on thermal expansion are extremely limited. On the basis of the available data, it seems likely that fly ash will slightly reduce thermal expansion at high cement replacement percentages /D 34, G 13, M 24, R 33/.

3.3 Frost Resistance (Prepared by H. Grube, S. Nagataki and H. Ohga)

3.3.1 Frost Attack

The use of fly ash has an effect on the frost resistance of concrete, dependent on

- the granulometry and the chemical composition of the fly ash
- the type and strength of the cement,
- the mixing ratios of all components of the concrete, especially the percentage of fly ash in relation to the cement and the relationship of water content to the content of cement and fly ash,
- the percentage of artificial air voids,
- the curing conditions, the age and the strength of the concrete up to testing,
- the methods of testing and the criteria on which assessment of frost resistance is based.

The tests which have been evaluated differ in terms of their significant influencing parameters, allowing various inferences to be drawn: the frost resistance of types of mortars or concrete in which a certain percentage by weight of cement has been replaced by fly ash is equal to or higher than that for the same type of mortar or concrete without fly ash /C 14, F 6, G 6, H 11, J 23, L 5, L 27, M 17, S 10, S 24, V 58/; no significant difference is observable /C 2, G 5, R 28, W 12, Y 3/; or the freeze-thaw resistance drops with increasing fly-ash substitution /B 16, E 23, G 6, H 10, H 30, K 16, L 4, L 5, S 42, V 2, W 4, W 5/. Even where a single exchange rate for the fly ash (25 % of the weight of the cement) was chosen, other investigators /B 6, K 23/ concluded that the frost resistance of a mortar with fly ash is lower than with a pure cement mortar.

3.3.1.1 Quality of Fly Ash

The effect of the composition of the types of fly ash on the frost resistance was not investigated systematically. Loss on ignition was varied by /L 27/ (3 % and 9 %), /V 2/ (1.5 % to 7 %), /W 4/ (1 % to 19 %), all using fly ashes of similar composition with approx. 50 % SiO_2, 25 % Al_2O_3, 10 % Fe_2O_3 and less than 5 % CaO, respectively. No influence of the loss on ignition on the frost resistance was observed within the ranges investigated .

However, in tests on mortar prisms in which fly ash had been substituted for 0 to 20 wt.% of the cement, Lühr /L 14/ observed that frost resistance varied inversely with increasing fly-ash content, provided that there was a loss on ignition exceeding 2 %. The investigation included fly ashes with losses on ignition of 2 to 20 % (*Fig. 3.3.1*).

Clendenning et al. carried out rapid freeze-thaw tests and outdoor exposure tests on concretes containing fly ash and examined the effects of carbon content on frost resistance /C 7/. Results indicated that there was no correlation between durability and total carbon content, but that there was a distinct correlation between weight loss and total carbon content (*Figures 3.3.2 and 3.3.3*).

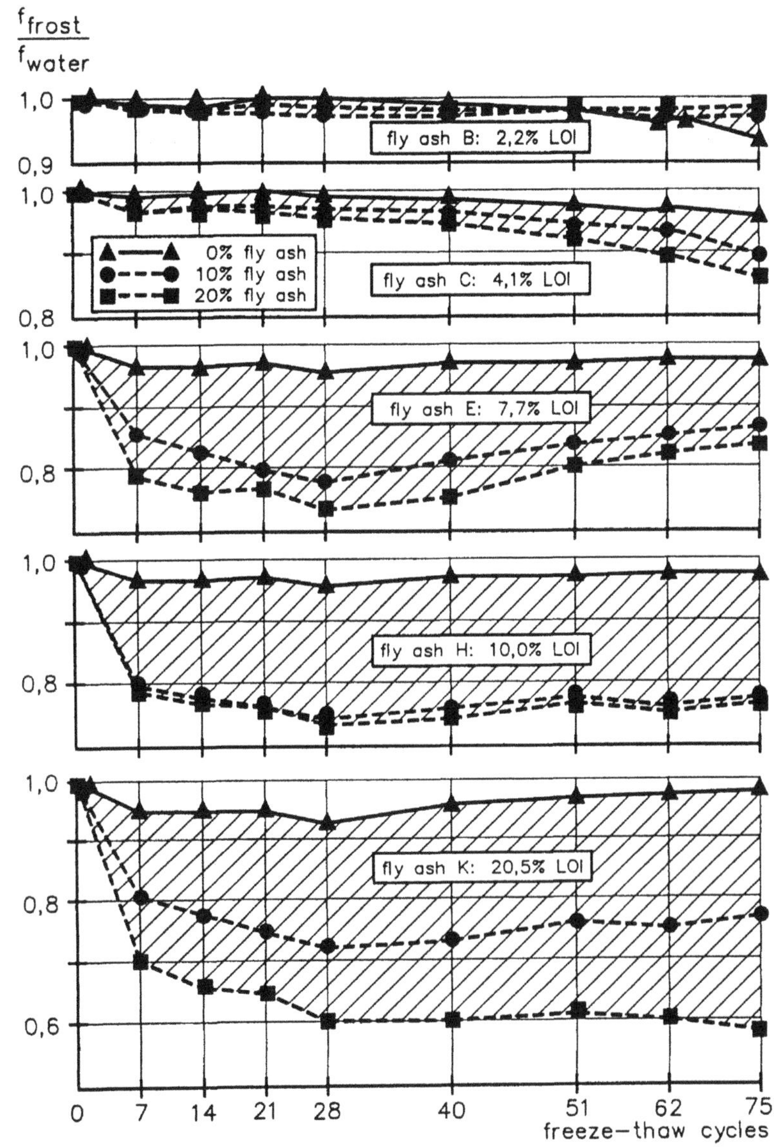

Fig. 3.3.1: Relative resonant frequency of fine-grained concrete prisms with 0,10 and 20% fly ash with graded loss on ignition and constant slump as a function of the number of freeze-thaw cycles (4 cm x 4cm x 16 cm).

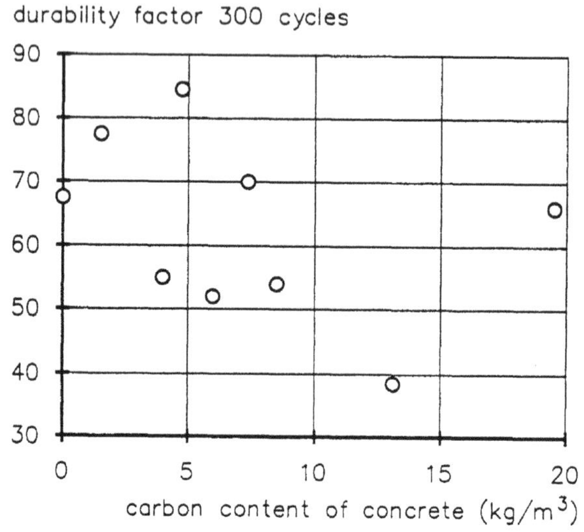

Fig. 3.3.2.: Relationship between freeze-thaw durability factor and carbon content of concrete

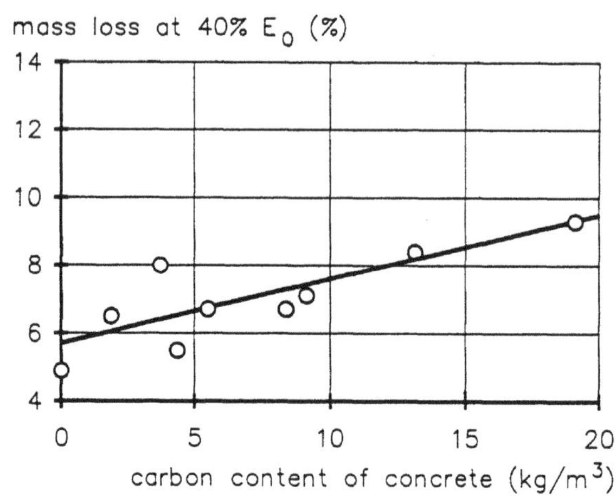

Fig. 3.3.3: Effect of carbon content on freezing and thawing mass loss of fly-ash concrete

It was also established that carbon in fly ash increased scaling of the surface of concrete but did not affect the frost resistance as long as the air content of the concrete was sufficient. A similar trend was evident in outdoor exposure tests /S 48/.

Hitherto, use of fly ash as a concrete additive in Germany has been restricted to coal fly ash, whereas in other countries brown coal fly ash with approx. 6.5 % CaO

/B 16, H 10, W 5/, fly ash of "sub-bituminous coals" with approx. 17 % CaO /I 22/ and with approx. 20 % CaO /H 30/ are, for example, in common use. /L 35/ summarizes papers which report on a type of fly ash whose chemical composition closely resembles that of granulated blast-furnace slag and which is therefore believed to possess not only pozzolanic but also latent hydraulic properties. A similar type of fly ash with 36 % SiO_2, 17 % Al_2O_3, 6 % Fe_2O_3 and 28 % CaO, with a loss on ignition of as little as 0.6 % is used in /S 50/ as the sole binding agent. Frost resistance according to the ASTM 666 procedure B was extremely low. A particular problem was the adjustment of setting, in this case accomplished by adding sodium borate. Fly ash with CaO percentages of 6 to 30 was also substituted for cement (15 to 25 %) in the tests presented in /C 14/. Good frost resistance with air voids is reported but not conclusively demonstrated.

Similar research results appear in /C 14/, obtained with batches containing only fly ash with a high CaO percentage, grate ash of up to 19 mm maximum grain size and water as well as a gypsum admixture of approx. 10 % of the fly-ash percentage. These batches, with strengths of approx. 35 MPa and an air void percentage of 5 to 6 %, are also reported to possess good durability when subjected to frost, but not without artificial air voids.

Because of these widely varying results for different types of fly ash and also in relation to air entraining agents, Lukas /L 32/ points out that extremely high quality standards must be imposed when using fly ash in concrete. Huber /H 10/ likewise stresses the importance of using "appropriate" types of fly ash.

3.3.1.2 Quality of Cement

The influences of the cement type and strength, as well as of storage conditions prior to freeze-thaw testing, were covered in the tests desribed in /L 8, S 10/. The specimens – with or without fly ash – generally exhibited reduced strengths after freeze-thaw tests. In /S 10/, using PZ 35 F and PZ 45 F, an increase in relative residual compressive strength with increasing fly ash content (peak value at a fly-ash content of approx. 25 % and 12 % respectively) was recorded, whereas with HOZ 35 L a decrease in relative residual compressive strength with increasing fly-ash content was observed. These differences in compressive strength with different types of cement were not matched by differences in splitting-tensile strengths in /L 27/.

3.3.1.3 Concrete Mix Ratios

The influence of cement content has been studied by Manz /M 5/ and by Teoreanu and Nicolescu /T 8/. Manz applied the ASTM freeze-thaw test to concrete containing four kinds of Class C fly ash. This test involves rapid freezing in air at -17,8 °C and rapid thawing in water at 4.4 °C. The results were inconclusive as to variation in freeze-thaw resistance with increasing cement plus fly ash /M 5/. In another experiment, Manz found hardly any effect of fly ash on the durability of lean mixes but, observed a fall in frost resistance with increasing cement replacement by fly ash as total cement content increased. Teoreanu and Nicolescu evaluated the freeze-thaw resistance of concrete with 50 % fly-ash replacement by measuring the loss of compressive strength after 25 freeze-thaw cycles. The freeze-thaw resistance decreased when the cement plus fly-ash content was increased from 460 kg/m^3 to 520 kg/m^3 /T 8/.

In freeze-thaw tests according to ASTM C666 Procedure A on concrete containing Class C fly ash, Yuan and Cook showed that the frost resistance of non-air-entrained concrete improves with increasing fly-ash content, although the durability factor falls below 60 % before 150 cycles have been completed. Conversely, air-entrained concrete shows the same frost resistance up to 400 cycles, irrespective of fly-ash content /Y 3/ (*Figures 3.3.4 and 3.3.5*). However, a decrease in frost resistance with increasing Class F fly-ash content is also reported /L 31/. Furthermore, as it has been reported that the influence of the fly-ash replacement ratio on frost resistance depends on the type of fly ash (four varieties of Class C fly ash) /M 5/, it would appear that the influence of fly-ash content on frost resistance also depends on the fly-ash type.

A comparison of the influences of lignite ash (Class C fly ash) and bituminous coal ash (Class F fly ash) on the frost resistance of concrete shows similar trends for the effect of the fly-ash replacement ratio and cement content. However, frost resistance is slightly higher with bituminous coal ash, as indicated in /G 6, M 5/.

The way in which cement and fly ash are blended together also affects frost resistance. Osborne and Nixon used two methods for mixing cement and fly ash – the intergrinding method (grinding cement clinker and fly ash together in a ball mill) and the blending method (blending cement ground in a ball mill with fly ash in a rotating drum) /O 12/. In order to study the frost resistance of the concretes, they then replaced 5 % and 20 % of the cement with three varieties of fly ash (Class F) exhibiting different levels of loss on ignition and fineness. They found that, whichever blending

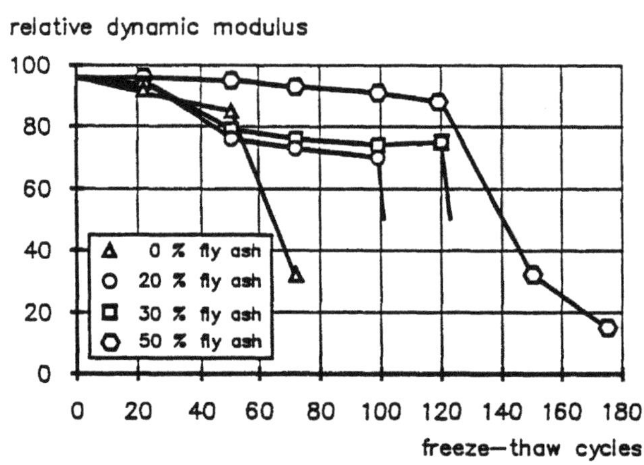

Fig. 3.3.4.: Relationship between relative dynamic modulus and number of freeze-thaw cycles for non-air-entrained concrete

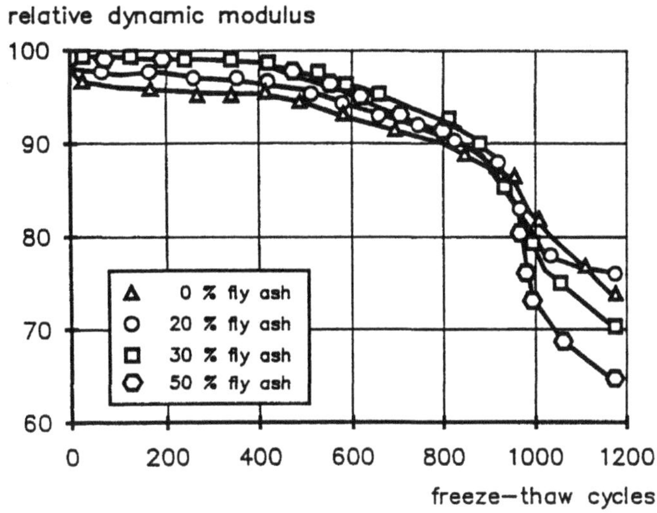

*Fig. 3.3.5.: **Relationship between relative dynamic modulus and number of freeze-thaw cycles for air entrained concrete***

method is employed, a coarse, high-carbon fly ash adversely affects not only early strength, but also frost resistance. The limiting replacement ratio is therefore 5 %, while even with a low replacement ratio, it is important to use a fine, low-carbon fly ash. In addition, the frost resistances of a concrete with 20 % admixture of a fine, low-carbon fly ash blended by the intergrinding method, and of a concrete with 5 % admixture by the blending method, were better than that of a concrete with no admixture.

3.3.1.4 Age, Strength and Curing of Concrete

The strength gain of concrete is generally slowed by fly ash, so that when freeze-thaw tests are begun at early ages, the frost resistance of fly-ash concrete tends to be lower than normal. However, Teoreanu and Nicolescu found that the freeze-thaw resistance of fly-ash concrete (measured in terms of loss in compressive strength after 50 freeze-thaw cycles) increased with maturity before testing /T 8/. Kovacs, Berry and Malhotra pointed out that when freeze-thaw tests were begun after prolonged curing, there were no apparent differences in the freeze-thaw resistance of concrete with and without fly ash /B 12, K 16/. However, there is also a report that when the age at which freeze-thaw tests were begun was extended from 14 to 90 days in order to assess the influence of the increase in compressive strength, the frost resistances for fly-ash concrete and normal concrete were identical /L 31/.

Crow and Dunstan /C 14/ emphasize the substantial effect of after-treatment on the frost resistance of concrete with or without fly ash. Concrete containing fly ash exhibited the same frost resistance as concrete without fly ash only where after-treatment took place in a humid environment and lasted more than 28 days before freezing occurred. Storage under humid conditions for 7 to 14 days with subsequent storage in fresh air or in a climatically controlled chamber (23 °C/50 % RH) entailed a substantial decrease in frost resistance (cf. *Fig. 3.3.6*).

3.3.1.5 Air Content

Several authors point out that it is useful or necessary to employ air-entraining agents in order to assure good frost resistance in fly-ash concrete /F 29, M 34, R 15, S 38/ Others even incorporate this in practice or in tests /B 16, C 14, H 10, H 30, S 42, V 3, W 4, W 5/, indicating that a high frost resistance can be achieved given a sufficient percentage of micro-air-voids. *Fig. 3.3.7* /B 26/ provides a particularly clear illustration. This reference is of major importance, because similar types of concrete have been utilized for concrete dams in high mountain regions.

Some papers report a similar frost resistance for both fly-ash and non-fly-ash concretes, provided that these contain artificially-entrained air voids /C 14, M 34/. Results also show very low w/(c + f) values of, for example, 0.45 /C 14/.

Capp and Spencer summarized a large number of references to fly-ash concretes. They found that air-entrained concretes showed equal frost resistance with or without fly ash, indicating that air content rather than fly-ash content is the determining factor /C 2, C 14, V 11/. Manz found that the frost resistance of a non-air-entrained concrete declined with increasing Class C fly-ash content, but that air-entrained concretes with high air contents had roughly equal durability regardless of their fly-ash or cement content /M 5/.

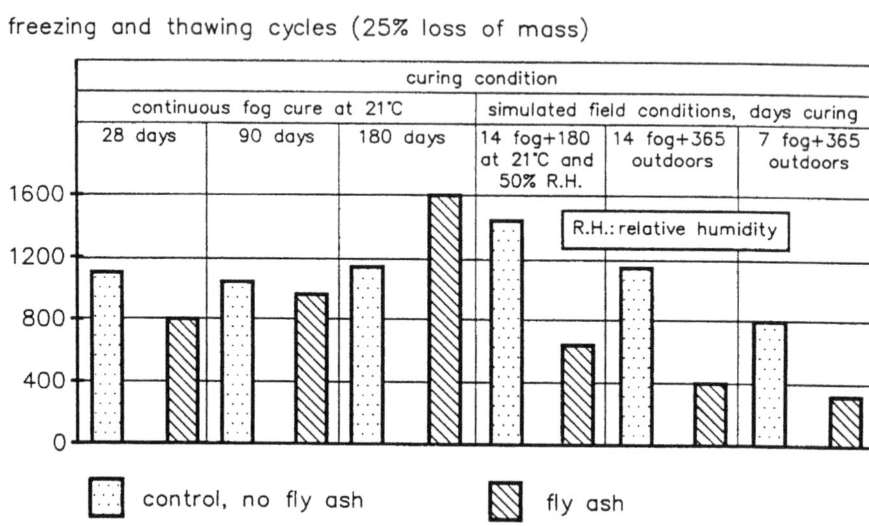

Fig. 3.3.6: Influence of curing conditions on the frost resistance of concrete with and without fly ash /C 14/

Virtanen evaluated the freezing expansion, frost-salt resistance, spacing factor, etc. of a concrete with a binder content of about 300 kg/m^3 and fly-ash (Class F) content of around 100 kg/m^3. His results indicated that if strength and air content can be kept at the same levels as for normal concrete, the fly ash has practically no influence on frost resistance /V 11/.

3.3.1.6 Testing Methods

The methods of frost attack used in tests are very different. There is no information on comparable tests with different frost attacks. The frost attacks used ranged from a single "fast freezing" /S 42/ via 14 freeze-thaw cycles with humid specimens wrapped in sheets /J 22/, 20 freeze-thaw cycles /S 42/, 25 freeze-thaw cycles /K 23/, 75 freeze-thaw cycles /L 14, S 10/, 100 to 200 freeze-thaw cycles /B 6, B 16, H 10, L 8, M 17, V 2, W 4, W 5/, 300 freeze-thaw cycles /H 30/ and 1000 freeze-thaw cycles with drilled cores /H 10/ to 1400 freeze-thaw cycles /V 3/. /C 14/ reports on 2600 to 3600 freeze-thaw cycles which lasted until 25 wt.% percent of the material spalled off. With the exception of /H 30/ and /J 22/, all specimens were stored in water, frozen in air and defrosted in water partially complying with standardized tests methods such as Austrian Standard 3303, DIN 52 252 or ASTM C 666 – procedure B. /H 30/ tested specimens in compliance with ASTM C 666 – procedure A (freezing and defrosting in water).

The following standards may be used for testing frost resistance:

- Frost attack /X 30 - X 32/,
- Frost plus de-icing agents /X 33/,
- Entrained air /X 1, X 3, X 34/.

Since the conditions in rapid freeze-thaw tests are vastly different those in the field, the relationship between laboratory test results and natural exposure has not yet been clarified, and few reports have been published on this topic. Legg studied the relationship between freeze-thaw tests in the laboratory and outdoor tests over an eight-year period /L 4/. The laboratory tests conformed to ASTM C290 (freezing in air at -17.8 °C and thawing in water at 4.4 °C). The rapid freeze thaw resistance of concrete with Class F fly ash was less than that of concrete without fly ash. By contrast, there was no deterioration of the concretes outdoors, as they were on a well-drained base and never became saturated.

Different criteria are used to assess freeze-thaw resistance. /J 23, K 23, L 27, M 17, S10/ concentrate chiefly on changes in the compressive strengths of specimens, whereas in other studies changes in the dynamic modulus of elasticity /B 5, B 16, H10, H 30/, changes in flexural tensile strength /B 6, W 4/, a loss of weight of 25 % due to pieces spalling off the specimen /C 14, E 23/ or strain /V 2/ are used as the test criteria. Evaluating all the results of tensile splitting tests in /L 27/ in the same manner as in /S 10/, it can be shown that the frost resistance tends to decrease if portland cement or blast-furnace cement are replaced by fly ash. This occurred despite utilization of

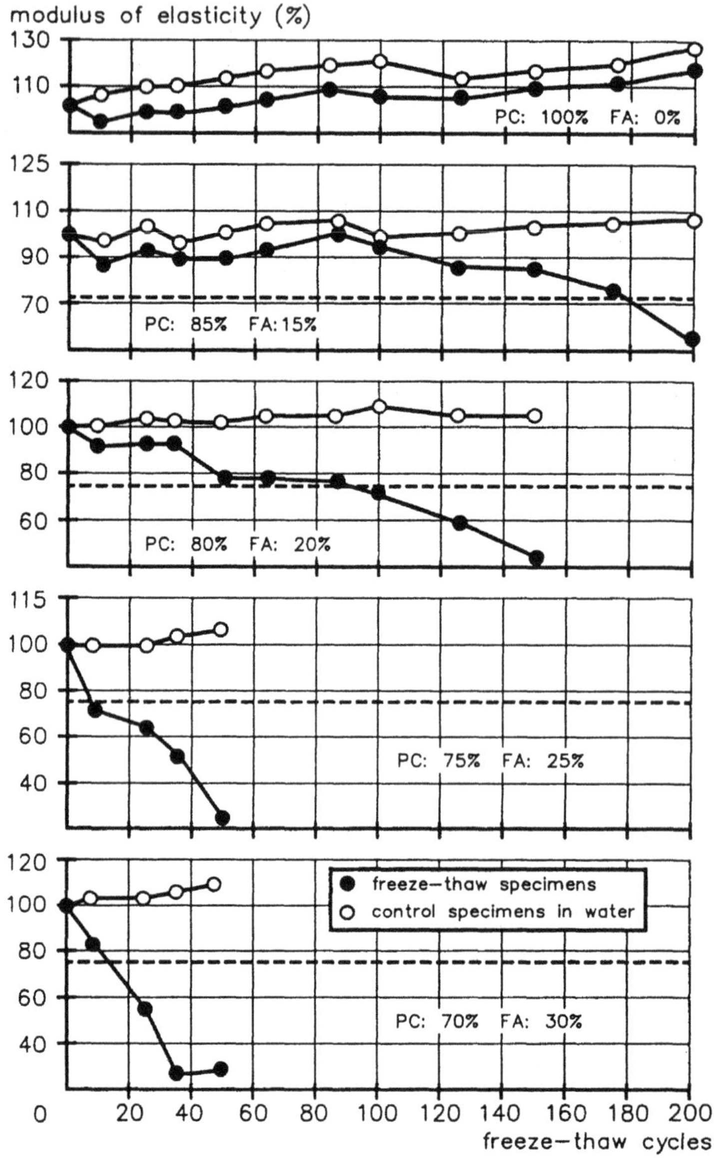

Fig. 3.3.7: Frost resistance of concretes with various fly-ash contents and without (left) and with air-entraining agent (right) /B 26/, cement and fly-ash content = 270 kg/m^3

the decreased water demand of the fly ash. It seems possible that the tests mentioned above reflect the strain-related influences of freezing conditions more distinctly than do compressive strength tests.

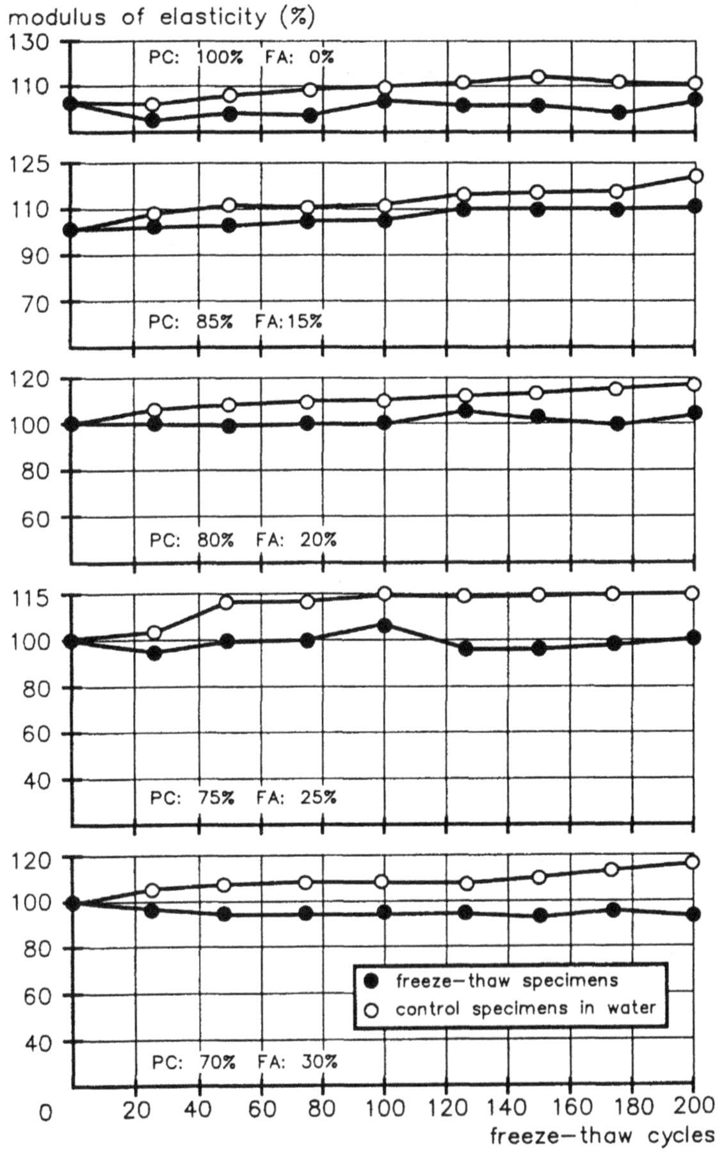

3.3.2 Frost Plus De-Icing Agents

One way to evaluate the resistance of concrete to de-icing agents is to use freeze-thaw tests in NaCl solution in the laboratory /L 4, V 11/. The other is to sprinkle de-icing agent repeatedly on specimens stored in the open /T 5/.

The resistance of fly-ash concrete to de-icing agents seems to depend on test methods and conditions. For example, in Virtanen's report /V 11/, specimens were 'frozen' in a saturated solution of sodium chloride at -15 °C and 'thawed' in pure water at 20 °C in the laboratory. The deterioration was measured as the change in volume after 25 freeze-thaw cycles. When the freeze-thaw test was started after curing in water for 7 days, the deterioration of non-air-entrained concrete was severe and the loss in volume of fly-ash concrete (Class F) was extremely large. However, when the concrete was air-entrained, it deteriorated much less and the influence of fly ash almost completely disappeared. It has also been reported that, when testing was started after 7 days of curing in water, followed by 28 days at 70 % RH, both non-air-entrained and air-entrained concretes showed the same increased resistance, and the fly ash made almost no difference.

On the other hand, Timms and Grieb studied the effect of fly ash (Class F) on the resistance of concrete to scaling due to cycles of freezing and de-icing with calcium chloride /T 5/. The specimens were cured in moist air for 30 days and placed outdoors on the ground. During autumn and winter, the top surface of the specimens was kept covered with water. When this froze, calcium chloride was applied to the surface at the rate of 1.3 kg/m^2. Once the ice was completely thawed, the surface was washed and fresh water left on the surface to await another freezing. Even air-entrained concrete shows a marked loss of resistance to deterioration if one-third of the cement has been replaced by fly ash. Fly ash should therefore be avoided when calcium chloride is to be used as a de-icing agent.

Legg also studied the effect of fly ash (Class F) on the resistance of concrete to scaling due to application of sodium chloride for de-icing /L 4/. Concrete containing fly ash (from 0 % to 35 % replacement) was frozen to -17.7 °C once daily in the laboratory, with 4 % sodium chloride solution ponded on the surface, and was frozen outdoors for eight years by natural weathering, with similar ponding for the first two years. In the open, resistance to scaling was very good, whether or not fly ash was present. Conversely, some scaling was observed on the fly-ash concrete in 150 cycles of the indoor tests.

Schorr /S 47/ reports on freezing and de-icing tests with paste matrix pats of PZ 35 F, EPZ 35 F, HOZ 45 L, HOZ 35 L and 2 types of fly ash with respective losses on ignition of 1.3 % and 4.7 %. The fly-ash percentages were 10 %, 18 %, 24 % and 30 % of the cement by weight. The water/solid matter relationship was $w/c = w/(c + f) = 0.28$. No air entraining agent was used. The after-treatment consisted of storing for 14 days under humid conditions and for several months under dry conditions. The specimens were then frozen in a 3 % NaCl-solution and defrosted. Results showed that the freezing-de-icing-resistance decreased with increasing percentages of granulated blast-furnace slag and fly ash. The damage assessment (splitting and destruction of specimens) was carried out on all specimens. 20 freeze-thaw cycles were sufficient. Schorr /S 47/ claims that these results are transferable to concrete.

Minnich /M 17/ reports on freeze-thaw tests with dewing salt on some concretes containing portland cement only and others with 20 or 30 % substitution of fly ash for cement. After 30 and 60 cycles no surface splitting was noted when using "dewing salt". It is not clear whether this concrete contained artificial air voids or not. When calcium chloride was substituted for "dewing salt" in the tests, substantial splitting was observed in portland cement concrete, but only minor splitting in concrete with added fly ash.

3.3.3 Entrained Air

The considerable advantages of artificially-entrained air voids have already been noted in Section 3.3.1.5. Generally, in /B 16, B 47, F 29, H 10, H 30, L 32, M 17, R 15, S 38, V 3, W 4, W 5/ it is stated that in order to attain a certain percentage of air voids, a larger amount of additives is necessary for fly-ash mortar or concrete than for non-fly-ash batches. Among other factors, this increased need for air-entraining agents is believed to depend on the percentage of fly ash /S 42/, or on certain properties of the fly ash such as percentage of activated carbon /L 32/, loss on ignition /C 14, S 38, W 4/ or fineness of the fly ash /F 29, H 10, V 3, W 5/.

In general, more air-entraining agent is required to entrain a specified volume of air in fly-ash as compared to non-fly-ash concrete /B 12, C 2, C 14, F 6, G 6, G 21, K 31, P 6, P 12, R 28, S 21, S 24, S 48, S 49, T 8, V 11/, and the cause of this is said to be the adsorption of air-entraining agent by unburnt carbon in the fly ash /S 21, S 48/. There is a good correlation between the carbon content of fly ash and the quantity of air-entraining agent needed to entrain a specified amount of air (Figure 3.3.8). A particular air content can hence be obtained by increasing the dosage of air-entraining agent. It has, however, been reported that some fly ashes make it difficult or impossible to entrain a specified amount of air /S 49/. Care will therefore be needed when using fly ash of high carbon content in a structure subjected to a severe environment.

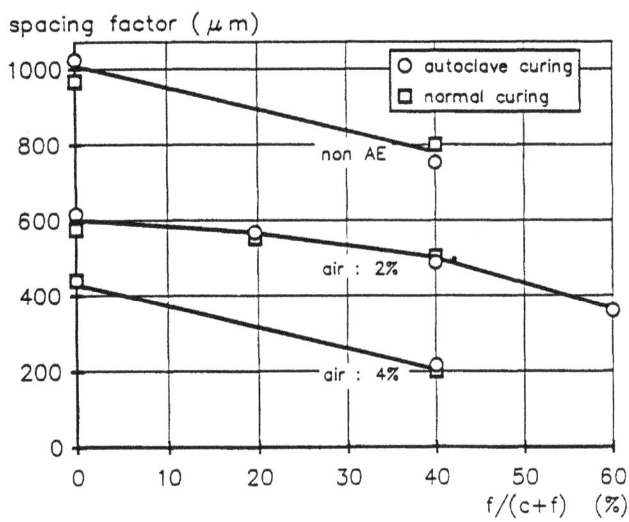

Fig. 3.3.8.: Relationship between spacing factor and replacement ratio of fly ash

If the carbon content of fly ash varies, the quantity of air-entraining agent to give a specified air content will also vary, making adjustment of air content extremely difficult /P 12, S 21/.

From studies of the air-entrainment properties of concretes with admixtures of 10 varieties of fly ash (Class C and Class F), Gebler and Klieger showed that more air-entraining agent is usually needed for a particular air content with Class F than with Class C fly ash /G 21/. If the quantity of organic substances, carbon content, and LOI of the fly ash increases, more air-entraining agent is required for a specified air content. It was, however, shown that when the total alkali in the fly ash is increased, less air-entraining agent is needed.

Sturrup, Hooton and Glendenning studied the effects of fly ash on the air content of concrete in the laboratory, and found that entrapped air is reduced by about 0.5 % by fly ash and that the total air content required to obtain adequate durability is reduced as compared to concrete containing no fly ash /S 48/. Consequently, even though total air content is lower with fly ash, the quantity of effective entrained air remains constant.

Wogrin /W 5/ points out that, given a constant AEA percentage, not every agent results in the same frost resistance. This problem is investigated by Pistilly /P 23/ without frost tests, analyzing the percentage of micro-air-voids and the distance factor. Results showed that Vinsol-resin (AEA type A) exhibited a larger distance factor at a constant AEA percentage and an increasing alkali content, irrespective of whether the alkalis originated from the cement or the fly ash or were added separately. Sulphated hydrocarbons (AEA type C) lead to a more favourable distance factor which is not dependent on the alkali content of the batch.

Virtanen studied the air-void content of hardened concrete with a fly-ash (Class F) replacement of 30 %, and found that the air content of hardened non-air-entrained concrete was less than that of fresh concrete, whereas the opposite was true for air-entrained concrete /V 11/. Further, it was shown that air voids are spaced more closely as air content increases, and that the relationship is the same whether or not fly ash is contained as an admixture.

However, Nagataki and Ohga /N 26/, in studies of the effect of Class F fly ash on the air entrainment of concrete, found that void spacing fell with increasing replacement ratio of the fly ash as well as with air content (*Fig. 3.3.8*). Soretz /S 42/ likewise determined smaller distance factors at constant air void contents for mortars containing fly ash than for mortar without fly ash.

3.3.4 Conclusions

The results of freezing and thawing resistance tests on fly-ash concrete differ considerably depending on the age at which freezing and thawing tests are started, the curing conditions, the characteristics of the fly ash and cement, etc.

There are no apparent difference in the freezing and thawing resistance of concretes with and without fly ash if these are of equal strength and air content.

Care needs to be exercised where high-carbon-content fly ash is used in a structure exposed to a severe environment, since it is sometimes difficult to entrain the specified amount of air.

The resistance of concrete to de-icing agents is not improved by the use of fly ash.

A larger amount of air-entraining agent is required to entrain the specified volume of air in fly ash as opposed to non-fly-ash concrete.

3.4 Chemical Resistance (Prepared by M. A. Ward)

3.4.1 Sulphate Attack

3.4.1.1 Phenomenological Examinations of the Effect of Fly Ash on Sulphate Attack

Since the advent of portland cement, it has been recognised that some chemicals inherent in the environment can undermine the integrity of concrete structures. Vicat /T 17/, for example, recognised the importance of protecting concrete from magnesium sulphates which could react with lime in the cement. The use of pozzolanas to counteract attack was also considered at an early stage; in Smeaton's time, it was common knowledge that the addition of pozzolanas made a durable mortar. Michaelis recommended the addition of pozzolanas to bind the free lime of the hydrated cement and so prevent reaction with the magnesium sulphate in seawater.

When considering modern concretes, the factors which determine the extent of attack are (a) the type and amount of sulphate in the water or soil in which the concrete rests, (b) the water table level and fluctuations in its movement, and (c) the chemical and physical properties of the concrete.

For a given structure, it is usually impossible to modify (a) and (b) significantly and the properties of the concrete hence become crucial. The importance of the concrete quality in determining the extent of sulphate attack is reflected in the number of phenomenological examinations which have been reported.

In the first half of this century, while researchers such as Thorvaldson et al. /T 12/ and Bogue /B 71/ grappled with the mechanisms and chemistry of sulphate attack, others examined the durability to sulphates of practical concrete mixes with and without the addition or replacement of pozzolanas. Davis, Carlson, Kelly and Davis /D 35/, for example, tested five low-calcium fly ashes at 20 % replacement level. Concrete cylinders with a diameter of 76 mm and a length of 152 mm were immersed in 10 % sodium sulphate solution. In terms of strength, all five fly-ash concretes were superior to plain concretes after five months' immersion.

Price /P 26/ performed a full-scale test programme on 5000 specimens exposed to the natural environment, correlating his results with laboratory tests using 2 % and 5 % mixed solutions of sodium and magnesium sulphates. He found that expansion of the fly-ash concretes was generally much less than that of the plain concrete; improvement due to ash addition was more pronounced when Type I rather than Type V cement was used. Price also obtained interesting results which indicated that sulphates may attack ash-concretes by mechanisms different from those affecting plain concretes; although the fly-ash concretes underwent greater surface erosion in the 5 % sulphate solution, their expansion was much lower (by a factor of 5) than that of the plain concrete. Other early phenomenological examinations confirm that the inclusion of fly ash in concrete or mortar increases resistance to sulphate attack /B 16, E 6, G 6, K 3, K 18, P 4, P 27, Z 3/.

Scholz and Scholz /S 53/ report on concrete specimens (100 x 150 x 350 mm^3) which were exposed to dump-waste water with a sulphate content of 1300 mg/l for 5 1/2 years. No ettringite or gypsum formed immediately at the surface of fly-ash concrete specimens (20 % replacement). In non-fly-ash concrete (Type I cement), SEM analysis revealed a large amount of ettringite and destruction of cement gel even in the core of the specimen.

Dikeou /D 36/ found that fly ash from bituminous coal markedly improved the sulphate resistance of concrete in the 20−35 % replacement range. In general, the order of resistance for different blends appears to be

(a) Type V cement with ash (best),
(b) Type II cement with ash,
(c) Type V cement alone,
(d) Type II cement alone,
(e) Type I cement with ash and
(f) Type I cement alone (the least resistance).

Scholz /S 7/ found that, for high-quality ash, replacement levels of 40−45 % in opc/pfa mortars resulted in about the same level of resistance as for a mortar made solely with sulphate-resisting portland cement. Scholz also noted that the "quality" of the ash (determined by its pozzolanic index, particle-size distribution, glass content and surface area) is important to the durability obtained. Bradbury /B 20/ examined the behavior of blended cement and noted that, if the blend contained more than 25 % ash, it achieved a performance similar to that of sulphate-resisting portland cement. On the other hand, Derdecka-Gryzmek /D 3/ noted substantial improvement in sulphate durability at the 15 % level, but much better performance at a 45 % replacement level.

Kalousek, Porter and Benton /K 1/ tested 34 pozzolana-concrete mixes under continuous soaking and also in an accelerated exposure test (cycles of 16 hours soaking in sulphate solution, followed by 8 hours drying at 54 °C). Although six of the pozzolanas failed to improve sulphate resistance, all of the 11 fly-ash concretes which they tested did show improved performance in sulphates. In the last 10 years, the beneficial aspects of fly ash have been confirmed by other workers /B 1, B 39, B 72,

L 23, L 41, M 60/. Wesche and Schubert /W 12/ replaced cement with ash to the 50 % level and observed increased resistance, especially when a cement of low sulphate resistance was used. Tyndall and Munn /T 20/ soaked specimens in 3 % sodium sulphate solution; pozzolanic replacement invariably improved resistance, and the greater the proportion of ash the better the performance. Elfert /E 23/ noted that many pozzolanas can effectively double the service life of concrete exposed to sulphate attack; although details of the pozzolanic reaction are not clear, it appears that low-calcium ashes perform best in sulphate environments. Sturrup, Hooton and Clendenning /S 48/ indicated that ash can be blended with Type II cement (30 % replacement) to produce a concrete whose resistance is greater than that of one made solely with Type V cement alone; for 30 months exposure in sulphate solution, mortar bars made with ash and Type II cement expanded by 0.026 %, while those made with Type V expanded by 0.152 %.

It would appear that there are no problems with sulphate resistance of ash-concretes in Australia. Samarin, Munn and Ashby /S 49/ reported that direct replacement of part of the cement with ash always improves resistance, the maximum improvement being in those which would be most susceptible without the ash. Good sulphate resistance is obtained with a moderate C_3A Type I cement and a good quality ash. In well-controlled experiments, Schubert and Lühr /S 12/ showed that the extent of improvement due to an ash is largely dependent upon the C_3A content of the cement.

The research reported above would suggest that it is important to include fly ash in concrete exposed to sulphate attack. Conversely, there is a considerable amount of research which indicates that fly ash is not always beneficial /M 28/. Kondo /K 47/, for example, cured specimens in water for seven days and then immersed them in 2 % magnesium sulphate solution; he found that pozzolana cements did not always perform as well as normal portland cements.

Hansen /H 39/ suggested that the benefits of pozzolanic addition are questionable. Much of the C-H in concrete may be in an amorphous form, intimately interspersed among the C-S-H hydrates; this form of C-H may have good cementing properties. Thus, the removal of C-H by the pozzolanic reaction may undermine the strength of the material and promote the detrimental effects of sulphate attack. Also, the fact that the C-H has reacted with the ash does not necessarily mean that it cannot still be leached from the pore structure. Hansen also points out that a highly siliceous pozzolana may result in increased porosity due to reaction rather than to the expected decreased porosity. The ash should have a high alumina content if total porosity is to decrease.

One of the reasons suggested for the beneficial action of fly ash in sulphates is the resultant decrease in pH in the pore solution due to removal of calcium hydroxide; if the pH is reduced below about 10, ettringite is unstable and thus the large expansions associated with its formation cannot occur. Hansen /H 39/ points out that, even given substantial pozzolanic reaction, the pH may still not decrease to the level which prevents ettringite formation; the beneficial action of pozzolanas in this respect may be negligible.

The pros and cons of the use of fly ash in sulphate environments have been compounded, most significantly in North America, by the increase in the use of so-called Type C (high lime) ashes. As compared to Type F ashes, Type C ashes, generally have a high lime content, a low silica aluminate iron oxide content, a high proportion of reactive crystalline alumina compounds and more minor water-soluble impurities. Several studies indicate that Type C, or high-lime, ashes may not be suitable if sulphate attack is prevalent.

A significant amount of work by Dunstan et al /C 14, D 37, D 38/ showed that sulphate resistance may be significantly reduced in concretes containing lignite or sub-bituminous ashes as compared to concretes with bituminous (Type F) ashes. This was observed even though the concretes made from the Type C ashes had adequate compressive strength, reduced drying shrinkage and satisfactory freeze-thaw durability. Emphasis was placed on the chemical composition of the fly ash; a sulphate resistance factor R = (%CaO-5)/%Fe_2O_3, inversely proportional to the sulphate resistance which a specific ash can contribute, was defined. The ratio R is an indication of the amount of sulphate-reactive alumina present in the ash; they suggested that a value for R of 1.5 or less would ensure that the ash has a beneficial effect on the concrete. It was also pointed out that the proportion of ash used is likewise important to sulphate resistance. Whereas Type F ashes increase resistance in all proportions, Type C ashes do not. At low proportions, the Type C ash reduces resistance; at higher proportions the trend reverses and the Type C ash tends to provide more resistance than the plain concrete mix.

This may be one of the reasons why investigations at the Prairie Farm Rehabilitation Administration (PFRA) /P 26-29/ indicated that concretes made with a high-lime lignite ash behaved well in concentrated solutions of sodium and magnesium sulphate. It should also be noted that the C_3A content of Dunstan's cement was 8 % while that of the PFRA cement was 2.6 %. Furthermore, the PFRA used a total cementitious content of 360 kg/m^3. The low C_3A and high cement content at the PFRA would, in any case, tend to favour adequate sulphate resistance.

Dunstan's examinations are by no means alone in indicating the poor sulphate resistance of concretes made with Type C ashes. Mehta /M 61/ used an accelerated test method in which 12.7 mm paste cubes were exposed to a 4 % sodium sulphate solution held at a constant pH of 7. He found that some ashes had questionable behaviour. Particularly where an ash with 30 % alumina is used, the strength falls considerably after 28 days' storage in the sulphate solution. In this regard, Mehta pointed out that the lack of performance tests is a major reason for the slow growth rate of the blended cement industry. Malhotra et al /M 60/, recognizing that all fly ashes differ from one another, strongly support the need for performance testing.

Mather's /M 62/ systematic approach to experimentation pointedly demonstrates the highly variable nature of fly ash and the subsequent inability to predict the behaviour of a given ash-concrete when exposed to sulphate attack. Mather mixed various mortars with variable water/cementitious ratios to yield constant flow of the fresh mix. When specimens reached a specified strength level, they were immersed in sulphate solution. The depletion of ion concentration in the solutions was examined and the solutions were exchanged frequently to maintain the most constant possible solution concentration.

Mather found that some ashes prevented serious sulphate attack while others made the situation worse. Of the eight different ashes examined, the three sub-bituminous ashes showed the best resistance, the single bituminous ash produced an intermediate resistance, while the four lignite ashes provided the worst response. Mather concluded that the most effective ashes were those which had high fineness and high silica content and were highly amorphous.

3.4.1.2 Mechanisms of Sulphate Attack and the Role of Fly Ash

The processes by which sulphates attack concrete can be classified as:

(a) diffusion of the attacking ions into the pore structure of the material,

(b) expansion and softening reactions occurring between the ions and the cement component once the ions have penetrated the pore structure and

(c) chemical reactions between magnesium, CSH and calcium hydroxide (when the attack is by magnesium sulphate) at the surface of the concrete.

The latter case represents a minor effect, but can lead to surface softening and results in a significant loss in strength through effective reduction of the cross-sectional area.

The principal reaction mode is therefore of a dual nature, with permeation and/or diffusion followed by chemical reaction; both the physical and chemical properties of the concrete are important to sulphate attack /K 14, W 21/. Mehta /M 73/ accordingly attributes improvements in the sulphate resistance of fly-ash concretes to two factors:

(1) reduction in the free lime content due to the chemical pozzolanic reaction and

(2) reduction in permeability due to pore refinement by the extra hydration product deposited by the fly ash.

Much existing research substantiates the assumption that these two mechanisms are primarily responsible for marked improvements in sulphate resistance due to fly ash /D 9, G 10, J 9, K 16, O6, O9/.

As far as permeation and diffusion are concerned, Bakker /B 51/ notes that differences in resistance to sulphate attack are mainly due to differences in permeability to ions and water. Other researchers recognized that it is the ion permeation rate, rather than permeability, which fundamentally affects the rate of attack /B 34, M 11, O 13/.

In designing concretes to withstand sulphate attack, great emphasis needs to be placed on the need for a dense concrete of low porosity /T 17/. Specifically, it is the volume of pores larger than 100μm which is significant for durability /M 38/; in fly-ash concretes, strength and durability are improved by the transformation of large pores into finer pores due to the pozzolanic reaction, although Hansen /H 39/ argues that this is not always the case – pozzolanic reaction can sometimes result in an increase in porosity. Recent results would, however, tend to confirm that pozzolanic reaction generally results in a pronounced refinement of the pore structure /F 31,M 63/.

Since the decrease in permeability of concrete is an important benefit from the use of fly ash, such concretes must be allowed to cure properly /M 38, W 12/. Caution must be observed when high fly-ash contents are used. Venuat /V 2, V 3/ demonstrated that although 20 % and 40 % replacement levels gave good performance in magnesium sulphate solutions, poor performance was observed at 70 and 90 % replacement levels owing to the extremely high porosity of these materials.

Another aspect of the benefit of fly ash in this respect is the reduced water demand. The ability to reduce the water/cementitious ratio for the same slump results in a less permeable concrete /B 39,B 72,S 48/.

Although there has been much controversy as to whether ettringite is formed by topochemical means or through solution /C 32, H 39, M 64/, it is now generally recognised that deterioration of concrete due to expansion and subsequent cracking is caused by the formation of ettringite and also, to some extent, by the formation of gypsum /B 2/. The expansive action of ettringite formation depends, however, on the nature of the pore solution within the material; Regourd et al. /R 45/ noted that expansion is not always proportional to the quantity of ettringite formed. Whether or not the ettringite expands depends on its crystalline form, the granularity of the C_3A from which it is formed and the nature of the interaction between C_3A and C_3S during hydration. Regourd noted that amorphous ettringite may also form; this may be expansive.

The explanation that there are different forms of ettringite was suggested by Kalousek and Benton /K 48/, who classified ettringite into two types: F-poor which is strongly crystallized and exhibits large expansions, and F-rich which has a gel-like nature (i.e. the amorphous form suggested by Regourd). Mehta /M 64/ defines the two forms differently: Type I has a large lath-like morphology which yields high strength and no expansion; this is the type found, for example, in expansive cements. Type II has a small rod-like form which can either be expansive or a source of strength, depending on the environment. Under the right conditions, Type II may adsorb large amounts of water, with subsequent swelling. This agrees with Thorvaldson's /T 17/ earlier suggestion that volume changes in mortars attacked by sulphate are caused not by crystal pressure, but by "osmotic forces concerned with the swelling and shrinkage of gels".

With respect to this second kind of ettringite formation, Mehta /M 64/ outlined the mechanism of sulphate attack: Type II ettringite first forms in hardened concrete under high pH; when the supply of aluminate is exhausted, ettringite formation ceases and gypsum formation begins, resulting in CH depletion which eventually causes CSH to lose strength and stiffness. The weakened CSH and the now favourable conditions for expansive action of Type II ettringite lead to substantial expansion and cracking.

The principal role played by fly ash in reducing expansion and cracking is to reduce the amount of free CH present, preventing the formation of large amounts of gypsum with subsequent loss in strength and stiffness of the CSH /C 29, M 65, R 46/. Reduction of the amount of CH in the so called "transition zone" between aggregate and paste matrix is also important /M 64/; this lime-rich zone is a source of weakness which may be detrimental to both strength (an initiation area for cracks) and impermeability. Pozzolanic reaction within this region depletes the weak CH component and improves homogeneity.

The second role of fly-ash replacement is to reduce the total amount of C_3A in the concrete. The amount of ettringite which forms is directly related to the amount of C_3A present; it is generally recognised that a cement with more than 8 – 10 % C_3A will be susceptible to sulphate attack /G 22, H 40, H 41, K 49/. Yet it may not be entirely correct to assume direct proportionality; Mehta /M 66/ found that 0 % C_3A cements were not necessarily more durable to sulphates under specific conditions (low pH in particular). Mehta also found that fly-ash/opc cements were not resistant to sulphate attack when the pH was controlled at 7 in a 4 % sodium sulphate solution.

It appears that care must be taken in choosing the correct fly ash. Mehta /M 61/ used an accelerated test method and showed that a fly ash of low (15 %) alumina content achieved superior performance in sulphates as compared to a fly ash of high (30 %) alumina content. There is little point in diluting a high C_3A content cement with a fly ash which contains significant amounts of reactive alumina. It is interesting to note that the glass content of the ash may be an important factor in determining whether the aluminates present in the ash are reactive or not. Bogue /B 71/ noted that crystalline C_3A is less resistant to sulphate attack than a glass rich in C_3A. In support, Parker /P 29/ tested pairs of cement clinkers with the same calculated compound composition, but with different glass contents; the high-glass cements were consistently more resistant to sulphate action. The correlation between C_3A and resistance applied solely to the low-glass cements.

3.4.2 Attack by Other Salts and Acids

Some past review papers have dealt with the resistance of concrete to attack by aggressive agents /C 30/, while others have paid special attention to the use of fly ash /B 12, F 2, J 6/. Nevertheless, there appears to be a distinct lack of published data on the general mode of attack. In contrast, there is a great deal of information on specific attack modes, such as those due to sulphates, carbonation and frost action. This section reviews attack by "other aggressors".

3.4.2.1 Sea-Water Attack

The mechanism of sea-water attack on concrete is very complex and has been reviewed in great detail by Calleja /C 30/. A shorter, but no less valuable review has been presented by Mehta /M 70/. The attack is a combination of the physical effect of salt crystallization with carbonation, chemical attack and mechanical, erosional and frost actions. The particular combination of these factors will depend on local circumstances and the location of the section of concrete with respect to tidal movements.

Composition of Sea-Water

A typical sea-water composition, in terms of ionic concentrations, is 18 g/l Cl, 12 g/l Na^+, 2.6 g/l $(SO_4)^{2-}$, 1.4 g/l Mg^{2+} and 0.5 g/l Ca^{2+}. In terms of salt concentration this becomes 2.7 g/l NaCl, 0.32 g/l $MgCl_2$, 0.22 $MgSO_4$ and 0.13 g/l $CaSO_4$. The pH of sea-water is typically about 8, but where there is an unusually high amount of dissolved carbon dioxide, this value can fall to less than 7. At this level it becomes quite aggressive to the hydration products of portland cement /M 70/. The synthetic sea-water used by Regourd in her tests, described below, contained 28.9 g/l NaCl, 2.7 g/l $MgCl_2$, $MgSo_4$, 1.2 g/l $CaSO_4$ and 0.2 g/l $KHCO_3$.

Mechanism of Attack

Chemical attack, which is the predominant form of attack in submerged concrete, is due to a combination of magnesium, chloride and sulphate ions and to dissolved carbon dioxide. Whereas the presence of chloride ions influences the corrosion of reinforcement, it is the magnesium salts that are most harmful to the cement paste. Attack due to the sulphate ions in sea-water is less than would be the case in the sulphate solution alone. This is due to the presence of the chloride ions /C 30/. A detailed study of the effect of the combined action of magnesium and chloride ions has recently been published /F 34/. The mechanism of chemical attack on cement paste may, however, be summarized as follows: At the concrete surface the lime is leached away by the action of the dissolved carbon dioxide. Lime removal will also occur due to substitution of calcium by magnesium. Once the hydroxyl ion (OH^-) supply to the surface is depleted, penetration of magnesium ions takes place: magnesium attack on the calcium silicate hydrates can then occur by conversion to the non-hydraulic magnesium silicate hydrates /F 34, K 48/. The reaction of sulphate ions is slow, and is secondary to the other modes of attack, but leads to the formation of gypsum and ettringite. Thus the aggressive ions, with the exception of chloride, are removed from solution by reaction with the paste. Chloride ions, however, penetrate deep into the concrete and threaten the reinforcement. Although in the absence of carbonation the penetration of sodium chloride has virtually no effect on the pH of concrete, it is effective in destroying the passive (gamma Fe_2O_3) layer of the steel /M 75/. The mechanism, however, is not yet fully understood.

The decisive factor in resistance to sea-water attack seems to lie not in chemical phenomena, but in the porosity, permeability and compactness of the concrete. It is these factors which influence the degree of penetration of the aggressive agents /C 30, R 48/. A cement with a low intrinsic chemical resistance may, for example, behave better than a second cement with a higher intrinsic chemical resistance, if the former has superior impenetrability.

Microcracks exist in concrete at the aggregate-paste and steel-paste interfaces. When these are limited in size and number they are discontinuous and do not affect durability. If, however, they are allowed to grow – by chemical reaction of the concrete constituents and sea-water, or through impact with floating objects – they facilitate transport of aggressive ions to the embedded steel /M 75/. Low permeability concrete will have a reduced tendency for enlargement of these microcracks and will thus have improved durability in the critical region just above the high-water level.

Effect of Fly Ash

The presence of fly ash in a concrete should substantially reduce sea-water attack due to reduced permeability and a reduced quantity of free lime, and to hydration products with a low C/S ratio /F 34/.

The following results emerge from studies both in the laboratory and on in-service concrete: A long-term investigation into the effect of fly ash on resistance to sea water attack is under way at Trent Island on the Eastern coast of the USA in Maine /M 60/. A total of 51 concrete mixes utilizing fly ash and/or slag are under test. Specimens were initially moist-cured for 28 days, followed by at least 30 days at 21 °C and 50 % relative humidity. Inspection by pulse velocity measurement after one year's exposure indicated no major deterioration of any of the samples.

The linear expansion of ISO mortar bars (20 x 20 x 160 mm^3), containing 20 % of fly ash, soaked in artificial sea-water (described above) was monitored by Regourd /R 48/. The bars were stored in fresh water for 28 days before exposure. After one year the fly-ash mortar had expanded by only 0.081 %, whereas the control mortar had expanded by 0.113 %. After three years, however, the fly-ash and plain mortars showed very similar expansions, at 0.188 % and 0.178 % respectively.

A French study /G 22/ tested 20 · 20 · 100 mm^3 mortar bars by half-immersion in a re-constituted sea-water solution of unspecified chemical composition. After two years' exposure, compression tests were performed on the immersed and emergent ends, and the results compared to control specimens which had been stored in fresh water. Ten of the eleven samples containing 15 % fly ash were found to have retained between 50 and 80 % of the strength of the control specimens. Their performance was similar to that of the French cement designed for use in sea-water (equivalent to ASTM Type II) and of cements containing similar amounts of various natural pozzolanas. The performance of the cements containing fly ash was superior to that of the ordinary French cement (equivalent to ASTM Type I). Here, 12 out of 15 specimens fell to less than 60 % of the strength of the control specimens.

Nicholescu /N 25/ showed that a combination of 70 % fly-ash and 30 % blast furnace cement, containing up to 50 % ground blast furnace slag, was able to resist attack only if properly cured. When immersed in artificial sea-water containing twice the salt content of the Black Sea, 141 mm cubes cured for only 72 hours declined in strength to only 95 % of the control strength after one year. Specimens allowed to cure for 28 days (7 days in water and 21 days in air) possessed 99 % of the control strength after one year. These results indicate extremely high resistance to attack, considering the low amount of cement present, which restricts the extent of possible fly-ash reaction. The possibility of a reaction between the fly ash and slag, in the presence of an activator, was suggested.

In China, reinforced concrete specimens containing 36,6 % fly ash were subjected to ten-year exposure tests in different harbour conditions /C 31/. One site was in sea-water containing 2.26 % NaCl, the second combined a 2.1 - 2.9 % NaCl content with an average of 82 annual freeze-thaw cycles and the third was in fresh water. The corrosion was found to be worst in the 1.5 m region above the mean high-water level in sea-water. No corrosion was found in the tidal zone or underwater. After 10 years the corrosion in the reinforcement was found to be worse in the concrete containing fly ash than in the plain cement concrete; the average corrosion rates were 1.7 and 0.4 mg/dm^2 per day respectively. There were no significant differences in any of the underwater concretes or above the high-water level in the fresh water harbour.

General Comments

Results show that the effects of using fly ash vary. Ten-year exposure studies in China showed reinforcement to be more susceptible to corrosion in fly-ash concrete /C 31/. Other tests, however, showed fly-ash mortars to be durable to sea-water attack /M 75, R 48/ and at least equivalent in performance to conventional materials /G 22/.

Although the amount of free lime is reduced by the inclusion of fly ash, the consequent reduction in leachability is probably overshadowed by the greater ease with which an impermeable concrete can be produced. This is especially true in the most vulnerable section of the concrete, that is, just above the high-water level.

3.4.2.2 Acid Attack

Probably the most common acid encountered by concrete structures is sulphuric acid. It can be formed by production from sewage, from the sulphur dioxide present in the atmosphere of industrial cities – especially in the form of acid rain – and even from the oxidation of iron sulphides sometimes present in shales, slags and coals /G 28/. Concrete may encounter other mineral or organic acids in industrial or agricultural applications.

Mechanism of Attack

Most acids attack concrete by a process of dissolution and leaching. Formation of water-soluble products, by reaction of the acid with the components of the cement paste, initially occurs at the surface but subsequently progresses inwards /C 30/. The deterioration is a long-term process but manifests itself as a gradual loosening and softening of the cement paste, eventually causing loss of aggregate particles and finally reducing the concrete to rubble.

Acid strength and concentration (pH) are the parameters governing pure acid attack. Calcium hydroxide is the least stable component and its solubility is increased at lower pH values. If the reaction products from the acidic action on the bases and basic salts in the hydrated pastes are soluble, and if leaching takes place, then deterioration will occur. With organic acids, the order of aggressiveness is defined largely by the solubility of the salt formed.

Deterioration of concrete due to sulphuric acid attack is, however, caused by complex reactions with the hydrated lime and other basic hydrated compounds in the cement paste. It is a two-step process in which the initial formation of gypsum is followed by the formation of expansive ettringite.

Effect of Fly Ash

All components of the paste matrix of concrete are susceptible to acid attack if the pH is less than about 4.5 /C 30/. Substantial attack will accordingly occur even if fly ash is included. The initial rate of attack may, however, be slower, as less highly vulnerable calcium hydroxide is present /T 19/. It should also be noted that the permeability of fly-ash concretes may be very low /M 47/ and that diffusion, controlling the rate of ingress of aggressive ions and removal of soluble reaction products, will be reduced. Most of the attack will, therefore, necessarily take place from the surface – a process which is likely to be fairly gradual.

Fattuhi and Hughes /F 35/ reported an investigation of the resistance of concretes containing various admixtures, including 25 % cement replacement by fly ash, in a continuously flowing sulphuric acid solution. The 102 mm cube specimens were initially cured for 28 days before immersion in the sulphuric acid for 172 days. During this time weight loss was monitored.

The weight loss of the fly-ash concrete was about 14 % after 172 days whereas the plain portland cement control lost approximately 10 %. This indicates that the behaviour of the fly-ash concrete is inferior despite a 4 % water reduction compared to the control and equal 28-day strength (60 MPa). It is interesting that the weight loss of the fly-ash concrete increased from 7.5 % to 14 % during the final 90 days of immersion, whereas that of plain concrete increased only from 7 % to 10 % over the same period.

Examination of the specimens showed the acid attack to be essentially limited to the surface in both cases, proceeding towards the interior of the concrete over time. Steel bars embedded centrally in some of the specimens showed no evidence of corrosion in either type of concrete. There was also no clear evidence of secondary sulphate attack under the test conditions, which entailed constant renewal of sulphuric acid. This was true even in the case of other concretes which exhibited severe acid attack.

Acid attack in both plain and fly-ash concretes was characterized by a white surface layer 1 to 1 mm thick. The pH of this layer was found to be 7.4 as compared to 11.5 for the rest of the concrete. The pH changed sharply at the edge of the coloured layer.

Nicholescu /N 25/ immersed $40 \cdot 40 \cdot 160\ mm^3$ mortar bars of a mixture of blast furnace cement containing up to 50 % ground blast furnace slag and 70 % fly ash in a solution of hydrochloric acid at a pH of 3.0. After 28 days'initial curing in moist air, specimens were immersed for one year. They were then tested in compression and the results compared to those for identical specimens stored in fresh water. The acid-immersed specimens suffered a decrease in strength of only 7 % as compared to the control. This performance was regarded as satisfactory considering the small amount of cement present. The attack manifested itself as surface damage in the form of sand granule detachment.

3.4.2.3 Chloride Attack

Attack on concrete by chloride solutions, usually manifesting itself in the form of reinforcement corrosion but sometimes also as softening of the concrete, is a subject of great interest and importance. The attack may be due to exposure to seawater (sodium and magnesium chlorides), to de-icing salts (sodium and calcium chlorides), to mine shaft water (calcium, magnesium and sodium chlorides) or to other environmental and industrial conditions. It has been the subject of a large volume of research: recent reviews are available /C 30, H 45, S 68/.

3.4.2.3.1 Mechanism of Attack

The influence of chloride salts on concrete may take two forms. One type of attack is always present and is due to the chloride ion; the other depends on the type of chloride salt which is present, and is due to the cation.

3.4.2.3.1.1 Effect of Chloride Ions

Some penetration by ions will always occur when a concrete is exposed to a chloride solution. Nevertheless, corrosion will occur only if the chloride level reaches a critical level at the steel. The depth of this penetration will depend on at least three factors /S 68/:

- chloride binding capacity,
- chloride ion diffusion,
- conveyance by water penetration.

Chloride Binding Capacity

Chlorides in concrete may exist in various forms, such as:

- free chloride ions,
- chlorides bound to the surfaces of the hydration products,
- chlorides chemically bound in the structure of the hydration products,
- chlorides strongly bound in other reaction products.

The relative proportions of the differently-bound chlorides will vary according to the chloride binding capacity of the cement paste. A high binding capacity will mean that only a small part of the total chloride content is dissolved in the pore water and is thus potentially aggressive to the reinforcing steel. The penetration rate of free chloride ions is also reduced in a high binding capacity paste, even if the diffusion rate of chloride ions is not reduced.

Chloride ions do not react with the complex calcium aluminate salt hydrates C_4AH_{13}, C_2AH_8,, C_3AH_6, $C_3A \cdot CaSO_4 \cdot 12H_2O$ or ettringite already formed in the paste /M 71/. They do, however, bind by reaction with the C_3A in the unhydrated cement grains and by adsorption and incorporation into the structure of CSH /R 48/. This interaction causes a change in the CSH from a fibrous to a reticulated morphology.

Up to 0.4 % of chloride (by weight of the cement) may be bound by reaction with the C_3A to form "Friedall Salt" (calcium monochloroaluminate hydrate – $C_3A \cdot CaCl_2 \cdot 10H_2O$) /S 68/. This chemically bound chloride is harmless in terms of reinforcement corrosion. Nevertheless, some free chloride ions will still exist in the pore water in equilibrium with the salt, even at concentrations less than 0.4 % by weight of cement. Midgley and Illston /M 71/ have shown that the quantity of Friedall salt formed is independent of the chloride concentration in the paste; it must therefore be dependent upon the C_3A in the unhydrated grains. They also showed, however, that the chloride ion concentration in the paste is dependent upon the concentration of the solution in which it is immersed.

Chloride Ion Diffusion

The coefficient of diffusion for chloride ions in hardened cement paste is influenced by many factors, including temperature, the type of salt, water/cement ratio, cement type and, in particular, pore size distribution. It is far greater than that of free cations, due to the electropositive character of hardened cement paste when considered as a semi-permeable membrane /P 33/, and between 10 and 100 times greater than that of free sulphate ions. The diffusion coefficients of chloride ions associated with various cations rank in the following order of decreasing magnitude:

- magnesium chloride,
- calcium chloride,
- lithium chloride,
- potassium chloride and
- sodium chloride.

It may generally be true that the greater the water/cement ratio, the greater the depth of penetration of chloride ions /H 45, M 7/. Nevertheless, Gjorv and Vennesland found this to be true only for surface layer penetration, and concluded that time is the major factor governing penetration into the mass /G 29/.

It is generally agreed /H 45, M 71, P 33, S 68/ that the most important factor influencing diffusion of free chloride ions is the pore size distribution, which, in turn, is influenced by the type of cement and the water/cement ratio. It is especially interesting to note that the pore size distribution is shifted towards smaller pores by the very penetration of chloride ions /M 71/. The greater the amount of chloride present, the smaller the pores.

In general, diffusion is reported to obey Fick's Law /G 30, S 68/. This law states that the rate of diffusion of matter across a plane is proportional to the negative of the rate of change of concentration of the diffusing substance in the direction perpendicular to the plane.

Conveyance by Water Penetration

This also obeys Fick's Law, but is likely to convey greater amounts than does pure chloride diffusion /S 68/, especially if influenced by capillary suction or high hydrostatic pressure /H 45/. A high chloride penetration rate will result in a high permeability concrete. This high permeability may be the result of improper mix design, poor concreting practice or cracks in the concrete. The effect of the type of cement is secondary if any of these conditions were present and resulted in a concrete that is porous, inhomogeneous or cracked.

In a dense, homogeneous and uncracked concrete, conveyance by water penetration will depend upon the permeability of the concrete itself, which is determined by the pore structure. Apart from this, penetration by chlorides depends on the chloride-binding capacity of the cement paste. The penetration of chloride reduces the permeability of the cement paste, by displacing the pore size distribution towards smaller pores /M 71/. To some extent, chloride penetration may be self-diminishing.

Midgley and Illston /M 71/ have shown that the penetration of chlorides into hardened cement follows the power relationship

$$C = k \cdot d^{-m} \qquad (3.4.1)$$

where
C is the concentration of chloride,
d is the distance into the paste and
k and m are constants depending upon time and permeability.

3.4.2.3.1.2 Effect of Cations

Much attention has been paid to the importance of the chloride ion, but it is now recognised that cations are always involved in the deterioration process /F 36/. The order of rates of diffusion of chloride ions into concrete was given above. This ranking does not, however, coincide exactly with the order of importance of the destructive effects. The difference is due to the influence of the cation. In decreasing order of their destructive effects /C 30/, chloride cations may be ranked as follows:

- magnesium chloride,
- calcium chloride
- lithium chloride,
- sodium chloride and
- potassium chloride.

The results of these differing degrees of aggressiveness may be illustrated by comparing the effects of two common de-icing salts: calcium chloride and sodium chloride. Partly owing to its low pH, calcium chloride causes leaching of calcium hydroxide and loss of strength of the concrete; sodium chloride does not. Calcium chloride attack is highly dependent on temperature. This is due to the formation of an expansive reaction product which is stable at around 5 ^{0}C and causes disruption, but is not found at 20 ^{0}C or 40°C /C 30/.

Solutions of magnesium chloride are particularly aggressive. Magnesium ions may react with calcium hydroxide to form magnesium hydroxide, and may form oxychloride ($Mg_2(OH)3Cl \cdot 4H_2O$) in the presence of chloride ions . Calcium silicate hydrate may also be broken down gradually. The presence of sodium and calcium chlorides enhances the solubility of many of these reaction products /M 70/, adding to the complexity of the attack mechanism when the solution is a mixture of chlorides. Leaching of reaction products leads to a softening of the structure, loss of strength and increased porosity. This in turn leaves the concrete vulnerable to further attack.

Once the chloride concentration at the reinforcement has reached the critical level, corrosion of the steel will occur due to breakdown of the passive layer. Steel corrosion is expansive and results in cracking of the concrete. This cracking increases the rate of ingress of the aggressive media. This process, showing cracking to be the most important factor in the acceleration of reinforcement corrosion, has been summarized by Katawaki /K 50/ in the form of a flow chart:

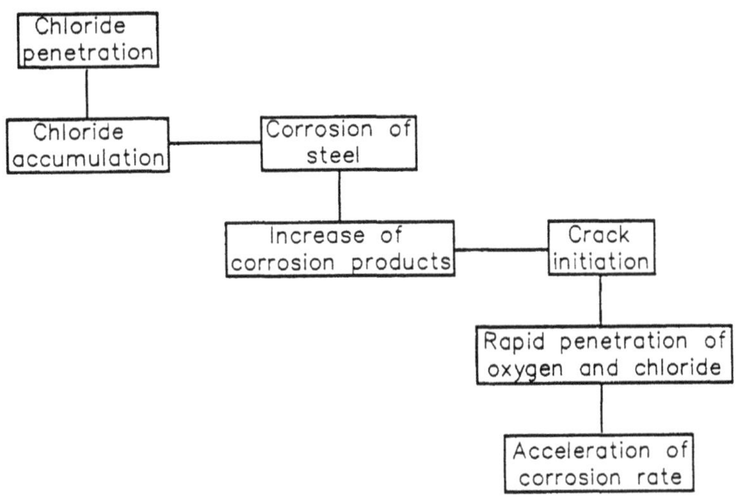

3.4.2.3.2 Effect of Fly Ash

Page et al. /P 33/ used 3 mm thick circular discs of hardened cement paste to measure the diffusion coefficient of sodium chloride. Comparative tests were made for specimens of plain portland cement, portland cement with 30 % fly-ash replacement, portland cement with 65 % blast furnace slag replacement, and sulphate-resisting portland cement. All the pastes were cast at a water/solid ratio of 0.5 and were cured in calcium hydroxide solution at 22 ^{0}C for 60 days before testing.

The effective diffusivity of chloride ions, at 25 ^{0}C was found to be:

	$D(m^2/s \times 10^6)$
Ordinary portland cement	4.47
Ordinary portland cement + 30 % fly ash	1.47
Ordinary portland cement + 65 % blast furnace slag	0.41
Sulphate-resisting portland cement	10.0

The effective diffusivity of the paste containing fly ash is only one third that of the ordinary portland cement, despite its higher porosity. Page et al. suggest that the reason blended cements are more effective in limiting chloride ion diffusion is to be sought in differing mechanisms of diffusion rate control. The difference is probably due to the substantially reduced permeability resulting from pore blocking /M 72/. Another factor causing a lower effective diffusivity, especially in the paste containing blast furnace slag, is an increase in chloride-binding capacity /H 45/.

Feldman and Remachandram /F 37/ investigated the effect of a mixture of salts in solution on the durability of hardened cement paste specimens. The solution comprised 27.5 % calcium chloride, 3.9 % magnesium chloride, 1.8 % sodium chloride and 0.1 % sodium bicarbonate; this corresponds to mine shaft water. Cement mortar discs, 6.4 mm thick, containing 20 % and 35 % of fly ash in sulphate resisting cement, were compared to plain sulphate-resisting cement and Type I cement mortars. The specimens were cured for periods of 15 or 140 days before testing.

Durability was assessed by measuring non-destructive deflection under constant load and was compared to that of control specimens stored in water. The small size of the specimens meant that any deterioration of the material would be observed much earlier than in real structures. The results showed that after 15 days' curing, the sulphate-resisting cement mortar failed first, followed by the two fly-ash mortars, then the Type I mortar. After 240 days' curing, however, the sulphate-resisting cement mortar still failed first but the fly-ash mortars were much more durable than the Type I mortar. The performance of the 35 % fly-ash mortar was slightly better than that of the 20 % fly-ash mortar under both curing conditions. The results tended to confirm that the susceptibility of the paste increases with an increasing quantity of free calcium hydroxide. They also show that prolonged curing is necessary in order for fly ash to be effective. This may, however, present problems in practice.

An examination of the salt-water channels at Munmorah power station in Australia after 18 years of service /S 49/ showed no detectable difference between plain and fly-ash concretes. These concretes were manufactured using fly-ash replacement for cement on a weight-for-weight basis.

3.4.2.3.3 General Comments

It is quite clear that the production of a high-quality, impermeable concrete is of the greatest importance in guarding against attack by chlorides. If the concrete is permeable or cracked, then the composition of the cementitious component is of little significance. The use of fly ash in a made, impermeable concrete will, however, reduce chloride and cation penetration due to changes in the pore structure and chloride binding capacity of the cement paste fraction.

3.4.2.4 Carbonic Acid Attack

Carbonic acid attack on concrete can occur when it is exposed to almost pure water such as the soft, natural acid-waters in mountain and moorland streams and water from melting snow. The carbonic acid in natural waters is essentially formed by organic processes or volcanic activities. Enriched carbonic acid may also be present in wastewaters along with other aggressive substances /E 24/.

The conditions governing carbonic acid attack are detailed by Calleja /C 30/ but the degree of aggressiveness generally depends on the pH, on lime content (hardness) and on the amount of excess carbon dioxide. At a given carbon dioxide content, aggressiveness decreases with increasing calcium content. Thus, a low pH water containing much dissolved calcium may be less aggressive than a higher-pH water which is less mineralized or not mineralized at all. The excess carbon dioxide is that component which is not used to maintain a saturated solution of calcium bicarbonate. All excess carbon dioxide may be considered aggressive. In its natural tendency toward neutrality, the carbonic acid attacks the alkaline hardened cement paste in mortar and concrete. Initially, it reacts with the calcium hydroxide to form calcium carbonate. The carbonate is precipitated in the pores and may, in fact, have a beneficial effect, even stopping the reaction. If, however, there is excess carbon dioxide in the solution, this then reacts with the carbonate to form the more soluble calcium bicarbonate. The attack is not, however, restricted to the calcium hydroxide but extends to the calcium aluminate and silicate hydrates.

In the Federal Republic of Germany, DIN Standard 4030 /X 28/ defines the classes of attack according to the carbon dioxide content of the solution:

- "low concentration" ≤ 30 mg CO_2/l,
- "high concentration " ≤ 60 mg CO_2/l,
- "very high concentration" 60 mg CO_2/l.

According to Calleja /C 30/, the very high concentration solution will lead to serious attack. Mlodecki /M 69/ states, however, that concrete is not resistant to an aggressive carbon dioxide content greater than 15–20 mg/l.

The quality requirements of the concrete, especially the density requirement, will depend upon the degree of attack to which the concrete will be exposed /F 33/. The extent of the attack will depend on the thickness of the member and the duration of attack. It can generally be stated, however, that the corrosion resistance of concrete to carbonic acid can be increased by improving concrete quality. The very low permeability obtainable only with a properly cured fly-ash concrete can thus be expected to increase the durability of this material to a value above that of plain portland cement concrete.

Resistance of concretes to carbonic acid attack has been tested in several ways. Bertacchi /B 76/ immersed specimens in tanks of distilled water through which carbon dioxide was bubbled. The water was changed periodically over a period of seven years because of the build-up of dissolved lime. The degree of attack was assessed in terms of weight loss, amount of leached lime and loss of strength. Ferric-pozzolanic cement (composition unstated) lost 172 g/kg of concrete from a $40 \cdot 40 \cdot 160$ mm^3 specimen over seven years. A plain portland cement concrete suffered a weight loss of 287 g/kg over the same period. The amount of leached lime (77,4 g/kg) was also appreciably less than for the plain concrete (116,6 g/kg). The loss of strength in compression was less than for the plain concrete, at 80,6 % as against 92,6 %; both values are, however, extremely high. At three years, the strength losses were 63,4 % and 72,4 % respectively.

Mlodecki /M 69/ proposed two test methods: an accelerated stationary test and a flow method. The former method has the advantage of allowing quantitative analyses of the chemical reactions between the alkaline components of cement and the acid to be assessed. In other words, the degree of damage to the concrete can be related to the quantity of the aggressive solution causing the problem. In this method, a sample is immersed in an acid solution and the time taken for the solution to be neutralized is measured. This procedure is then repeated until resistance to attack can be assessed by measuring strength, weight loss and the change in the aggressive solution. The flow method is an attempt to simulate the natural flow of acid waters and, as such, is useful only for comparative studies of different materials.

In a study of the effect of the use of slag in concrete, Efes /E 24/ soaked mortar bars in a solution containing 100 mg/l of CO_2 for five years. He found that degradation was preceded by considerable leaching of calcium oxide. This eliminated the alkalinity of the external zone of the mortar. X-ray diffraction analysis of this corroded zone showed that all the free lime (calcium hydroxide) had been removed, together with some of the combined lime (calcium aluminate and silicate hydrates).

Friede /F 33/ has deduced a calculatory method for estimating the corrosion resistance of concretes. The estimate is based on measured mass changes in laboratory specimens and on the depth of the corroded zone in the eroding and still existing component. This method also allows long-term predictions of the corroded layer thickness.

Friede also pointed out that corrosion causes changes in bulk density, dynamic elastic moduli and strength. Eventually, dimensional changes will occur, possibly followed by complete destruction of the concrete. The depth of the corroded zone will increase gradually until dimensional changes occur. The corrosion rate will, however, decrease as the path length for inward diffusion of carbonic acid and the outward diffusion of the dissolved corrosion products increase. This rate will again increase once dimensional changes take place.

None of these methods has, however, been applied to the study of concretes containing fly ash. Data on this subject are concequently very sparse. Pozzolanic cements are generally considered to be more resistant than portland cements to leaching of calcium hydroxide caused by soft water, especially if the latter are rich in C_3S. Plain portland cements are not considered adequate for very pure water or for water containing aggressive carbon dioxide. In one of the few published studies, however, the use of Romanian fly ash from brown coal and bituminous coal as a cement replacement has been found to be beneficial in terms of carbonic acid resistance /T 8/; unfortunately, however, no actual test results are given in the publication. This also applies to the use of other pozzolanic materials /B 76/. The consumption of free lime by fly ash to form insoluble aluminates and silicates is suggested as the reason for the beneficial effect. Comparative tests showed fly ash to be more effective in controlling carbonic acid attack than either natural pozzolanic material or inert filler (at the 50 % replacement level) /K 5/. The consumption of the free lime by the fly ash makes the cement paste, and hence the concrete, less vulnerable to the "lime-hungry" water /T 19/. It should, however, be remembered that this enhanced resistance may not be obtained if exposure is allowed before significant pozzolanic reaction has occurred.

Concrete containing 33 % by weight of the minimum 350 kg/m^3 cementitious content, to obtain a characteristic strength of 40 MPa at 90 days, was used for the tunnel linings of a recently constructed power station in Wales /C 10/. This concrete, at a water/cementitious ratio of 0.43, was specified to resist the attack of the mildly acidic, soft moorland water. This water contained 6.3 ppm of free carbon dioxide and, at a pH of 5.6, was regarded as aggressive to plain portland cement concrete. No problems were experienced in the first two years, however. No long-term results are yet available, as construction was so recent.

3.4.3 General Comments on Attack by Aggressive Agents

The most important characteristic determining long-term durability in aggressive environments is the permeability of the concrete, which itself depends on the permeability of the paste. Chemical attack occurs only if the concrete is permeable. Permeability is fundamental to the rate of chemical attack because of the primary role played by the movement of aggressive ions into, or of dissolved reaction products out of the concrete. Partial replacement of cement by fly ash can, under suitable conditions, reduce the permeability of paste by reducing the volume of large pores and, more importantly, by blocking pores.

The use of fly ash changes the chemical composition of the cementitious material and hence of the hydration products: this has a pronounced effect on durability, especially with respect to chloride penetration. The combination of chemical composition and physical properties (notably fineness), which determines the rate at which hydration proceeds, is also affected by the use of fly ash. This must affect permeability, at least at early ages, when the ash is inert and the effect is similar to that of a reduction in cement content. At later ages, however, the ash contributes to the cementitious components, but in a way which alters the proportions of the usual hydration products. The stabilization of free lime through the formation of less reactive calcium silicates and aluminates is generally considered beneficial in terms of resistance to aggressive environments. It is clear, however, that sufficient time should be allowed for significant pozzolanic reaction to occur before enhanced performance can be expected from the use of fly ash. Premature contact with aggressive media should therefore be avoided, in order to prevent rapid, intense chemical attack. The performance of fly ash concrete improves with the length of curing.

3.4.4 Alkali-Aggregate Reaction

Before 1940, aggregates were generally assumed to be an essentially inert and chemically unreactive component of concrete. It is now agreed that all aggregates are reactive to a greater or lesser degree. Some reactions may be of a beneficial nature, but others may result in serious damage to the concrete, owing to abnormal expansion with the accompanying cracking and loss of strength /W 22/.

Stanton /S 69/ was one of the earliest researchers to identify the deleterious effects which could result from a chemical reaction between certain hydroxyl ions in the pore water of the concrete and the poorly ordered forms of silica present in some aggregates. This reaction was originally referred to as "alkali-aggregate reaction" (AAR), but has since been more properly designated as "alkali-silica reaction" (ASR) /X 29/.

Stanton's early work triggered considerable research until the early 1950s when the number of articles decreased, interest reviving only in the mid-nineteen-sixties /R 47/. The reasons for this renewed interest include:

- Production of portland cements with higher alkali contents due to energy considerations and pollution controls: Brotschi and Mehta /B 73/ cite data for a dry process suspension preheater plant showing that the amount of energy required to reduce the alkali content of the cement from 0.73 to 0,58 % almost tripled in terms of bypass heat and dust per ton of clinker.
- The use of higher cement contents in modern concretes, increasing the total alkali content per unit volume and possibly triggering unexpected reactions.
- Depletion of sources of high-quality aggregates, entailing the use of larger quantities of marginal aggregates and increasing the risk of AAR.

The magnitude of the problem in specific areas has been demonstrated by Semmelink /S 64/, who notes that in the Cape Town region of South Africa more than 50 % of the concrete structures built in the last 15 years have exhibited some form of deterioration due to the use of reactive aggregates and the use of a moderately high-alkali cement.

3.4.4.1. Types of Alkali-Aggregate Reaction

Most of the known reactions involving aggregates in concrete are due to the alkaline character of the pore solution and are designated "alkali-aggregate reactions"

(AAR)/D 39/. These reactions are known to occur with at least two distinct kinds of aggregates, and a third category has been suggested /G 25/. To date, little is known about this reaction /D 40/ and the category has consequently not found universal acceptance /M 67/.

Alkali-Silica Reaction

The alkali-silica reaction (ASR) involves aggregates containing more then 94 % silica /R 47/. The most important alkali-reaction aggregates are forms of opal, chert, dolomites, tuffs, shales, phyllites and microcrystalline or strained quartz /X 29/. Quartz is relatively unreactive owing to its ordered structure of Si-O tetrahedra; conversely, reactive silica has a random network of tetrahedra with irregular spacing, which entails a high surface area readily attacked by the alkalis present in pore water.

The most generally accepted expansion theory /D 39/ suggests that silica aggregate reacts "in situ" with the alkaline solution to form alkali-silica glasses or relatively dry gels. These gels may vary from a high Ca-alkaline non-expanding gel to a high alkali-silica gel with substantial expansive properties /R 47/. If free water is available, it is absorbed by the gel, which then expands, generating sufficient expansive forces to damage the surrounding cement paste matrix. A variety of reactions have been observed: some aggregates maintain their rigidity and expand while others convert to gelatinous products in situ /D 39/.

Alkali-Carbonate Reaction

While some carbonate aggregate reactions may be beneficial in that they enhance the bond between cement paste and aggregate, certain carbonate aggregates produce the typical map cracking associated with expansive aggregates /H 42/. Reactive aggregates are thought to exhibit the same general features, i.e. dolomite rhombs in a fine-grained calcite matrix with finely divided clay and calcite. Expansive carbonate rocks contain small areas ranging in composition from 40 to 90 % dolomite as a percentage of the total carbonate, 5 to 49 % of acid insoluble residue and equal volumes of calcite and dolomite, and often containing illite and chloride clay minerals /W 23/.

This type of reaction does not generate the same reaction products as ASR. The expansion mechanism is attributed to a "dedolomitization reaction", in which the alkalis attack the dolomite constituent of the carbonate rock /H 43, S 65/. Hadley /H 43/ suggested that the alkali carbonate reaction product could react with the $Ca(OH)_2$ from the cement hydration to produce NaOH, which further attacks the dolomite crystals until all the dolomite has reacted or all the alkalis have been used up. These reactions are unlikely to be responsible for the expansion, as the volume of solid products from the dedolomitization reaction is less than the dolomite replaced /K 3, D 35/. Swenson and Gillott /S 66/ suggest that the dedolomitization reaction is necessary to allow access of moisture to a previously unswelled included clay (illite and chloride). The moisture absorbed by the clay is responsible for the expansion. This mechanism is supported by results from Feldman and Sereda /F 32/.

Alkali-Silicate Reaction

Gillott /G 25/ proposed this category to draw a distinction between reaction of various forms of silica (those included under ASR) and reactions involving complex rock types in which other silicate minerals form the active component (greywackes, phyllites, siltstones, etc.). This category is not clearly defined, as many of these rocks include fine quartz plus silicates, and some may contain dolomites /D 39/.

3.4.4.2 Factors Other than Fly Ash Affecting the Alkali-Aggregate Reaction

Alkalis

Alkalis in cement are derived primarily from the raw materials and the fuel used in the manufacturing process, and may be present in a soluble or insoluble form. The soluble form is present largely as sulphate while the insoluble alkalis are present in the C_2S, C_3S and C_4AF components. The alkalis in the paste pore solutions are usually present as hydroxides, the hydroxyl being derived from lime formed as one of the hydration products /G 25/. This accounts for the high pH of the system. Both sodium and potassium hydroxide cause AAR expansion, sodium hydroxide being more expansive /S 65/. The limit of 0.6 % Na_2O for low-alkali cements was established very early /S 69/. However, this assumes that both sodium and potassium produce the same effect, which cannot be substantiated. It has been suggested that the absence of expansion in concretes with low alkali contents may be due to retention of the alkalis by calcium silicate hydrate, lower calcium/silicate ratios being more effective than higher ones, since they lead to higher $Ca(OH)_2$ contents /G 26/.

The most common method of preventing AAR is to reduce the alkali contents. This is not always successful because of :

- the presence of highly reactive aggregates,
- the presence of additional alkalis through ground water or deicing salts and
- the use of higher cement contents /G 26/.

When using reactive aggregates, it is, however, important to limit the total alkalis in the concrete, not just in the cement. Moisture migration has been shown to concentrate alkali ions near the surface from which moisture is evaporating /N 24/.

Pettifer and Nixon /P 30/ have suggested that sulphate attack on concrete by the sulphates of alkali metals can additionally promote AAR on susceptible aggregate.

Moisture

Moisture has been shown to be a requirement for all classes of AAR /G 25/. Ludwig /L 42/ states that the critical humidity required to prevent expansion is 85 %. Vivian /V 15/ found that mortar bars containing reactive silica at a low water/cement ratio do not expand and that storage of specimens with normal ratios at low RH reduces expansion. The converse has been noted for the alkali-carbonate reaction, where a lower ratio may lead to greater expansion /S 65/. In some instances, a drying period during testing results in a reduced rate of expansion /G 25/.

Temperature

It has been suggested that an increase in temperature increases the AAR. However, there is disagreement about the temperature at which maximum expansion occurs. For example, Duncan et al. /D 41/ showed that for most of the aggregates tested the expansion increased with temperature. Ludwig /L 42/, however, found more severe damage at room temperatures.

Admixtures

McCoy and Caldwell /M 74/ have proposed the use of lithium compounds to control expansion, while Jensen et al. /J 27/ introduced an air void system in an attempt to reduce the expansion due to ASR. Results showed that the voids were filled with gel, 4 % of air reducing expansion by 40 %. However, the freeze-thaw resistance may be reduced as a result of void filling.

3.4.4.3 Effect of Fly Ash on the Alkali-Aggregate Reaction

Stanton /S 67/ may have been the first to recognize the beneficial effects of pozzolana in reducing the expansion due to AAR. Early tests by Blanks /B 13/ indicated that fly ash was more effective in reducing expansion at later ages.

Numerous other investigators (e.g. Pepper and Mather /P 31/) have found that the addition of a sufficient quantity of a fine reactive material to a potentially expansive concrete mix will inhibit expansion. The usual level of replacement is 20–30 % by weight of the cement; Owens /O 5/ and Sutton /S 24/ recommend a minimum replacement level of 25 %. Brink and Halstead /B 19/ found that the 20 % replacement level showed varying effectiveness between ashes, while a 10 % replacement produced no significant reduction in expansion. Crow and Dunstan /C 14/ reported similar findings. Bradbury /B20/ found that at least 30 % replacement is required to reduce expansion to a level similar to that produced by low-alkali cement. Johnston /J 28/ demonstrated the effectiveness of 20–30 % fly-ash replacement of cement in ensuring dimensional stability even when waste glass is used as a coarse aggregate.

Results of tests carried out by Kordina and Schwick /K 25/ on concrete bars also indicate that a minimum of 20 % replacement of cement by fly ash is needed in order to reduce ASR. Beyond a certain minimum, the expansion may increase if the pessimum conditions for the silica/alkali ratio have not yet been attained in the specific concrete without fly ash, but are brought into being through the addition of SiO_2 with the fly ash.

The mechanism by which a pozzolana reduces the AAR is not fully understood. To date, fly ash and ground granulated blast furnace slag are the most widely used materials. Silica fume is now being utilized more extensively, because of its faster reactivity.

Factors influencing the effectiveness of pozzolanas are the change in hydration (reaction) products, the change in the reaction rate, decreased permeability and the change in concentration of the pore water solution.

Powers and Steinour /P 32/ suggested that, given the presence of enough pozzolanas, a non-expanding lime-alkali-silica complex forms in preference to the expansive water-absorbing alkali-silica gel. Gratton-Bellew /G 26/ states that the reactive components in pozzolanas are silicates which react with the $Ca(OH)_2$ to form a calcium-silicate-hydrate similar to that formed by cement. Together with the reduction of porous $Ca(OH)_2$, formation of this extra gel decreases the permeability of the paste. The removal of lime was suggested as a significant factor in the ASR-reducing role of pozzolana.

The significance of the pore-solution chemistry has already been noted. Ming Shu et al. /M 68/ suggest that the lower the basicity of the reaction products, the more alkalis they might retain. It was also felt that mineral admixtures absorb alkalis from the pore solution during the initial stages of hydration, even if the alkali content of the admixture is high. Admixture of larger amounts of minerals may absorb the alkalis, preventing their release to the pore solution. This effect may be attributable to the fineness/surface area of the pozzolana.

Crow and Dunstan /C 14/ were unable to correlate the alkali contents of the ash with the AAR, although they did feel there was some minimum replacement level for fly ash in respect to AAR. Powers and Steinour /P 32/ suggested that lower expansion is due to reduction of the alkali content of the pore solution through reaction at the large surface area of the pozzolana. Oberholster and Westra /O 19/ also felt that the surface area of the pozzolana may be significant in ASR. Hobbs /H 25/ suggests that fineness may be important in reducing ASR; test results indicated that the coarsest fly ash was correlated with the highest expansion.

Gratton-Bellew /G 26/ stated that when the pozzolana reacts with the $Ca(OH)_2$, the gel which is formed incorporates the Na^+ and K^+ ions, reducing the concentration in the pore solution and hence the AAR. Butler /B 34/ also felt that the beneficial action of fly ash in reducing the ASR is due to entrapment of the Na^+ and K^+ ions in the CSH gel formed by the pozzolanic reaction. In addition, the movement of the hydroxyl ions associated with the alkali metal ions is severely inhibited by formation of the additional CSH gel. While investigating the ways in which blast furnace slag cement affects AAR, Bakker /B 74/ studied the effect of the alkali content of the pore solution on the expansion of mortar bars. Test results indicated that even when Na^+ is added to a mix, expansion is reduced, negating the concept that pozzolanas are effective in diluting the alkalinity of the pore solution. Ludwig /L 42/ investigated the effect of alkali salt admixtures on the ASR. The pore solution had a higher alkali concentration and the rate of deterioration of these mortar bars was accelerated significantly. Diamond /D 6/ tested two Danish fly ashes and found that neither contributed alkalis to the pore solution; one was essentially inert while the other extracted a small amount of alkali.

Blackie /B 26/ points out that, although much discussion has centred on the availability of sodium and potassium ions in the constituents of the concrete, only the water-soluble alkalis are significant.

Diamond and Lopez-Flores /D 20/ studied the alkali metal content of the ash to establish the extent to which they appear as alkali hydroxides in the pore solution. Both low and high lime ashes were studied, with measurements of the pore solution concentrations of K^+, Na^+, Ca^{2+} and OH^-. Results indicated that the Ca^{2+} concentration in the pore water is low, falling to zero at 90 days, and that the sum of the cations is effectively equal to the concentration of OH^- anions. After seven days, the pore solutions are essentially solutions of sodium and potassium hydroxide. The high sodium contents of the two high-calcium fly ashes represent alkali readily mobilized into the pore solution, substantially increasing the long-term concentration of alkali hydroxide. The high potassium contents of the low-calcium ashes are not mobilized in the pore solution.

Rayment /R 25/ investigated the reduction in the C/S molar ratio of CSH when fly ash is used as a partial replacement for portland cement, and attempted to establish whether this reduction would cause an increase in retained alkalis. While there was a small reduction in the C/S molar ratio, accompanied by a small increase in retained alkalis in the hydrate, it was not clear whether this sufficed to explain the effectiveness of fly ash in reducing the AAR.

Diamond et al. /D 7/ have observed "duplex films" which develop rapidly in hydrating cement systems around all exposed grains including fly ash. The film is a continuous layer of calcium hydroxide plus a thin layer of CSH gel particles which has been seen to form on any inert surface in contact with solutions developing in hydrating cement paste. Long-term pozzolanic reactions must take place through this film, presumably by diffusion; moreover, the need for alkalis to diffuse through this "skin" may influence the AAR-reducing potential of fly ash.

A number of researchers – including Ramachandram et al. /R 47/, Hobbs /H 25/ and Idorn /I 3/ – have suggested that the replacement of portland cement by pozzolana diminishes the total amount of alkali in the mix and that this "dilution effect" reduces or prevents AAR. Other researchers have disputed this scenario. Gaze and Nixon /G 20/ studied mortar bars similar to Hobbs /H25/ and found that the reduction in expansions was too substantial to be explained by simple dilution. They also studied specimens where the alkali level was held constant; fly-ash specimens still exhibited a reduction in expansion when compared to portland cement bars. Gutt and Nixon /G 12/ noted that fly ashes with high alkali contents are significantly less effective in reducing AAR.

It should be re-emphasized that some pozzolanas contain significant amounts of alkalis. Gratton-Bellew /G 26/ notes that the total or acid-soluble alkali contents may vary from 0.48 % for silica fume to 3.78 % for fly ash, these values being based on limited data and so unlikely to represent the upper limit for fly ash. Hobbs /H 18/ stated that the amount of cracking is related to the quality of reactive aggregate present and the total amount of alkali available in solution. The amount of available alkali depends on the mix proportions, on the water-soluble alkali content of cement and aggregate and possibly on the diffusion of alkalis from external sources. Test results indicate that cracking occurs in mortar bars with total water-soluble alkali contents (equivalent Na_2O) in excess of 2.5 kg/m^3. This value includes all sources of alkali.

Nixon and Gaze /N 14/ found little correlation with total or available alkali content and expansion. One fly ash exceeded 1.5 % available alkali but was very effective in reducing expansion.

Using a mortar bar test, Ming Shu et al /M 68/ investigated the effects of tuff, fly ash and ground granulated blast furnace slag on the ASR, establishing that their effectiveness in reducing ASR is ranked in the above order. They suggested a reduction mechanism based primarily on the acidity of the admixture and the basicity of the cement. Results indicate that the higher the acid oxides of the admixture, the greater will be its effect in preventing expansion. The addition of CaO to the cement increased expansion, so low basicity is required to reduce expansion. Basicity is defined as the ratio of basic oxides (CaO) to acidic oxides (SiO_2 + Al_2O_3 + Fe_2O_3). Comparison of cements with the same alkali contents showed that those with a lower basicity underwent less ASR.

The effect of the hydroxyl ions present in the pore solution has also been investigated. Fly ash may lower the hydroxyl concentration to a point at which the alkali silica reaction is reduced. Hobbs /H 18/ stated that his results showed no evidence that hydroxyl ions are depleted by fly ash. Butler et al. /B 75/ suggest that the reaction of fly ash with $Ca(OH)_2$ thickens the gel produced by hydration, reducing the mobility of the hydroxyl ions.

Various researchers (e.g. Hobbs /H 44/) have noted that there is often a "pessimum content" of reactive aggregate for a given mix and cement alkali content. This is the content which produces maximum expansion; as the content of reactive aggregate increases beyond this point, expansion decreases. Nixon and Gaze /N 14/ found that the type of aggregate affected the pessimum value. If the beltane opal content is increased at a constant alkali level, a pessimum value is obtained, whereas with Pyrex aggregate the expansion is proportional to the aggregate content. The introduction of fly ash reduces the expansion, but behavior is similar. Gaze and Nixon /G 20/ also noted an increase in expansion when the alkali content was reduced from 1.18 % to 1.02 % (equivalent Na_2O), again indicating a pessimum content. These results indicate the existence of a critical alkali/silica ratio for mixes with opal aggregate.

3.4.4.4 Test Methods

Although the accelerated test methods currently in use do not simulate in situ conditions, they do provide an estimate of an aggregate's potential performance. The optimum test method depends on the type of AAR. An excellent review of the test methods and their evaluation is given by Dollar-Mantuani /D 42/ and a very brief review of current test methods is provided below.

Mortar Bar Test (ASTM C227) /X 23/

This test method measures the expansion of mortar bars containing specific aggregate-cement combinations, and is one of the most reliable techniques /X 29/.

However, instances are cited /D 4/ in which the reaction takes much longer to reach a limit, usually a problem when dealing with potential alkali-silicate reactions /R 47/. Swenson and Gillott /S 65/ have indicated that this test cannot be used to assess potential alkali-carbonate reactivity.

Concrete Prism Test (ASTM C157) /X 24/

This test is similar to ASTM 227 but uses larger specimens. Gratton-Bellew / G 27/ notes that this test procedure is best for alkali-carbonate reactions and for slowly expanding siliceous aggregates.

Rock Cylinder Test (ASTM 586) /X 25/

This test was designed to study the alkali-carbonate reaction. It does not appear to be a good indicator of ASR, owing to breakdown of the specimens /G 27/.

Chemical Method (ASTM C289) /X 26/

This is a rapid test which produces results in three days. The test measures the amount of silica dissolved, and is used primarily to detect ASR.

Mortar Bar Test (ASTM C441) /X 27/

This test method is designed to determine the effectiveness of mineral admixtures at preventing expansion due to AAR. The procedure is similar to that in ASTM C227, but in this case a Pyrex aggregate is specified as the reactive material. This choice has been criticized by Hobbs /H 25/, who argues that the use of Pyrex is unsatisfactory. It is too dissimilar from natural aggregates, since it is non-porous and contains significant amounts of alkalis.

3.5 Carbonation (Prepared by P. Schubert)

3.5.1. Definition

Carbonation is the reaction of carbon dioxide in air with the calcareous components of hardened cement paste. Initially, this means that the $Ca(OH)_2$ formed by the hydration of portland cement clinker reacts to form $CaCO_3$. Carbonation considerably reduces the alkalinity of the pore water in the hardened cement paste, from pH > 12.6 to < 9. As a result, the active corrosion protection of the steel reinforcement in the concrete is lost and the steel can corrode in the presence of oxygen and water. The rate of carbonation is influenced mainly by the density of and by the quantity of lime in the hardened cement paste, together with the atmospheric conditions (humidity and CO_2 content of the air). The use of fly ash in concrete affects its density and lime content and may therefore be assumed to influence its carbonation behaviour.

3.5.2 Alkalinity of the Pore Water

Before carbonation, the alkalinity of pore water in the hardened cement paste is determined by the content of soluble alkalis and their effect upon hydration. Fly ashes (mainly coal fly ashes) react with $Ca(OH)_2$ to form similar CSH phases as the cement hydrates. As $Ca(OH)_2$ is consumed, the alkalinity of the pore water falls. The consumption depends on the amount of reactive silica (SiO_2) in the fly ash and hence on the duration of hydration.

The alkalinity of the pore water is determined not only by $Ca(OH)_2$ but also by the alkalis Na_2O and K_2O. These continue to have an important effect, unlike $Ca(OH)_2$, which is of importance only in the initial stage. Alkalinity in relation to the alkalis is, however, maintained by the much higher supply of CaO in the hardened cement paste.

Investigations made on water-coal-fly-ash dispersions /C 8, R 22/ have shown that the pH at first falls rapidly to 4 or 5 and later rises quickly or slowly to a value above 7 (*Figures 3.5.1, 3.5.2*). The low initial value is attributed to the absorption of

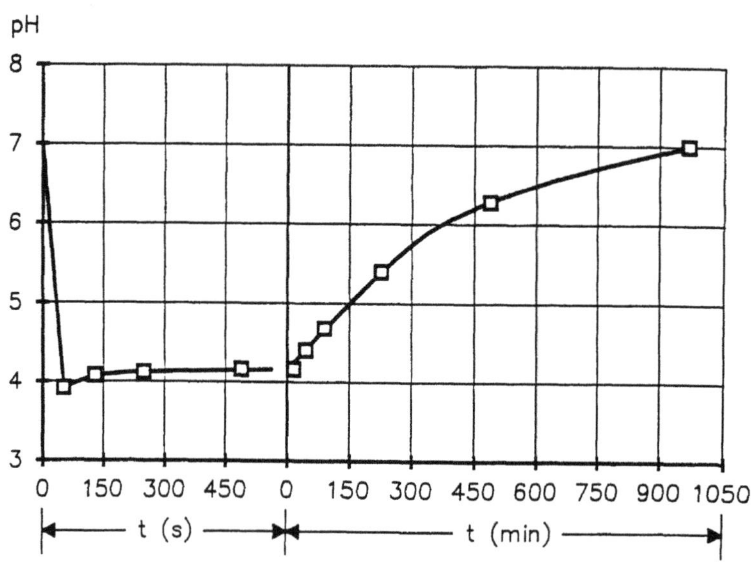

Fig. 3.5.1: Variation of pH with time t for a water/ash dispersion /C 8/

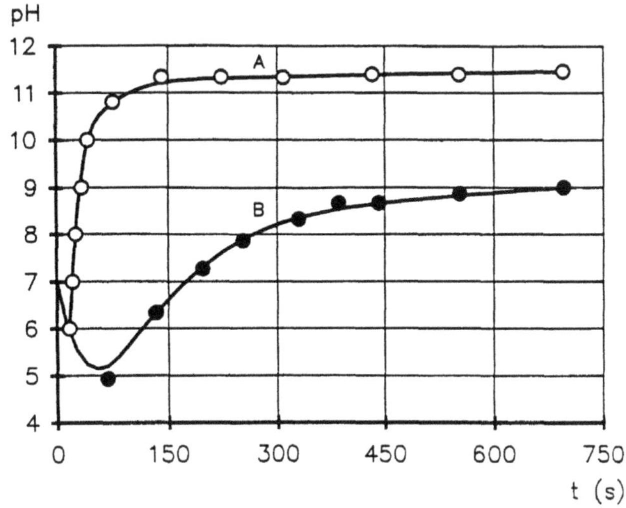

Fig. 3.5.2: Variation of pH with time t for 5 % ash slurries /R 22/
A — highly alkaline ash
B — feebly alkaline ash

dissolving sulphates – SO_2, SO_3 – on the surface of fly-ash particles. Later, calcium and other alkalis also go into solution, raising the pH. The two lines in *Figure 3.5.2* represent the limits for English fly ashes. Assuming that the pH was measured after 24 h, similar results are to be seen in /B 19/ for 34 different coal fly ashes. The pH ranged from 8.5 to 12.3. No obvious connection between the pH value and the amount of soluble alkalis was found, though high pH values corresponded to large $Ca(OH)_2$ contents and *vice versa*.

In /B 26/, it is indicated that a fall in pH from 12.5 to 9.5 implies that 99.9 % (by weight) of the $Ca(OH)_2$ present must be bound to the fly ash, if alkalinity is affected solely by $Ca(OH)_2$. Calculations in /B 34/ show that the huge decrease in $Ca(OH)_2$ caused by a high fly-ash content and high fly-ash reactivity could lead to a critically low pH. The value could, however, remain sufficiently high because of the Na^+ ions and K^+ ions released during the pozzolanic reaction of the fly ash. /B 21/ also notes that a drop in pH to a level critical for corrosion would appear unlikely to result from the pozzolanic reaction. The reason is believed to lie in the failure of fly ash to react completely with $Ca(OH)_2$ even at higher temperatures (60 °C) and in the effect of the reaction products in blocking the diffusion of $Ca(OH)_2$.

According to tests described in /K 15/, performed on hardened cement paste made from different fly-ash cements with fly-ash contents up to 40 % by weight, the $Ca(OH)_2$ content increased up to 28 days at 20 % (by weight) of fly ash. At 40 wt.% of fly ash it increased up to only 7 days, and stayed constant thereafter.

In /M 36/, hardened cement paste specimens with (f/c = 0.28) and without coal fly ashes were moist-stored for 7, 28 and 90 days. The $Ca(OH)_2$ contents were 15.0,

16.9 and 18.9 wt.% for the portland cement specimens without fly ash, and 11.8, 13.2 and 13.4 wt.% for those with fly ash.

Tests on pore water squeezed from hardened cement pastes with and without two different fly ashes /D 6/ showed that the pH was only 0.2 lower for the fly-ash paste (*Fig. 3.5.3*). Although the Na_2O contents of the fly ashes were different, the alkali content of the pore water was virtually the same. Other tests /D 21/ have shown that lignite coal fly ashes with a high lime and Na_2O content (but an extremely low K_2O content) released Na into the pore solution, whereas coal fly ashes with a high K_2O content and negligible Na_2O content released no K into the solution.

Only small reductions (0.7) in maximum pH as compared to non-fly-ash concretes were observed /B 26/ in three-year-old fly-ash concretes (f/c = 0.25). Even hardened cement paste with f/c up to 0.68 exhibited only a small or zero fall after three years (minimum pH = 11.9).

Similarly, /R 40/ reported little difference in the pH of concretes made with and without fly ash after 50 cycles of water storage followed by drying at 110 °C.

Examination of concrete in structures between 10 and 30 years old made from portland cement and from portland cement plus fly ash /N 20/ revealed little difference in pH, either in the mass concrete or in the carbonated zone. The pH ranged from 10.4 to 12.4. To some extent, this reflects the variation in free lime content (0.14 to 6.11 wt.%) which, as expected, increased from the outside inwards.

On the basis of tests on cement mortars containing two fly ashes, f/c = 0.4 has been suggested /S 13/ as a limit for reinforcement corrosion inhibition (pH falling) and accelerated carbonation: at f/c = 1.0, the $Ca(OH)_2$ content was high enough to assure passivity of the steel. Heat treatment accelerated the pH drop.

In tests in which 80 % of the cement had been replaced by fly ash /P 26/, the pH was 12.5 to 13.0, even though 90 % of the $Ca(OH)_2$ had been consumed by the coal fly ash. pH values in excess of 12.5 have also consistently been found in a number of other tests.

3.5.3 Mechanism of Carbonation

In the tests reported in /K 15/, hardened cement paste specimens with different fly-ash cements (made from three different fly ashes with CaO up to 19 wt.% and three different portland cements), containing up to 40 % fly ash, were variously moist-stored, stored in CO_2 and dried. It was found to be easier to attain carbonation of the "secondary" hydrate phases from the reaction between fly ash and hydrated lime; the reaction was more vigorous than that with the hydrates from pure portland cement. Carbonation was greatest at 40 % fly-ash content, and there was a considerable increase in strength. $Ca(OH)_2$ could no longer be detected.

3.5.4 Rate of Carbonation

At constant temperature, humidity, atmospheric CO_2 content and air flow, the carbonation depth d_c increases parabolically with time t (*Fig. 3.5.4*):

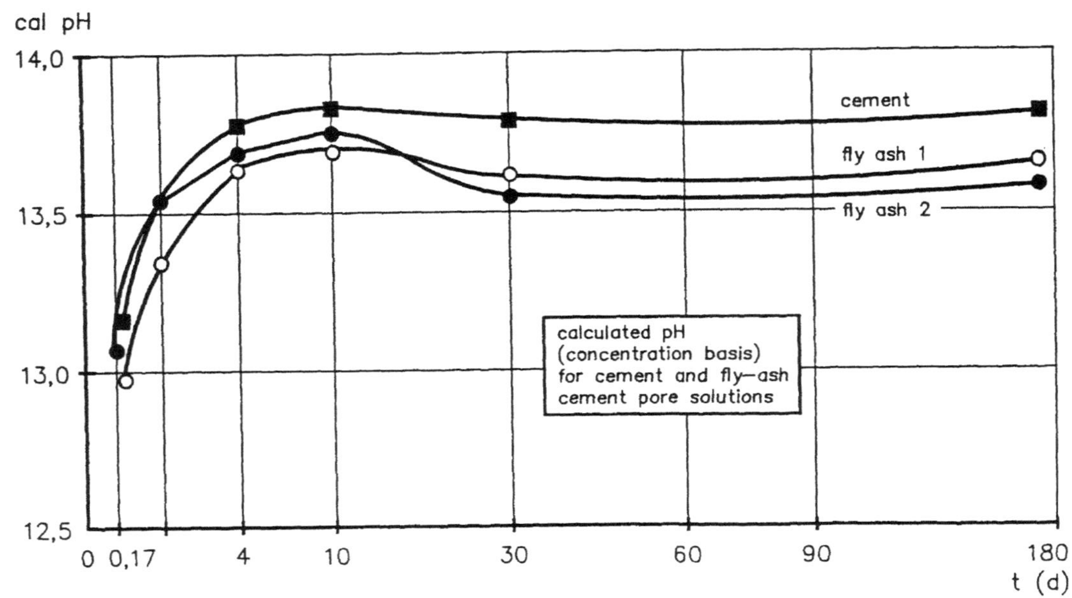

Fig. 3.5.3: Variation of calculated pH values cal pH with time t for pore solutions expressed from cement-bearing and fly-ash-bearing pastes /D 6/

$$d_c = t^n \qquad (3.5.1)$$

This approximates to:

$$d_c = c_o + v_c t^{0,5} \qquad (3.5.2)$$

In other words, the so-called root-t-law applies /S 10/ (cf. bottom of *Fig. 3.5.4*).

The slope of the line v_c is a measure of the depth of carbonation with time, and is referred to as the rate of carbonation /S 10/. The value of c_o depends primarily on the water content of the specimen at the start of carbonation curing, i. e. from the time when drying out has progressed sufficiently for carbonation to begin. However, c_o is very small in comparison with $v_c t^{-2}$ and may safely be neglected /B 47/. The rate of carbonation therefore depends principally on v_c.

In /S 13/, tests on cement mortars with f/c up to 1.0 in a 20/60 environment (i.e. 20 °C and 60 % relative humidity) indicated a considerable increase in carbonation depth at f/c > 0.4 after about a year's storage. In other words, v_c was roughly constant up to f/c > 0.4, and increased substantially with higher f/c.

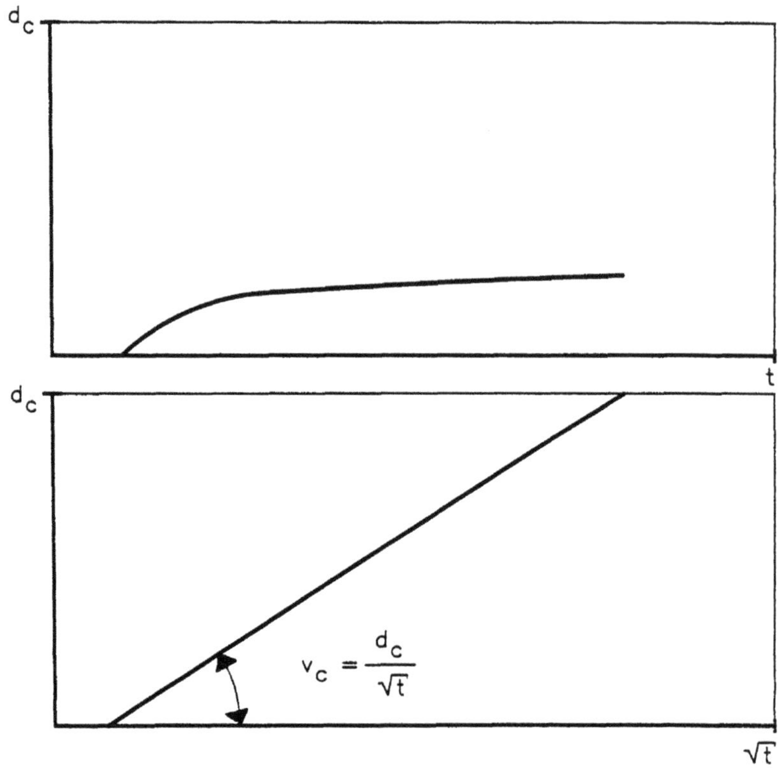

Fig. 3.5.4: Relationship between depth of carbonation dc and time t

According to tests in /K 40/, carried out on mortars with three fly-ash cements first stored in water up to 7 days, and thereafter in a 20/70 environment, the carbonation depth increased very rapidly from 7 and 14 days up to three months, and then more slowly up to six months. The rate of carbonation was higher with fly-ash cement mortars than with portland cement mortars, and was also accelerated by additional fly-ash content. The shorter the initial water storage, the faster the carbonation.

In /R 30/, carbonation tests were carried out on mortars with two fly-ash cements containing 0–30 wt.% of fly ash. Following moist storage up to 28 days, the mortars were cured in four different environments in the laboratory and outdoors. The relationship for the 70-day storage period so far completed is approximately:

$$\log d_c = \log c_o + \log t \qquad (3.5.3)$$

An effect of fly-ash content was determined only in tests with 5 % CO_2 in the air, water/cement ratio = 0,80 and 30 % fly-ash content, where a much higher rate of carbonation was observed.

According to /L 32/, fly-ash content should not exceed 20 % by weight of the cement if a high rate of carbonation is to be avoided.

3.5.5 Factors Affecting Carbonation

If the rate of carbonation v_c is taken as the primary characterizing variable for carbonation, the following influencing factors should be taken into account:

- the density, and in particular the water/cement ratio and the compressive strength,
- the components in the mortar and concrete involved in hardening, i. e. matrix and fly ash,
- initial storage and curing conditions before exposure to atmospheric CO_2,
- ambient conditions, including relative humidity, CO_2 content, rain, snow or frost, etc.

In /B 22/, pastes were made using portland cement with and without fly ash (f/c = 0.33) with different water/cement ratios and roughly equal 28-day strength. The specimens were moist-cured for one or 28 days and then stored in pure CO_2. In some of the tests, the pastes were treated with $CaCl_2$ before CO_2 storage. As expected, v_c increased according to the amount of initial drying out. The longer the hydration period before the CO_2, the lower were d_c and v_c. The hardened cement pastes containing fly ash had lower drying and carbonation rates than those without. However, the $CaCl_2$ did not affect drying. The results show that the CSH phases must also have reacted with the CO_2.

Results from various sources cited in /W 16/ show that v_c increases with increasing f/c. Concretes of equal strength with and without fly ash have equal v_c values.

In /S 13/ cement mortars were moist-stored for 28 days and subsequently stored in CO_2 in a 20/60 environment. They had a water/cement ratio of 0.60 and f/c values of 0.11, 0.43 and 1.00, with some substitution of fly ash for the sand. Several mortars were also heat-cured. Polarization lines were determined to establish $Ca(OH)_2$ contents, d_c values and pore size distribution, revealing a marked effect of f/c on v_c. v_c increased considerably with f/c values somewhat above 0.4. Mortars with f/c = 0.43 and especially those with f/c = 1.00 had more capillary pores than those with f/c = 0 and 0.11. The rate of carbonation depends on the number of capillary pores and the $Ca(OH)_2$ content.

In /K43/, portland cement concretes with and without fly ash (f/c = 0.43) were tested at 13 different institutes. After 14 days' water storage, the reinforced test cylinders were stored in the open and d_c was measured after two and five years. Results indicated an almost linear increase in d_c with water/cement ratio. At an equal water/cement ratio, the d_c of the fly-ash concretes was lower than that of the non-fly-ash concretes, owing to the pozzolanic reaction (*Fig. 3.5.5*). There was a linear relationship between 28-day compressive strength and d_c, almost regardless of whether or not the concrete contained fly ash (*Fig. 3.5.6*). Additional tests on concrete specimens from port installations confirmed these results.

Further tests on 7-year-old concrete specimens with and without fly ash showed that the $Ca(OH)_2$ content in the non-carbonated zone for concrete without fly ash was high. By contrast, it was low in concretes with fly ash, possibly because of the pozzolanic reaction. With further carbonation, CO_2 will react with other hydrates in the fly-ash concretes. As expected, total porosity increased with an increasing water/cement ratio, whether or not fly ash was present. Contrary to other published data, accelerated carbonation entailed much higher total porosity in the carbonated than in the non-carbonated zone. It is concluded in /K 43/ that d_c depends not on the presence of fly ash, but on the concrete quality.

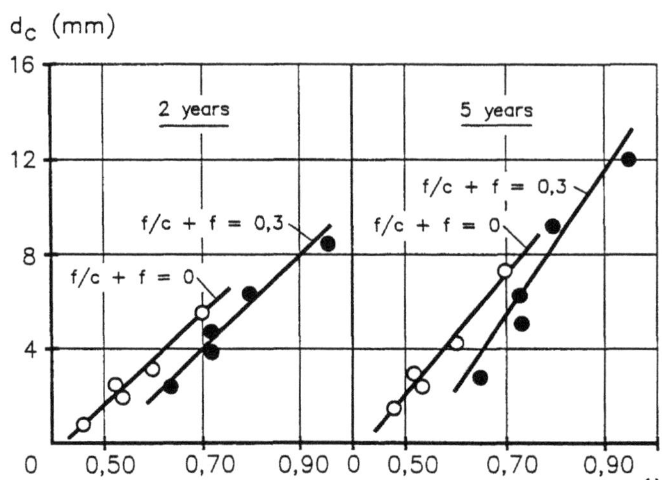

Fig. 3.5.5: Relationship in concrete between water cement ratio ω and carbonation depth d_c /K 43/

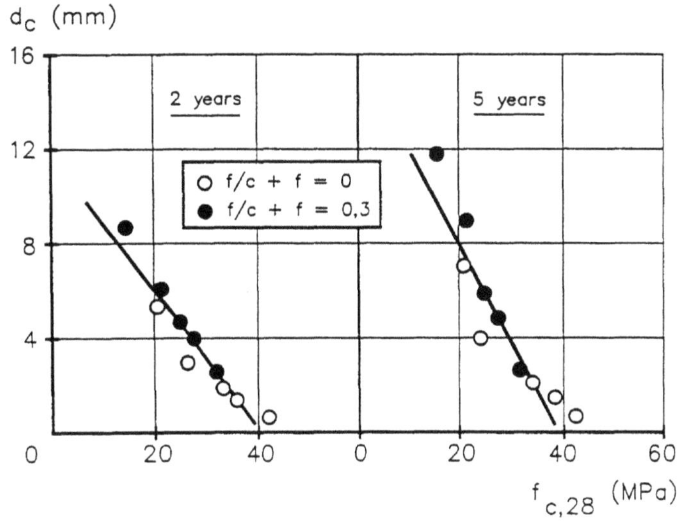

Fig. 3.5.6: Relationship in concrete between compressive strength $f_{c,28}$ and carbonation depth d_c /K 43/

Carbonation tests on various fly-ash and non-fly-ash concretes with identical workability are reported in /M 53/. Two fly ashes, essentially differentiated by their 45-micron sieve residue, and two portland cements, one ordinary and the other sulphate-resisting, were used. The concrete mixes were 225 kg cement-plus-fly ash per m^3, f/c = 0, 0.37 to 0.38 and 335 kg cement-plus-fly ash per m^3, f/c = 0, 0.32 to 0.33. Compressive strengths and carbonation depths were determined on cubes stored in a 20/65 environment and on prisms stored in water for 90 days and thereafter in the open. In addition, compressive strength tests were carried out on cubes demoulded after one day and then stored in water. The carbonation depths of the cubes were measured after 10 years and those of the prisms after seven to eight years.

If the carbonation depths are compared with the compressive strength at 28 days (assuming that the compressive strength of a prism is identical to that of a water-stored cube), no significant difference was found between concretes of equal strength with and without fly ash. However, when 28-day compressive strength is below some 30 to 35 MPa, the concrete containing fly ash behaves slightly worse, i. e. the depth of carbonation is slightly greater (*Fig. 3.5.7*).

In relation to compressive strength after 6 to 8 years (under the same storage condition as for compressive strength and carbonation tests), the mean depth of carbonation of prisms and cubes is about 5 mm greater for concretes containing fly ash (*Fig. 3.5.8*). This may be due to the delayed strength development of the fly-ash concretes.

No difference was found in the effects of the two different fly ashes. Nor was there any substantial difference in carbonation behaviour between the concretes with and

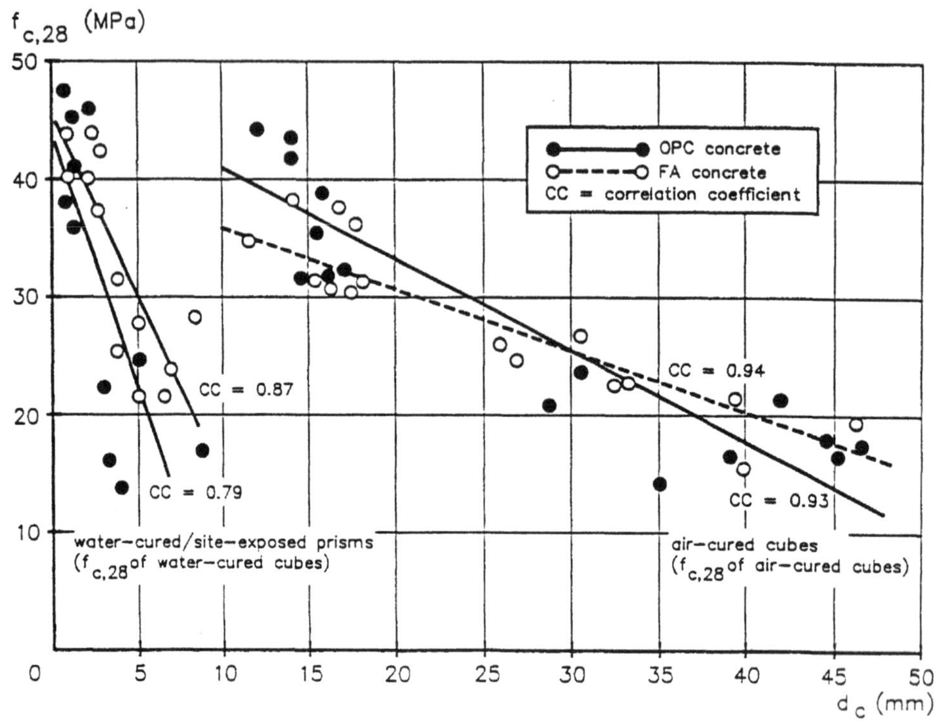

Fig. 3.5.7: Depth of carbonation d_c versus 28-day compressive strength $f_{c,28}$ /M 53/

without fly ash after storage for one day at air temperature and after water storage for 90 days at air temperature.

In /S 10, W 12, W 20/, various tests on mortars and concretes containing fly ash ($f/c \leq 0.25$) and with identical water/cement ratios were evaluated. Initially, the mortars and concretes were either moist-stored or immersed in water for various periods and then kept in a 20/65 atmosphere with a normal CO_2 concentration. *Fig. 3.5.9* /W 12/ shows a linear increase in v_c with f/c. The shorter the storage period in water, the greater the rate of increase. At equal compressive strength, v_c was almost the same for fly-ash and non-fly-ash mortars (*Figures 3.5.10 and 3.5.11*).

/R 40/ reports on Indian investigations on reinforced concrete prisms with and without fly ash in various environments (natural climate, alternate water storage and drying) over several years. In all cases, f/c was 0.34 and the compressive strengths of comparable concretes with and without fly ash were identical. At various stages, embedded steel bars were examined for corrosion, and the depth of carbonation was determined. There was virtually no difference between the two concretes. This supports the results given in /N 20/ for tests on structures ranging in age from 10 to 30 years made from concrete with and without fly ash.

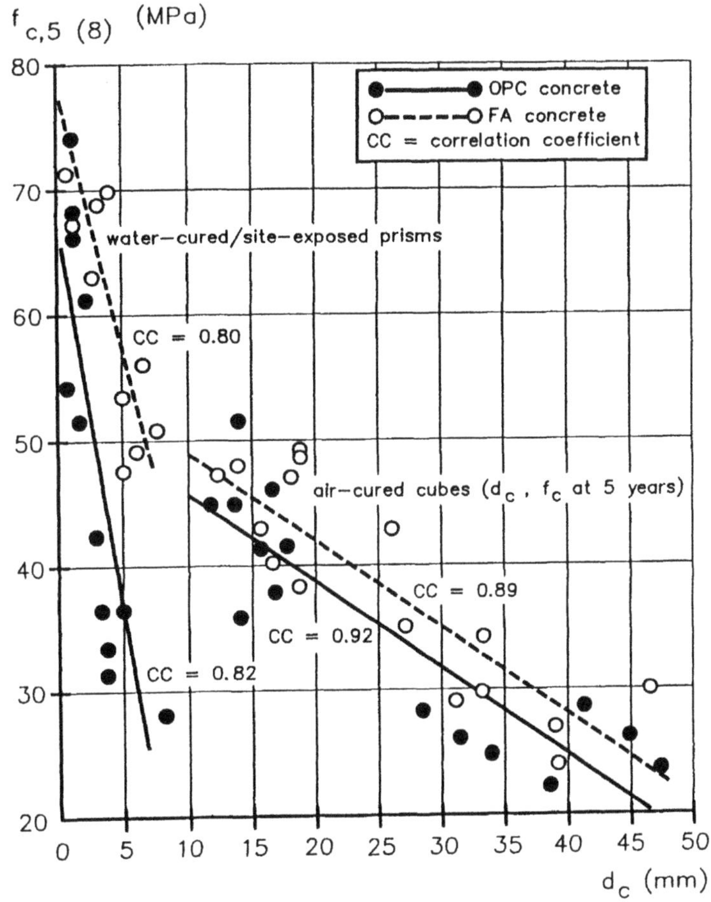

Fig. 3.5.8: Depth of carbonation d_c versus compressive strength at 5 ... 8 years /M 53/

3.5.6 Calculating Carbonation

According to /S 10, W 12, W 20/, Equation (3.5.2) can be used to approximate the depth of carbonation of fly-ash mortars and concretes in the same way as for non-fly-ash mortars and concretes:

$$d_c = c_o + v_c\, t^{0,5} \qquad (3.5.2)$$

Since c_o and especially v_c depend upon the composition of the mortar and concrete – particularly the water/cement ratio, type of binder, content of fly ash and general storage conditions – their validity is not universal, but is limited to special cases. Given test values of d_c and t covering a period of at least six months, c_o and v_c can be

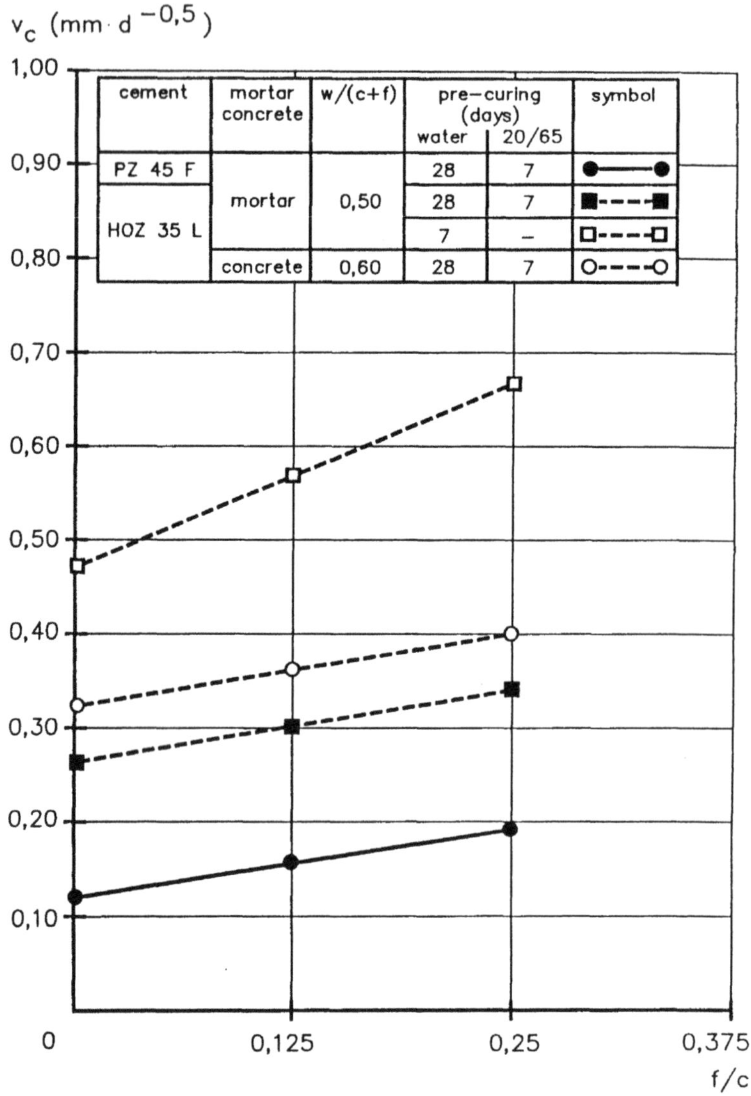

Fig. 3.5.9: Rate of carbonation v_c as a function of fly-ash content f/c /W 12/

determined by successive approximation. However, since concrete is usually designed for a particular compressive strength – normally the 28-day value – the carbonation behaviour of concrete with and without fly ash should be calculated as a function of v_c and f_c (see *Figures 3.5.10 and 3.5.11*).

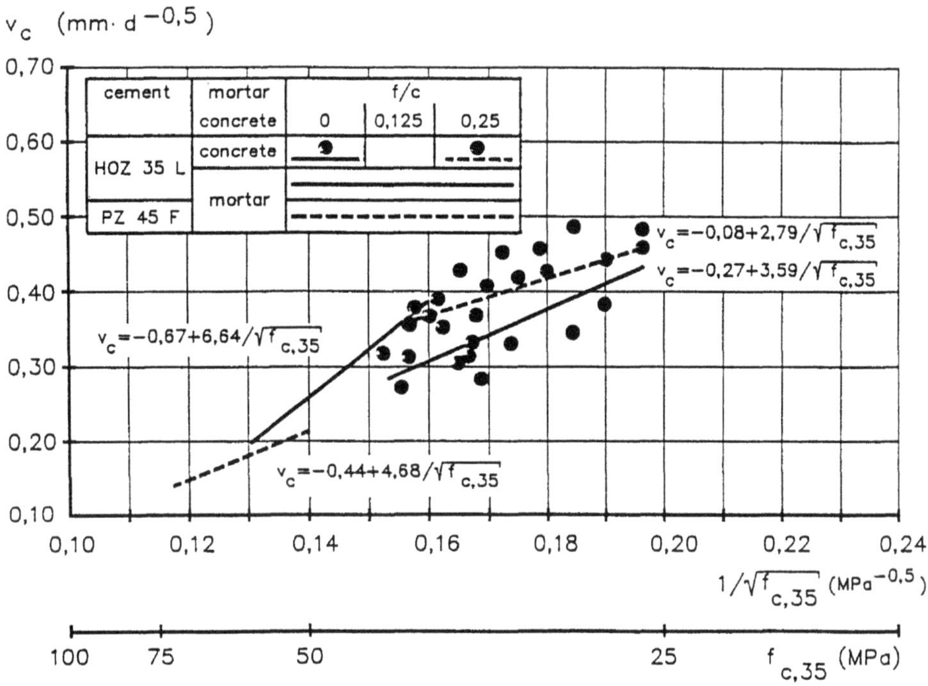

Fig. 3.5.10: Relationship between rate of carbonation v_c and 35-day compressive strength $f_{c,35}$ /W 12/

The relationship can be approximated in linear form:

$$v_c = a + b/f_c^{0,5} \tag{3.5.4}$$

and, neglecting C_o in equation (3.5.2),

$$d_c = at^{0,5} + b(t/f_c)^{0,5} \tag{3.5.5}$$

3.5.7 Summary

The alkalinity of pore water in the hardened cement paste depends on the proportions of the alkalis Na_2O, K_2O and $Ca(OH)_2$. Coal fly ash consumes $Ca(OH)_2$ during the pozzolanic reaction. However, a further reduction in the pH of the pore water is possible only when there is extremely high consumption of $Ca(OH)_2$. In all cases, various laboratory and other investigations on existing concrete structures have shown only a small decrease in the pH of cement paste, concrete and mortar containing fly ash in f/c proportions up to 4.

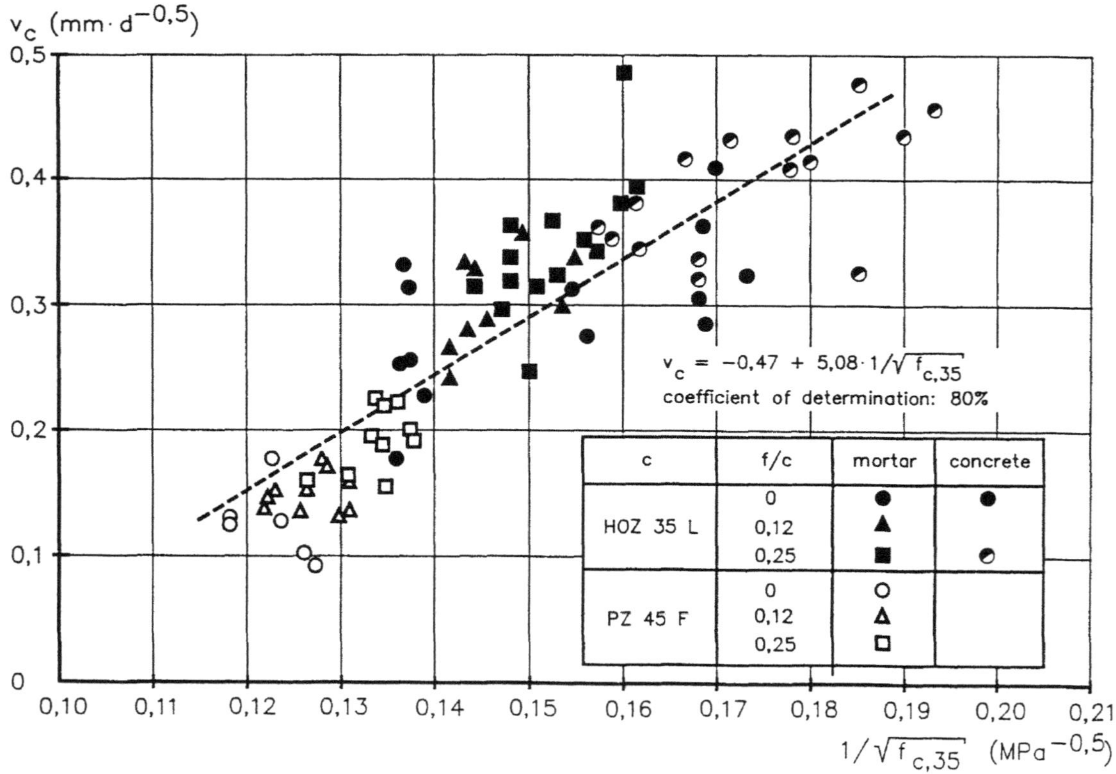

Fig. 3.5.11: Relationship between rate of carbonation v_c and 35-day compressive strength $f_{c,35}$ for various cements and fly-ash contents f/c /S 10/

Atmospheric CO_2 reacts with $Ca(OH)_2$, alkalis and CSH to produce carbonates. Carbonation lowers the alkalinity of the pore water, leaving the steel reinforcement unprotected against corrosion in the presence of oxygen.

In most tests, the depth of carbonation increased with time and was higher in fly-ash than in non-fly-ash mortars and concretes. The rate of carbonation increases with the f/c ratio, the most marked increase occurring above f/c = 4. Conversely, no significant difference was observed in the carbonation behaviour of fly-ash and non-fly-ash mortars and concretes at a given compressive strength, provided that compressive strength was above 30 MPa and f/c was no greater than 0.4.

Several investigators have shown that the carbonation behaviour of fly-ash and non-fly-ash mortars and concretes can be expressed by the following equations:

$$d_c = c_o + v_c t^{-2} \qquad (3.5.2)$$

$$v_c = a + b/f_c^{-2} \qquad (3.5.4)$$

$$d_c = at^{-2} + b(t/f_c)^{-2} \text{ neglecting } c_o \qquad (3.5.5)$$

where
d_c is the depth of carbonation,
v_c is the rate of carbonation,
t is the duration of carbonation,
f_c is the compressive strength and
c_o, a and b are parameters.

The validity of the equations can be improved if other substantial factors such as the type of binder are differentiated.

3.6 Chloride Attack on Steel Reinforcement (Prepared by J. Bijen)

Steel reinforcement embedded in concrete normally exhibits good long-term durability. This is generally attributed to passivation of the steel in the presence of the highly alkaline pore solution in the concrete. Nevertheless, corrosion can occur when depassivating ions are present or penetrate through the cover to the steel. Chloride is the most important of these aggressive ions. It can be present from the start or as a result of penetration from the environment.

In most standards for concrete, chloride is either prohibited or restricted to extremely low levels. Experience in the West suggests that the risk of corrosion in concrete made with ordinary portland cement is small when the chloride concentration at the surface of the steel is less than 0.4 % by weight /T 21/. It is now generally agreed that this critical level depends on the ratio between the concentrations of chloride and inhibitive hydroxyl ions: the higher the ratio, the greater risk of corrosion. The rule of thumb relying on a critical total chloride concentration relative to the cement content can be misleading, because it ignores the chloride concentration in the pore solution, i. e. it does not distinguish between bonded and unbonded chloride ions in the pore solution and the hydroxyl concentration.

Different types of cement have different bonding capacities for chloride ions and various hydroxyl concentrations. For example, it is well known that portland cements with a high C_3A content have a higher bonding capacity than those with a low C_3A content. In the latter case, therefore, the critical chloride concentration expressed as total chloride versus cement mass is lower.

The effect of fly ash on the chloride/hydroxyl ratio has not been extensively studied. *Table 3.6.1* presents some results from Holden et al. /H 46/. It shows that, if 30 % of the ordinary portland cement is replaced, the chloride concentration in the pore solution hardly changes, although the hydroxyl level falls. As a result, the chloride/hydroxyl ratio increases but the difference is small in relation to other factors such as the C_3A content of the cement.

Table 3.6.1: The concentration of chloride and hydroxyl ions (mmole l^{-1}) in pore solution expressed from the hardened cement paste specimens /H 46/.

	Series (1) No additions			Series (2) +0.4% Cl			Series (3) +0.4% Cl, 1.5% SO_2		
Cement	Cl^-	OH^-	$\frac{Cl^-}{OH^-}$	Cl^-	OH^-	$\frac{Cl^-}{OH^-}$	Cl^-	OH^-	$\frac{Cl^-}{OH^-}$
OPC-A/C_3A 7,7%	2	589	0.003	83	741	0.112	215	1318	0.163
OPC-B/C_3A 14,3%	3	479	0.006	41	661	0.062	153	1047	0.146
OPC-B/30% PFA	5	355	0.014	28	457	0.061	147	741	0.198
OPC-B/65% BFS	2	347	0.006	110	501	0.220	257	1000	0.257
SRPC									

Table 3.6.2: Effective diffusivity of chloride ions at 25 °C in various cement pastes of w/c = 0.5 /H 46/.

Type of cement	Diffusivity (10^8 cm^2s^{-1})
OPC-A/C_3A 7,7%	3.14
OPC-B/C_3A 14,3%	4.47
OPC-B/30% PFA	1.47
OPC-B/65% BFS	0.41
SRPC	10.00

A much more significant effect, observed by several investigators, concerned the effective diffusion coefficient of chloride ions in hardened cement or concrete. *Table 3.6.2* shows results obtained by Holden et al. /H 46/. The effective diffusion coefficient is reduced by a factor of 2.5, which means that the initiation period of chloride-induced corrosion of the steel reinforcement is substantially extended.

3.7 Electrical Resistivity (Prepared by J. Bijen)

The electrical resistivity of concrete is a measure of the density of the cement gel structure and is crucial to the rate of corrosion once this has begun.

It is known that the electrical resistivity of concrete is strongly dependent on the type of cement. In general, blended cements have a higher resistivity than portland cement. Data on portland cement/fly-ash mixtures are limited, but it is likely that resistivity will be higher with than without fly ash /H 47/.

4 OTHER USES OF FLY ASH

J. BIJEN, J.P. SKALNY and E. VAZQUEZ

4.1 Cement (Prepared by J. Bijen)

Fly ash can be utilized in several ways in cement works. It can be incorporated in the raw mix for the production of portland clinker (low-quality fly ash), as a raw material and a fuel in the kiln (high-carbon fly ash), for clinker production, and as replacement for portland clinker for the production of portland fly-ash cement (low-carbon, high-quality fly ash) /B 77/.

In the case of portland clinker production, fly ash is used as an alternative to clay, shale etc. The silica, aluminium and iron oxide content are used to obtain the desired chemical/mineralogical composition of cement. In general, up to 8 % of the clinker can consist of fly ash. For this purpose the fly ash can be low quality, i. e. rather variable and with a fairly high coal content. As a raw mix component, the fly ash behaves less favourably in wet processes, because of its negative effect on the viscosity of the slurry. Fly ashes with a high coal content are used both as a fuel and as a clinker raw material by blowing the fly ash into the burning zone of the rotary kiln.

Portland fly-ash cement is a mixture of portland clinker, fly ash and gypsum anhydrite. In general, the fly-ash content is 30 % at the most. In principle, the effects of this fly ash on concrete properties are similar to those of the fly ash added to the mixer as a partial portland cement replacement. However, the following advantages are claimed in the literature:

- by treating the fly ash (screening, homogenization, etc.), the variability in its composition can be reduced;
- by grinding the clinker (and possibly also the fly ash) to a higher degree of fineness, it is possible to compensate for the loss of early strength development which occurs when only a part of the clinker is replaced by fly ash. This is illustrated in *Fig. 4.1*;
- gypsum anhydrite addition can be adjusted to give the desired setting time.

In general, these advantages should allow production of a cement with strength and other quality characteristics similar to those of portland cement.

This similarity in characteristics is one of the main reasons why the production of ordinary portland cement has been, or is being, totally or gradually replaced by production of portland fly-ash cement in countries such as Belgium, Denmark, Germany, the Netherlands, Norway and Sweden.

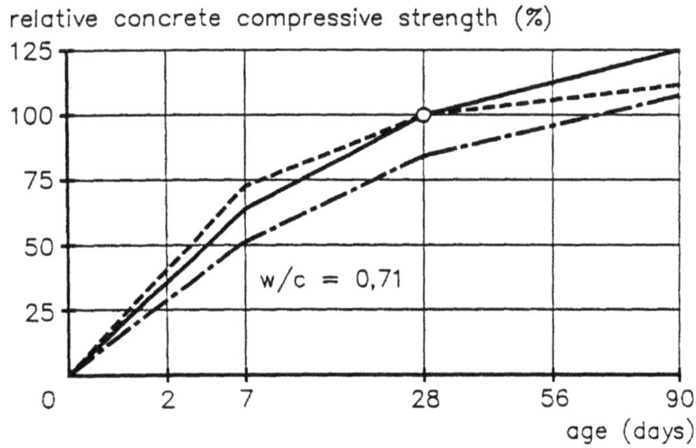

concrete with

------- 300 kg/m³ portland cement PZ 35 F w/c = 0,57

——— 300 kg/m³ portland cement FAZ 35 F w/c = 0,57

—·—·— 240 kg/m³ portland cement PZ 35 F + 60 kg/m³ fly ash* w/(c+f) = 0,57

* same fly ash as used in FAZ 35 F

Fig. 4.1: Compressive strengths at various ages of concrete made with portland cement PZ 35 F, with portland fly-ash cement FAZ 35 F and with a mixture of portland cement and fly ash /H 32/
Note: portland clinker in FAZ 35 F is ground finer than in PZ 35 F

The addition of fly ash to the mix is sometimes preferred to the use of fly-ash cement, for the simple reason that any loss in strength may be corrected by changing the fly-ash content.

There are a number of methods for producing fly-ash cement (*Fig. 4.2*). The system actually chosen by the cement manufacturer will depend on many factors, such as existing plant facilities, energy costs, fly-ash quality, and whether or not there is a steady demand for portland clinker and for fly ash.

Ternary cements made from portland clinker, fly ash and granulated blast-furnace slag are produced in France:

- portland blast-furnace and fly-ash cements falling into the category of portland composite cements, CPJ; these include no less than 65 % of portland clinker and no more than 35 % of slag + fly ash.
- slag and fly-ash cement (CLC) not belonging to the portland cement; their clinker content ranges between 25 and 60 %, whereas the proportion of slag + fly ash is between 20 and 45 %.

Although both fly ash and slag can be regarded as pozzolanas, their joint effects in this cement appear to be more complementary than competitive, probably because of the physical effect of fly ash in terms of improved workability.

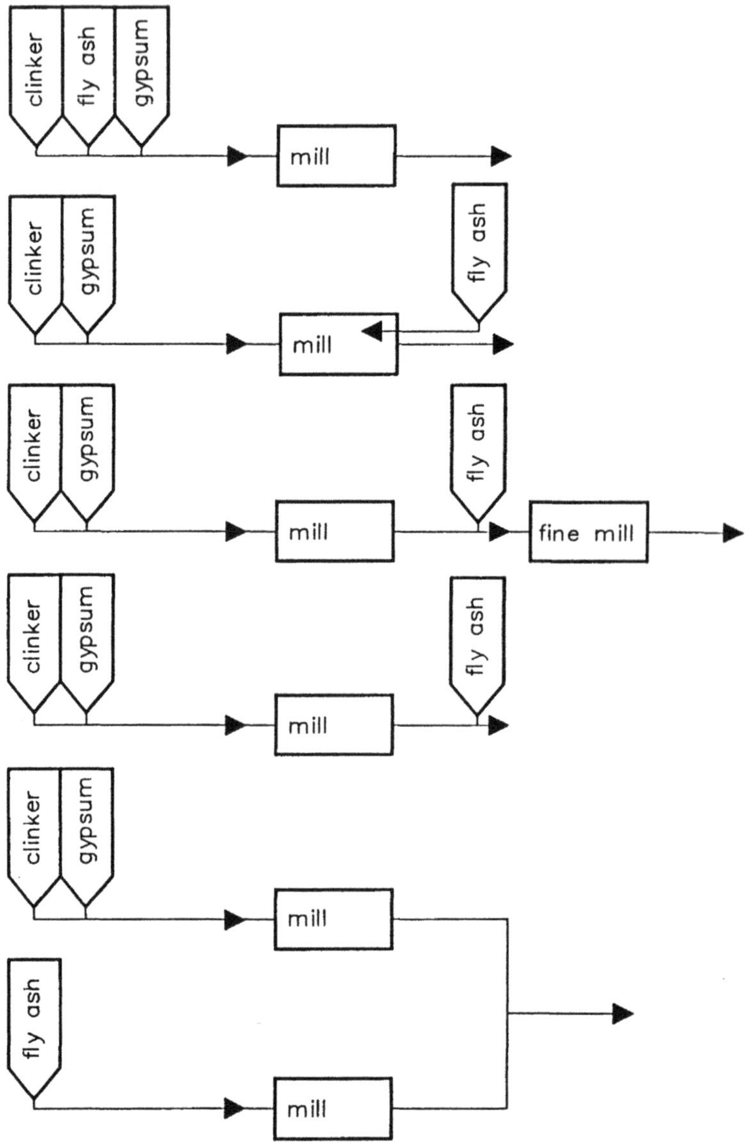

Fig.4.2a: Open circuit grinding/mixing processes for the production of portland fly-ash cement

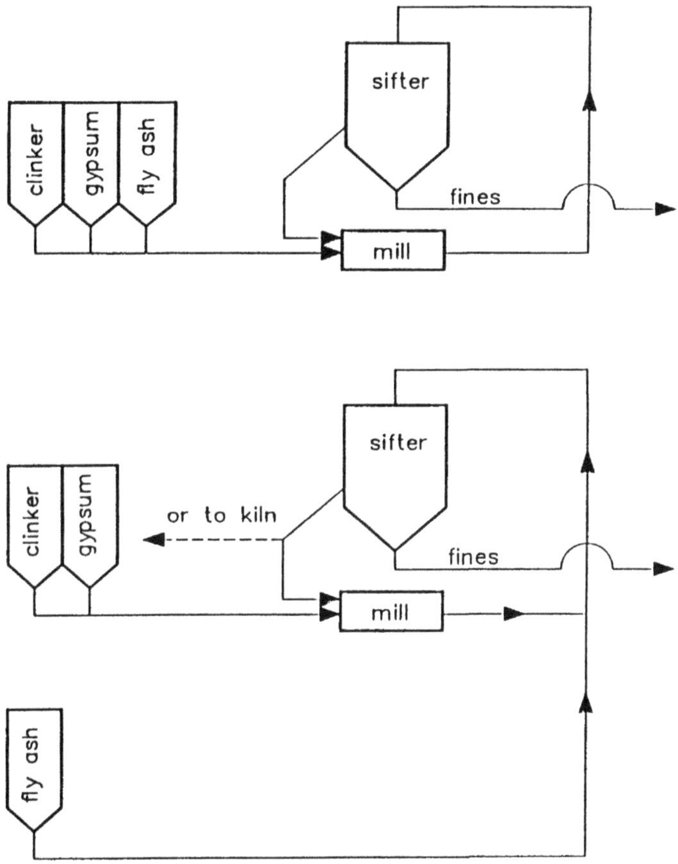

Fig.4.2b: Closed circuit ginding processes for the production of portland fly-ash cement

4.2 Binders with Fly Ash (Prepared by J. Bijen)

Apart from its use as a constituent in blended cements and concrete, fly ash is also used in other types of binder, which generally exploit the pozzolanic nature and the rounded particle shape of fly-ash. Fly ash is applied in these binders in combination with:

- cement; to distinguish this application from utilization in cement and concrete, cement/fly-ash binder is defined here as a mix containing more than 30 % of fly ash;
- lime;
- lime plus gypsum; the latter material also comprises spray dry desulphurization residues and wet desulphurization residues consisting of calcium sulphite, calcium sulphate and some lime /D 43/;

– slag and alkaline activators.

The binders are applied with or without additional fillers or aggregates.

The choice of cement, lime or lime plus gypsum depends on a number of factors, which may vary widely from area to area. One of the main criteria will, however, be strength development, particularly early strength. *Fig. 4.3* shows typical differences between cement, lime and lime plus gypsum.

High fineness, low coal content and rounded particle shape are, in general, favourable properties for use in cement and concrete. The amount of cement or lime or lime plus gypsum required to achieve a certain strength depends on the amount of free lime available in the fly ash. A Type C fly ash (according to ASTM) containing free CaO needs less free lime than a Type F. The literature draws attention to the destructive formation of expanding ettringite when lime plus gypsum is used. Although the data are difficult to compare, it can be stated that the use of a composition in the "safe area" depicted in *Fig. 4.4* will be trouble-free, while unacceptable expansion may occur with compositions in the "non-safe area", depending on the kind of application and the prevailing conditions. In the latter case, extensive practical testing is recommended. Sulphate/sulphite residues appear to be less expansive than pure sulphates in combination with fly ash. However, in the long term, expansion due to sulphite oxidation into sulphates needs to be taken into account.

These binders are used in a wide range of application, comprising:

- masonry mortars for brick walls,
- renderings (plastering) of walls,
- oil well cements,
- block production (discussed in Section 4.4),
- production of artificial aggregates (discussed in Section 4.5) and
- road construction (discussed in Section 4.6).

In general, most applications are low strength (5 – 20 MPa).

A special binder application investigated in recent years is the use of fly ash in alkali-activated slag cement. In this cement, ground granulated blast-furnace slag is activated by alkaline compounds such as sodium hydroxide and waterglass. Fly-ash contents up to 40 % of the total cement mass can be used, depending on the fineness and activity of the slag. Although the strength development of the binder may be similar to that of ordinary cements, some disadvantages have been reported, such as:

- a high rate of carbonation,
- loss of tensile strength in drying,
- irritation of the skin due to high alkalinity,
- increased danger of alkali-aggregate reaction and
- the need to add alkaline activator in the concrete factory rather than at the cement works.

More research will be needed in order to judge the merits of this binder.

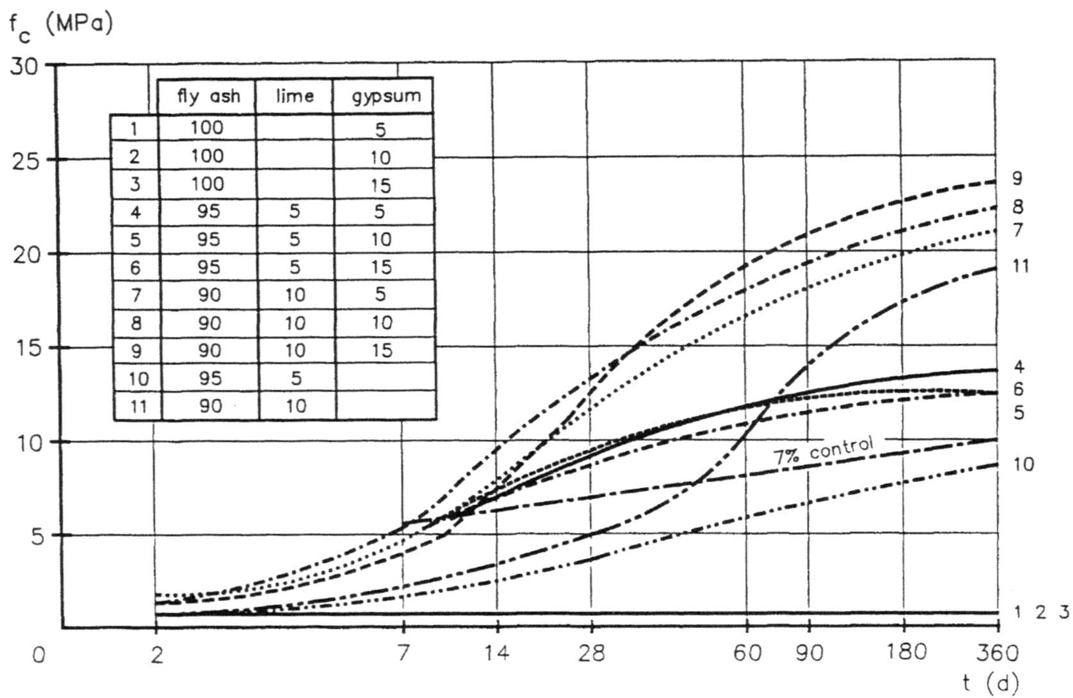

	fly ash	lime	gypsum
1	100		5
2	100		10
3	100		15
4	95	5	5
5	95	5	10
6	95	5	15
7	90	10	5
8	90	10	10
9	90	10	15
10	95	5	
11	90	10	

Fig. 4.3: Compressive strength development for fly-ash lime, fly-ash lime gypsum and fly-ash cement mixtures as a function of time t

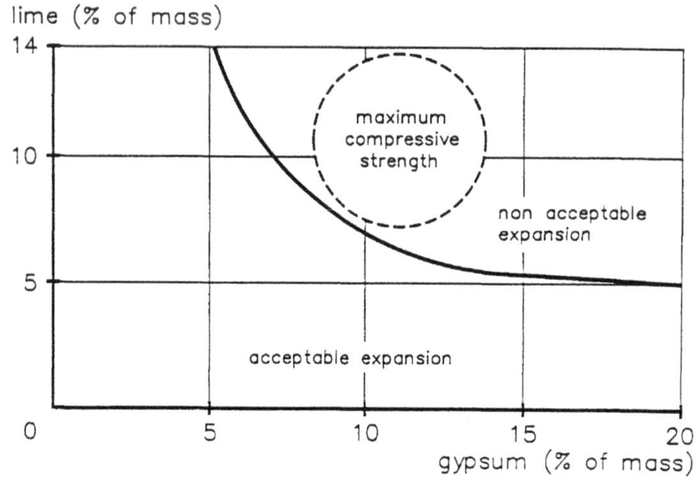

Fig.4.4: Composition area with acceptable and non-acceptable expansion respectively and with maximum compressive strength after 28 days of hardening at 20 °C and 99 % RH

4.3 Precast Concrete (Prepared by J. P. Skalny)

As one would expect, applications for fly ash have been found in the precast concrete industry /e. g., M 30, K 11, J 5, F 17, G 18/. Because factory-produced concrete products are usually heat treated (steam cured, autoclaved), the beneficial pozzolanic properties of fly ash are exploited at a higher rate, and material and labour savings may be achieved /C 15/.

Whereas little literature exists on the specifics of high-temperature curing of fly-ash concrete, limited information is available on autoclave curing /N 10, S 2/ as well as on low-pressure steam curing /e.g. R 24, R 10, S 11/. According to Schubert and Jaegermann /S 11/, short-time curing of fly-ash concrete may lead to decreased strength; however, adequate curing leads to improvements in both compressive and flexural strength. Understandably, results are influenced by a multitude of variables such as the amount of fly ash used and the curing conditions. Experience shows that proper curing is one of the most important factors in the production of durable concrete containing blended cements or supplementary materials.

Further information on the effect of temperature on fly-ash concrete products can be found elsewhere in this report.

4.4 Bricks and Blocks (Prepared by J. Bijen)

4.4.1 Aerated Concrete

In a number of countries, fly ash is used as a raw material for the production of autoclaved aerated concrete. It is used to replace ground quartz sand alone or ground quartz sand and binder (lime/portland cement) /B 78/.

In general, the requirements listed in *Table 4.1* are imposed on fly ash. These requirements are rather similar to those for concrete, except that there is a stipulation concerning the proportion of silica, the main component reacting with the lime.

Table 4.1: Recommendations regarding requirements on fly ash for autoclaved aerated concrete

Chemical composition (as a percentage of mass)		
LOI	%	6*
Sulphate (SO_3)	%	2.5
Magnesium oxide (MgO)	%	2
Silicium oxide (SiO_2)	%	40
Fineness		
Residue on sieve: 200 μm	%	< 10
90 μm	%	< 20
60 μm	%	< 30

* The percentage varies strongly between countries.

Under the usual autoclaving conditions, less well crystallized reaction products are formed when fly ash is used instead of quartz sand. Nevertheless, as in the case of quartz sand, tobermorite is found, together with a microcrystalline phase (CSH 1) and a hydrogarnet which can be described by the formula $C_3A_{1-n} \cdot F_nS_nH_{6-2x}$. The strength of autoclaved aerated concrete appears to be closely related to its tobermorite content, which depends partly on the silica content of the fly ash, especially the silica in the glass phase.

Given an appropriate fly ash, the strength of aerated fly-ash concrete is similar to that of concrete made with ground quartz sand, although the ratio of wet strength to dry strength appears to be more favourable with fly ash. Shrinkage and creep are very similar. One of the major advantages of using fly ash is that the heat flow resistance may be 15 – 40 % higher than with quartz sand. This is due to the amorphous character of the fly ash. Some of the gain in thermal insulation will be lost, however, because the moisture content of the aerated concrete in equilibrium with the environment is usually higher when fly ash is used.

The coefficient of linear expansion is almost the same, and no great difference has been found as regards durability aspects, e.g. freeze-thaw resistance. A corrosion-inhibiting effect of the fly ash has been reported with uncoated steel reinforcement. Steel reinforcement in aerated concrete is, however, usually coated.

It would appear to be possible to replace up to 30 % of the lime/cement binder without greatly altering its properties. In general, there appear to be no major technological drawbacks in the use of fly ash in autoclaved aerated concrete. In some countries such as the UK, this application is the most important in terms of total fly-ash utilization.

4.4.2 Foamed Concrete

Foamed concrete is very similar to autoclaved aerated concrete with respect to its constituents: cement, a filler (mostly α-quartz sand), water and air. But its properties are rather different. At the same apparent density, foamed concrete is less strong, has a very much higher drying shrinkage and creeps more. It hardens at ambient temperatures and can be placed *in situ*. It has acquired a market as a material for floors, roofs and sometimes walls, owing to its insulating properties. Owing to its lightweight properties, it is used in road foundations on low bearing-capacity soils and for filling disused pipes, oil tanks, etc. It covers a wide range of apparent densities, usually from 600 – 1400 kg/m^3.

Replacement of α-quartz sand by fly ash slightly increases the compressive strength (see *Fig. 4.5*). Drying shrinkage is not affected. The modulus of elasticity is diminished and tensile strain capacity is somewhat increased. Foamed concrete made with fly ash instead of quartz sand is therefore rather less prone to drying shrinkage cracking.

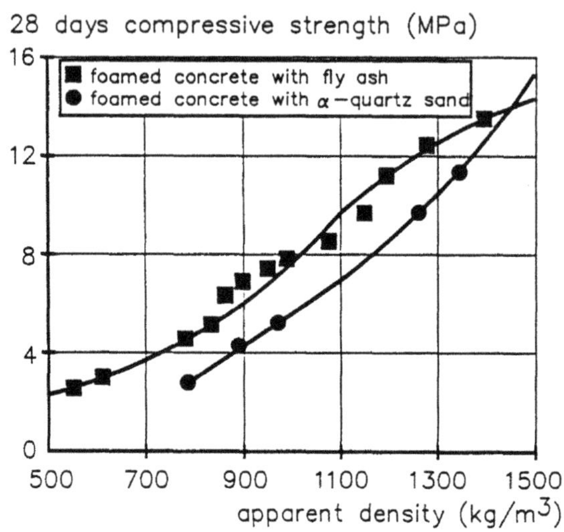

Fig.4.5: Compressive strength of dry foamed concrete with α-quartz sand and fly ash respectively as a function of the apparent density

The thermal conductivity for dry foamed concrete with fly ash is lower than with α-quartz sand (see *Fig. 4.6.*), evidently for the same reason as mentioned in Section 4.4.1. Under ambient climatical conditions the difference is less obvious.

Partial replacement of cement is also possible. Up to a percentage of 20 to 30 %, no very large differences in final properties are observed in comparison with the reference foamed concrete.

4.4.3 Lime-Silica Bricks

Lime-silica bricks are produced in considerable quantities in a number of countries. The silica used is mostly quartz sand, and the lime content is 8 to 10 % by weight. The sand can be replaced partially or totally by fly ash, the lime only to a limited extent. When a silica-lime mixture is autoclaved, the reactions are very similar to those in aerated concrete (see Section 4.4.1) /H 48/.

In general, the use of fly ash has the following deleterious effects on brick quality:

- The brick colour alters from the usual fairly white colour produced by the lime-quartz sand mixture to varying shades of grey.
- The brick is more vulnerable to efflorescence.
- Strength appears to be lower.
- The brick is less freeze-thaw resistant.

A lime-silica brick consisting of fly ash, lime and wet bottom boiler slag (Granusand) is produced in the Federal Republic of Germany. This lime-silica brick has the advantage of a high thermal insulation, owing to the amorphous character of both slag and fly ash.

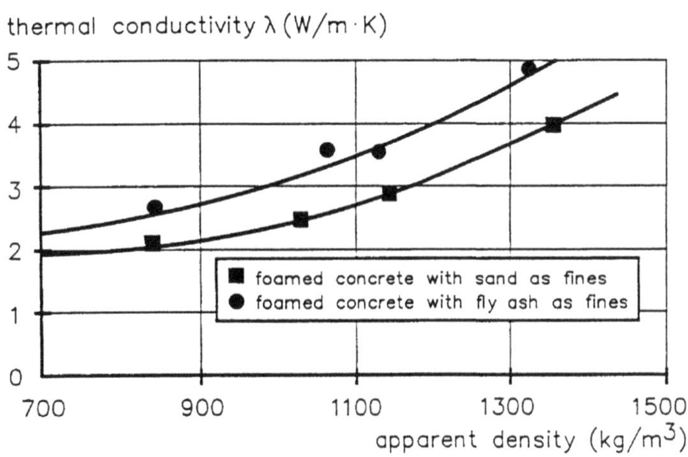

Fig.4.6: Thermal conductivity of dry foamed concrete with α-quartz sand and fly ash respectively as a function of apparent density

4.4.4 Ceramics

Ceramic building products such as bricks, tiles, pipes, etc. are usually made of clay but occasionally of other raw materials, such as ground shale. Fly ash has a chemical composition similar to that of clay, but its molecular structure and particle shape are quite different. Hence, its rheological properties are unlike those of clay; clay has a good plasticity while fly ash has not. Because of the similarities in chemical composition, the firing properties of the green products (such as fusion temperature, for example) are similar. However, the different rheological behaviour means that products have to be shaped differently from the conventional clay brick, unless clay/fly-ash mixes with a minor quantity of fly ash are used.

The uses of fly ash may be sub-classified as follows /A 9–11/:

- products made of clay/fly-ash mixes using conventional methods, which in Europe are generally the so-called wet processes;
- products with a high fly-ash content (e.g. 70 %), produced by means of so-called semi-dry processes using various binders to facilitate formation and drying of the green product;
- porous bricks exploiting the calorific value of the fly ash;
- ultra-lightweight ceramics: materials used for their insulating and refractory properties.

Wet Process

In Europe and some non-European countries, building bricks, etc. are often produced using so-called wet processes. The products are made from clay containing a substantial (30–40 %) proportion of water. In general, the wet clays can be shaped easily without applying high compaction pressures.

It has been shown that up to 40 % of the raw material for such processes can be replaced by fly ash. With clays that are too plastic, filler is often added to reduce the drying shrinkage of the products. Quartz sand is mostly used for this purpose. Fly ash can replace this filler and part of the clay without spoiling workability.

Incorporating fly ash has a number of effects:

- It changes the colour of the product, e.g. it is less easy to produce a red brick;
- There is a maximum permissible coal content, typically 2.5 %, for dense products. For a fly ash with 6 % coal content, this means that replacement must be limited to 40 %.
- The bricks are usually more porous;
- The fly ash may increase the content of soluble salts, increasing the probability of efflorescence more likely;
- The quantity of water can often be decreased, reducing both drying costs and drying time;
- Drying shrinkage will be lower;
- Firing can be faster, partly because there are no quartz transitions in the filler.

Semi-Dry Process

With high percentages of fly ash, hot processes are no longer suitable for ceramic building materials, but semi-dry processes – used mainly for refractory bricks and wall tiles – are.

In the semi-dry process, the mixed raw materials are compacted under pressure (e. g. 10 – 40 MPa), after which the green products are dried and subsequently fired.

Fly ash alone has insufficient plasticity to produce a green product strong enough to survive handling, drying and firing; a binder has to be added. This is generally clay, but starch solution, waterglass, lignin wastes, etc. can also be used; moisture content must be much lower than for the wet processes (e. g. 5 – 15 %).

The semi-dry process has the following advantages over wet processes:

- Less drying energy is required and drying is faster;
- Drying shrinkage is less, perhaps as little as a tenth of the value for wet processes;
- Firing is faster and less firing energy is generally required.

There are, however, certain disadvantages:

- Higher investment is required;
- Production is less flexible, especially in terms of brick appearance.

It has been found that, at very high fly-ash contents, fine ash provides much better density and strength than coarse ash; this is evidently due to better packing of the fine fly-ash particles.

Processes have been developed in the United States (pilot plant at West Virginia University) and development work is in progress in other countries (e.g. U.K. and Holland).

Porous Bricks

There are several processes for producing porous bricks using fly ash. All exploit the calorific value of fly ash; the porous structure allows the fuel in the solid to be used to advantage.

In Holland, a brick is produced from clay, fine colliery shale, fly ash and sawdust: its composition is shown in *Table 4.2*. The brick is produced by extrusion, dried and fired in a tunnel kiln in which process heat is fully maintained by the fuel in the brick itself.

In the United Kingdom, successful experiments have been made with the production of porous hand-moulded bricks fired on clamps. Fly ash has been used instead of coal slurry and town ash.

Table 4.2: Composition of Porous Bricks Containing Fly Ash Produced in the Netherlands (wt.%)

Clay	28 %
Fine colliery shale	44 %
Fly ash	20 %
Sawdust	8 %
Total dry solid	100 %
Moisture content	25 %

Ultra-Lightweight Ceramics

In the past decade, ultra-lightweight ceramics have attracted considerable attention. Fly ash can be used to produce these lightweight ceramics. One of the methods is to mix an artificial wet foam with fly ash and binder, dry the mixture and fire the dried foam.

Other techniques used include simple heating of an uncompacted heap of fly ash in a microwave kiln.

4.5 Lightweight Aggregates (Prepared by J. Bijen)

There are many ways of producing artificial aggregates from fly ash. Typical process steps for the manufacture of fly-ash aggregates are depicted in *Fig. 4.7*. The chief distinctions between the various processes are in the methods used for agglomeration and hardening /B 50/.

Agglomeration techniques may be sub-classified into methods

- without external compacting forces (agitation, granulation)
- with external compacting forces (compaction).

Hardening methods may be differentiated according to the hardening temperature employed, for example:

– Sintering processes	≥ 900 °C
– Hydrothermal processes	100 – 250 °C
– Cold-bonding processes	10 – 100 °C

Given the same apparent density, mechanical properties such as strength, drying shrinkage and pellet creep will generally decrease across the hardening spectrum of sintering, hydrothermal and cold-bonding processes. The loss in strength and other properties can, however, be totally or partially compensated by increasing the density of the pellets through compaction agglomeration.

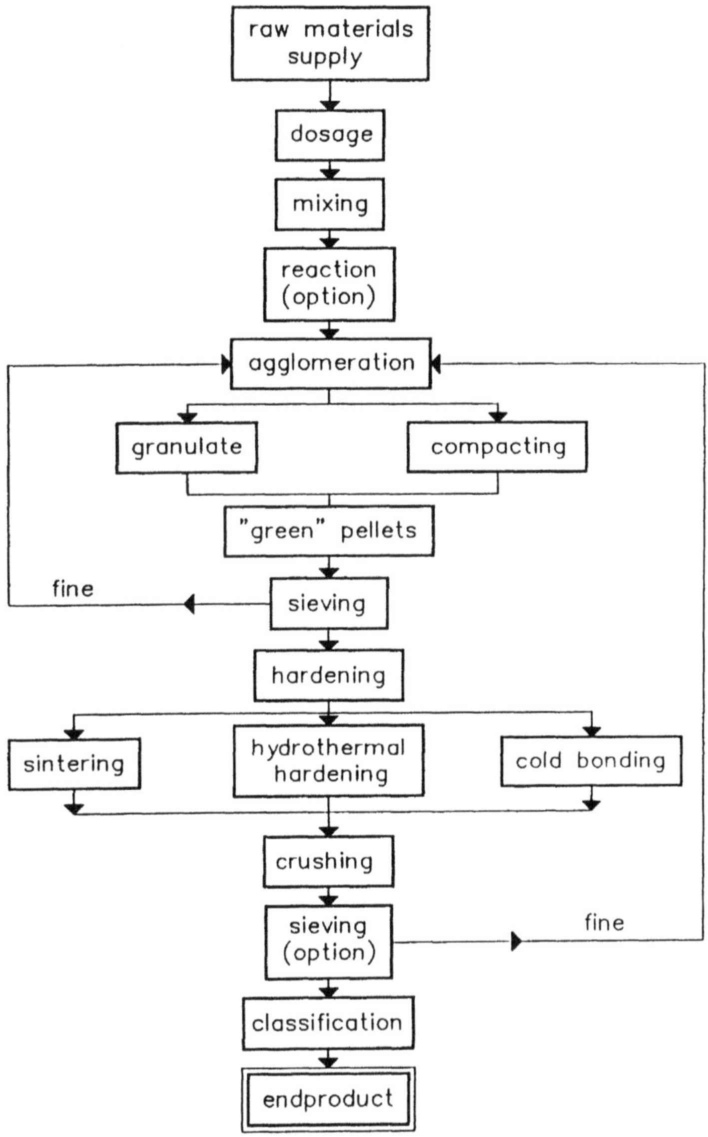

Fig.4.7: Process steps in fly-ash aggregate manufacture

Sintering processes have been known for many decades. Sintered aggregates are currently produced in a number of countries. The hydrothermal and cold-bonding processes have, however, recently attracted great attention, presumably because production costs are lower.

Fly ashes with a relatively high coal content are preferred for sintering processes. The ashes do not necessarily have to be pozzolanic and fine. For hydrothermal and cold-bonding processes, however, a low coal fly ash with high fineness and good pozzolanicity is preferred.

Properties of lightweight aggregates made from fly ash are similar to those of aggregates manufactured from other raw materials such as clay.

As compared to normal-weight aggregates, lightweight aggregates have the following disadvantages when used in concrete:

- More cement is needed to achieve the same characteristic concrete strength;
- The modulus of elasticity is lower;
- Creep and drying shrinkage are higher (although differences are small in the case of sintered pellets). It has been suggested that this is due to an expansive reaction at the fly-ash aggregate/cement interface;
- There is a decrease in workability during the first hours after mixing;
- Carbonation is faster, theoretically increasing the risk of reinforcement corrosion;
- Acoustic insulation properties are less favourable.

The following advantages are also reported:

- The apparent density is lower, reducing handling and transport costs;
- The relatively low deadweight of the pellets could be of major interest for high-quality conrete production;
- The heat flow resistance is relatively high; this effect is reinforced by the amorphous character of the fly ash;
- Internal stresses due to temperature gradients are lower.
- Fire resistance is improved.

Most fly-ash aggregates are not suitable for use in road construction, because their crushing value is too low, their bitumen absorption too high, their abrasion resistance inadequate and – if unbonded – their stability too low. Some of the newly-developed crushed compacted aggregates may, however, be able to meet road construction requirements.

4.6 Fly Ash in Road Construction (Prepared by E. Vazquez)

Fly ash has been quite extensively used as a road construction material /G 12, O 20, S 70/. It can be used as fill, sub-base and road base material, as a filler in bituminous mixtures and as an additive or partial substitute for portland cement in concrete. The total amount of fly ash employed in roadmaking is very high. In France, for example, 600 000 tonnes were utilized in just six months for construction of a section of the A2 motorway /A 2, A 12/. In England, 400 000 m^3 of fly ash have been employed in constructing embankments for the Trent Bridge works /C 30/.

Fly ash stabilized with lime or cement can be used as a sub-base and road-base. It may be considered as a replacement of the soil if this is impossible to stabilize. The amount of cement necessary to reach the minimum strength required (2.8 MPa in UK) is between 5 - 15 %. Cement increases the resistance of ash against frost.

Lime is also used to stabilize fly ashes, but the slower setting time of lime must be considered. Around 4 - 5 % of gypsum improves the strength at early ages.

Fly-ash-lime mixtures have been used to stabilize a wide range of materials. Sands mixed with fly ash, lime or cement can reach sufficient resistance to be employed as base. The optimum lime/fly-ash ratio is 1/4 and the dose of binder 20–30 %. The best compressive strength is reached when the mixture is compacted with a humidity slightly below the optimum Proctor. Mixtures of 91 % of fly ash, 5 % of phosphogypse and 4 % of quicklime have been used to overcome the slow setting problem. The compressive strength after one year is 15 MPa. With 75 % fly ashes, 15 % lime and 10 % phosphogypse a resistance of 35 MPa is reached after one year.

Mixtures of gravel, crushed stone and several types of slag have been used successfully: 85 % gravel 0/20, 13 % fly ash and 2 % quicklime is a very common proportioning. By replacing 30 % of sand by gravel in mixtures of sand, fly ash and lime, a higher compressive strength can be reached.

Good thermal insulating layers can be obtained by mixing fly ash, agglomerated ash-Agloporit and slags. The Agloporit coated with bitumen is more suitable than one stabilized by cement. If it is necessary to increase the strength characteristics, crushed stone can be added /M 76/.

4.7 Fly Ash in Soil Stabilization (Prepared by E. Vazquez)

Fly ash with lime or cement can be used to stabilize soils /C 33, J 29, T 22, V 16, V 17/. Pozzolanic reactions between fly ash lime and water give rise to cementitious products which bind the soil particles. In sandy or muddy soils, the fine particles of soil act as fillers, and the cementing products generated by pozzolanic reactions bind the soil. Soils containing clays require a larger lime/fly-ash ratio to ensure an adequate supply of lime for the lime-fly-ash reaction and the lime-clay reaction. There appears to be no optimum ratio of lime to fly ash for soil stabilization, since various proportions can provide satisfactory results. To achieve maximum compressive strength in clay soils, the lime content should be 5 to 9 % and the fly-ash content 10 to 25 % . For granular soils, the lime content should be between 3 and 6 % and the fly-ash content between 10 and 25 %.

Adding lime and fly ash decreases the maximum dry density and increases the optimum moisture content of the soil. The compressive strength of compacted soil-lime-fly ash is related to its density. Sandy soils will derive initial strengths from improved grading, and ultimate strengths from the lime-fly-ash reaction. The compressive strength may attain a value of 7 MPa.

Fly ashes possessing self-hardening properties can be used to stabilize sandy and clay soils without any other additive /M 77/. Self-hardening ashes provide cohesion improved grading, and ultimate strengths from the lime-fly-ash reaction. The compressive strength may attain a value of 7 MPa.

Fly ashes possessing self-hardening properties can be used to stabilize sandy and clay soils without any other additive /M 77/. Self-hardening ashes provide cohesion to sandy soils, reducing the plasticity index of plastic clays and increasing their compressive strength. High calcium fly ashes can be used to stabilize even organic clays. 10 % fly ash is generally sufficient for stabilization in sandy soils, and 15 % in clay . Some self-hardening fly ashes may have expansive effects.

The use of fly ash and the choice of the optimum content to mix with soils are very much conditioned by the degree of expansion and the available pore space of the compacted soil. Other self-hardening ashes containing fine free lime generate heat when mixed with the soil and water, but effectively stabilize sandy and clay soils when compacted after mixing. Strength development takes place rapidly up to 30 minutes. A small delay in compaction substantially reduces the soil-stabilizaing capacity of the fly ash. Salt retards the reaction between soil and fly ash.

4.8 Fly Ash as Asphalt-Filler (Prepared by E. Vazquez)

In 1939, fly ash was for the first time approved as a mineral filler for bituminous mixtures (by the Department of Public Works of the City of Detroit). The chemical composition of natural filler found in Trinidad asphalt does not differ materially from that of fly ash.

Fly ashes with no self-hardening properties shed water readily, reducing the tendency of moisture to strip the bitumen from the filler. They have good void-filling capacity and meet mineral filler specifications in terms of particle size distribution and moisture content. The beneficial effects of properly proportioned filler (max. 6 % by weight) are increased stability and better durability. Increased stability is attributed to the stiffened binder, while improved durability is a function of the character of the absorbed film.

Mixtures containing fly ash possess unconfined compressive strength at least equal to that of mixtures containing limestone dust. The flow determined by the Marshall Test shows no significant difference attributable to the fly ash, and resistance to water action is satisfactory tested by immersion-compression. The carbon content of the ashes has a negative effect on the stability of the bituminous mixtures: the maximum acceptable proportion seems to be 9 %.

Fine fly ashes with a high stabilizing effect require larger quantities of bitumen. If the quantity of such fly ash is reduced, asphalt mixes with bitumen quantities in the usual range can be obtained /A 13, O 20, Z 6/.

Fly ashes with self-hardening properties have been used with varying results. Experience reported by M. Feller /F 38/ and O. Manz /M 4/ indicated poor results, while Brama /B 79/ and Vasquez /V 18/ described the good properties of bituminous mixtures containing self-hardening fly ashes as a filler.

4.9 Fly Ash as Fill (Prepared by E. Vazquez)

Fly ash has been used as fill material in road construction, under buildings in old mine shafts and as land fill in general. The most important properties of fly ashes when used as fill are their particle size distribution, density, compactability, angle of internal friction and permeability /D 44, G 12, J 30, O 20, S 71, T 22/.

Dry maximum density is usually between 1 100 and 1 500 kg/m^3. This low density as compared with most other materials is advantageous for use in embankments constructed over compressible and weak bearing soils /J 4/. The optimum moisture content has been found to be 18 – 30 %. Vibratory compaction is best for fly-ash fills. Vibratory loads destroy the apparent cohesion in the fly ash by breaking the surface tension of the pore water. Steel rollers are not effective, because fly ash forms a wave in front of the forward roller which may bring it to a standstill.

Fly ash without self-hardening characteristics has no cohesion apart from that produced by capillary forces. Self-hardening fly ash may have a cohesion of up to 0,5 MPa. As a result of hardening, settlement within fly-ash fills is less than that in other materials. This makes it particularly useful as selected fill behind bridge abutments. Variations in self-hardening cementing properties due to variations in the free lime content and pozzolanic properties of the ash may cause difficulties.

The angle of internal friction in self-hardening fly ashes depends on density, and ranges from 29° to 46°; it increases with time. These ashes are incompressible as compared to a fly ash without self-hardening properties.

Permeability of fly ashes is low, but is much greater than that of clay. Water rises through ashes by capillarity. In some cases, it is advisable to include a 300 to 450 mm thick draining layer under the ashes to avoid the effects of frost.

4.10 Waste Neutralization and Stabilization (Prepared by J. Bijen)

Toxic waste materials are released from a large number of industrial processes /F 39, L 43/. One method of immunizing these wastes is to solidify and immobilize them. A relatively cheap method of solidification and immobilization appears to be to mix the waste with a mixture of water, fly ash and a bonding and activating agent such as lime, lime/gypsum or cement. Soluble silicates (waterglass, etc.) are often added to make specific ions insoluble (e. g. Cd^{2+}, Ca^{2+}, Ni^{2+} and Zn^{2+}).

After hardening, a stony material results. The rate at which toxic substances can be leached out of the material is greatly reduced. The technique is mostly applied to inorganic wastes.

Hazardous inorganic compounds are immobilized as a result of:

- encapsulation in the gelly matrix structure;
- a decrease in the solubility of heavy metal ions due to the prevailing high alkalinity and to the formation of insoluble silicates;
- physical adsorption and physico-chemical bonding; the reaction products of the activated fly ash and water have a very large specific surface, which promotes this type of bonding;
- low permeability and diffusivity of the solidified material to water and ions.

5 REFERENCES

/A1/ ALTNER, W. ; KOENIG, P. ; GRIEGER, C.: ZEMENTEINSPARUNG DURCH EINSATZ VON BRAUNKOHLENFILTERASCHE. IN: BAUSTOFFINDUSTRIE 16 (1973) AUSG. B, S. 11

/A2/ ANDRIEUX, P. ; COLOMBEL, J.-H.: UTILISATION DES CENDRES VOLANTES EN TECHNIQUE ROUTIERE : LES GRAVES-CENDRES VALANTES. IN: BULLETIN DE LIASON DES LABORATOIRES DES PONTS ET CHAUSSEES (1976) NR. 83, S. 73-90 ALSO IN: SILICAT INDUSTRIE 41 (1976), S. 227-244

/A3/ EL-AROUSY, H.T.: UEBER DIE STEINKOHLENFLUGASCHE UND IHRE WIRKUNG AUF DIE EIGENSCHAFTEN VON LEICHTBETON MIT GESCHLOSSENEM GEFUEGE IM FRISCHEN UND FESTEN ZUSTAND. BRAUNSCHWEIG : INSTITUT FUER BAUSTOFFKUNDE UND STAHLBETONBAU, TU BRAUNSCHWEIG (1973), NR. 24 (DISSERTATION)

/A4/ ABDUL-MAULA, S. ; ODLER, I.: HYDRATION REACTIONS IN FLY-ASH-PORTLAND CEMENTS. BOSTON : MATERIAL RESEARCH SOCIETY. - IN: EFFECTS OF FLY-ASH INCORPORATION IN CEMENT AND CONCRETE : PROCEEDINGS SYMPOSIUM ANNUAL MEETING (DIAMOND,S.(ED.)) BOSTON, NOVEMBER 16-18, 1981, S. 102-111

/A5/ ARCHITECTURAL INSTITUTE OF JAPAN: TENTATIVE GUIDE FOR MIX DESIGN AND WORKMANSHIP FOR CONCRETE USING FLY ASH CEMENT OR CONCRETE USING FLY ASH AS ADMIXTURE. ARCHITECTURAL INSTITUTE OF JAPAN

/A6/ ASHBY, J.B.: ANSWERS TO THE OBJECTIONS TO THE USE OF FLY ASH IN CONCRETE. IN: THE CHALLENGE OF CHANGE - SIXTH INTERNATIONAL ASH UTILISATION SYMPOSIUM PROCEEDINGS, DOE/METC/82-52(VOL.1), S. 246-258

/A7/ ABE, H. ; NAGATAKI, S. ; TSUKUJAMA, R.: WRITTEN DISCUSSION ON "FLY ASH AND FLY ASH CEMENT" BY M. KOKUBU. IN: DISCUSSION NO. IV 15, PROCEEDING: 5TH. INTERNATIONAL SYMPOSIUM ON CHEMISTRY OF CEMENT, V-ISCC, TOKYO 1968

/A8/ ALONSO, J.L.: ESTUDIO FISICO-QUIMICO Y TECNICO DE DIVERSOS TIPOS DE CENIZAS Y SU EMPLEO COMO MATERIAL DE CONSTRUCCION. LABOR. CENTRAL DEL MOPU. PUBLIC. (1966), NR. 199

/A9/ ANDERSON, M.: A NEW LOW-COST PFA BRICKMAKING PROCESS'. CONFERENCE ON ASH TECHNOLOGY AND MARKETING, LONDON, 1984

/A10/ ANDERSON, M.: PROCESSED HIGH CARBON PFA AS A DUAL ADDITIVE IN BRICKMAKING'. CONFERENCE ON ASH TECHNOLOGY AND MARKETING, LONDON, 1984

/A11/ ANDERSON, M. ; JACKSON, G.: THE BENEFICATION OF POWER STATION COAL ASH AND ITS USE IN HEAVY CLAY CERAMICS. NORTH DAKOTA : UNIVERSITY OF NORTH DAKOTA- IN: PROCEEDINGS OF THE UTILIZATION OF ASH WORKSHOP, MAY 13-15,

1987 ALSO IN TRANSACTION JOURNAL OF THE BRITISH CERAMIC SOCIETY 82 (1983), NR. 2, S. 50-55

/A12/ ANDRIEUX, P.: UTILISATION DES CENDRES VOLANTES EN TECHNIQUE ROUTIERE. PROCEEDINGS OF THE INTERNATIONAL CONFERENCE ON THE USE OF BY PROD. AND WASTE IN CIVIL ENG., PARIS, VOLUME II (1978), S. 333-339

/A13/ ANON ; N.N.: UNTERSUCHUNGEN UEBER DIE MOEGLICHKEITEN DER FLUG- ASCHEN-VERWERTUNG IM BITUMINOESEN STRASSENBAU. IN: BITUMEN- TEERE-ASPHALTE-PECHE 15 (1964), NR. 8, S. 363-373

/B1/ BABATCHEV, G.: UNTERSUCHUNGEN ZUR VERWERTUNG VON FLUGASCHEN AUS DEM KRAFTWERK REPUBLIKA FUER DIE HERSTELLUNG VON ZEMENT UND ZEMENT-BETONEN. IN: BAUSTOFFINDUSTRIE 23 (1980), S. 18-21

/B2/ BARBER, E.G.: THE UTILIZATION OF PULVERIZED-FUEL ASH. IN: JOURNAL OF INSTITUTE OF FUEL 43 (1970), S. 4-9

/B3/ BEAUDOIN, J.J. ; FELDMAN, R.F.: A STUDY OF MECHANICAL PROPERTIES OF AUTOCLAVED CALCIUM SILICATE SYSTEMS. IN: CEMENT AND CONCRETE RESEARCH 5 (1975), S. 103-118

/B4/ BEAUDOIN, J.J. ; MAC INNIS, C.: THE EFFECT OF ADMIXTURES ON LENGTH ANOMALIES DUE TO SLOW COOLING AND WARMING OF HARDENED CEMENT PASTE. IN: CEMENT AND CONCRETE RESEARCH 4 (1974), S. 347-356

/B5/ BENDER, R.J.: THE ASH UTILIZATION SYMPOSIUM. IN: POWER 117 (1970), S. 120-122

/B6/ BENNERT, E.M. ; GROSS, G.: BETONSCHAEDLICHE BESTANDTEILE IN FLUGASCHEN ERLAEU- TERUNGEN ZUM STANDARDENTWURF TGL 36859/04. IN: BETONTECHNIK 5 (1980), S. 13-15

/B7/ BERETKA, J.: UTILIZATION OF FLY IN AUSTRALIA. IN: PROCEEDINGS OF THE 4TH INT. MEET. ON MOD. CER. TECHN., SAINT VINCENT, ITALY 1979, S.335-346 ELSEVIER SCIENTIFIC PUBLISHING COMPANY

/B8/ BERETKA, J. ; BROWN, T.: UTILIZATION OF FLY ASH IN BUILDING. IN: JOURNAL OF THE AUSTRALIAN CERAMIC SOCIETY 13 (1977), S. 24-30

/B9/ BERETKA, J. ; MIDDLETON, L.W.: AERATED CONCRETE FROM FLY ASH. IN: CIA NEWS 6 (1980), S. 10-12

/B10/ BERG, VOM, W.: KOSTENORIENTIERTER BETONENTWURF FUER FLUGASCHEHALTIGE BETONE. IN: BETONWERK UND FERTIGTEILTECHNIK 47 (1981), S.401-407

/B11/ BERRY, E.E.: STRENGTH DEVELOPMENT OF SOME BLENDED-CEMENT MORTARS. IN: CEMEMENT AND CONCRETE RESEARCH 10 (1980), S. 1-11

/B12/ BERRY, E.E. ; MALHOTRA, V.M.: FLY ASH FOR USE IN CONCRETE : A CRITICAL REVIEW. IN: JOURNAL OF THE AMERICAN CONCRETE INSTITUTE 77 (1980), NR. 1, S. 59-73

/B13/ BLANKS, R.F.: FLY ASH AS A POZZOLAN. IN: JOURNAL OF THE AMERICAN CONCRETE INSTITUTE 21 (1950), S.701-707

/B14/ BLASCHKE, R. ; HOLZAPFEL, TH.: ANALYSE EINES FLUGSTAUBS UND SEINES EINFLUSSES AUF DIE GEFUEGEENTWICKLUNG IM ZEMENTSTEIN. IN: BEITR. ELEKTRONENMIKROSKOP. DIREKTABB. OBERFL. 12 (1979), S. 389-390

/B15/ BLASCHKE, R. ; SCHOLZ, H. ; BACH, H.: KRAFTWERKSSTAEUBE UND BETON. IN: BEITR. ELEK- TRONENMIKROSKOP. DIREKTABB. OBERFL. 14 (1981), S. 213-232

/B16/ BLUEMEL, O.W.: OESTERREICHISCHE FLUGASCHE ALS BINDEMITTELZUSAETZE. IN: MITTEILUNGEN VGB 39 (1959), S. 421-431

/B17/ BLUEMEL, O.W. ; LUKAS, W.: DER EINFLUSS DER FLUGASCHE AUF DIE FESTIGKEITSENTWICKLUNG UND DEN REAKTIONSABLAUF IN PORTLAND-ZEMENTGEMISCHEN. IN: MATERIAL UND TECHNIK 5 (1977), S. 44-50

/B18/ BRAUKMANN, B.: DIE KORNKENNLINIE DER FLUGASCHE DES KRAFTWERKS FORTUNA I UND IHRE ERMITTLUNG. IN: STAUB 14 (1954), S. 382-389

/B19/ BRINK, R.H. ; HALSTEAD, W.J.: STUDIES RELATING TO THE TESTING OF FLY ASH FOR USE IN CONCRETE. IN: PUBLIC ROADS 29 (1957), NR. 6, S. 121-141 ALSO IN: PROC. ASTM 56 (1956), S. 1161-1214

/B20/ BRADBURY, H.W.: THE USE OF FLY ASH IN PRE-BLENDED CEMENT. PROC.: INT. SYMPOS.: THE USE OF PFA IN CONCRETE, DEPT. CIV. ENG. LEEDS UNIV. 1982, VOL.1, S. 41-49

/B21/ BUTTLER, F.G. ; WALKER, E.J.: THE RATE AND EXTENT OF REACTION BETWEEN CALCIUM HYDROXIDE AND PULVERIZED FUEL ASH. PROC.: INT. SYMPOS.: THE USE OF PFA IN CONCRETE, DEPT. CIV. ENG. LEEDS UNIV. 1982, VOL.1, S. 71-81

/B22/ BUTTLER, F.G. ; DECTER, M.H. ; SMITH, G.R.: STUDIES OF THE RELATIONSHIP BETWEEN DESICCATION, CARBONATION AND COMPRESSIVE STRENGTH FOR SYSTEMS CONTAINING PORTLAND CEMENT AND PFA. PROC.: INT. SYMPOS.: THE USE OF PFA IN CONCRETE, DEPT. CIV. ENG. LEEDS UNIV. 1982, VOL.1, S. 101-110

/B23/ BROWN, J.H.: THE STRENGTH AND WORKABILITY OF CONCRETE WITH PFA SUBSTITUTION. PROC.: INT. SYMPOS.: THE USE OF PFA IN CONCRETE, DEPT. CIV. ENG. LEEDS UNIV. 1982, VOL.1, S. 151-159

/B24/ BANFILL, P.F.G.: AN EXPERIMENTAL STUDY OF THE EFFECT OF PFA ON THE RHEOLOGY OF FRESH CONCRETE AND CEMENT PASTE. PROC.: INT. SYMPOS.: THE USE OF PFA IN CONCRETE, DEPT. CIV. ENG. LEEDS UNIV. 1982, VOL.1, S. 161-171

/B25/ BROOKS, J.J. ; WAINWRIGHT, P.J. ; CRIPWELL, J.B.: TIME-DEPENDENT PROPERTIES OF CONCRETE CONTAINING PULVERIZED FUEL ASH AND SUPER PLASTICIZER. PROC.: INT. SYMPOS.: THE USE OF PFA IN CONCRETE, DEPT. CIV. ENG. LEEDS UNIV. 1982, VOL.1, S. 209-220

/B26/ BLACKIE, A.D.: THE INFLUENCE OF PFA ON THE PROPERTIES OF STRUCTURAL CONCRETE. PROC.: INT. SYMPOS.: THE USE OF PFA IN CONCRETE, DEPT. CIV. ENG. LEEDS UNIV. 1982, VOL.1, S. 289-299

/B27/ BRACKETT, C.E.: AVAILABILITY, QUALITY, AND PRESENT UTILIZATION OF FLY ASH. PROC.: EDISON ELECTRIC INSTITUTE, NATIONAL COAL ASSOC., BUREAU OF MINES SYMPOS., PITTSBURGH 1967, S. 16-36, COMPILED BY FABER, J.H. ; CAPP, P. ; SPENCER, J.D.

/B28/ BARTON, W.R.: RAW MATERIALS FOR MANUFACTURE OF CEMENT. PROC.: EDISON ELECTRIC INSTITUTE, NATIONAL COAL ASSOC., BUREAU OF MINES SYMPOS., PITTSBURGH 1967, S. 46-51, COMPILED BY FABER, J.H. ; CAPP, P. ; SPENCER, J.D.

/B29/ BELOT, J.R.: FLY ASH IN CONCRETE AND CONCRETE BLOCK MANUFACTURING. PROC.: EDISON ELECTRIC INSTITUTE, NATIONAL COAL ASSOC., BUREAU OF MINES SYMPOS., PITTSBURGH 1967, S. 101-106, COMPILED BY FABER, J.H. ; CAPP, P. ; SPENCER, J.D.

/B30/ BARENBERG, E.J.: LIME-FLY ASH-AGGREGATE MIXTURES. PROC.: EDISON ELECTRIC INSTITUTE, NATIONAL COAL ASSOC., BUREAU OF MINES SYMPOS., PITTSBURGH 1967, S. 111-134, COMPILED BY FABER, J.H. ; CAPP, P. ; SPENCER, J.D.

/B31/ BERGEMANN, G.O.JR.: USE OF FLY ASH IN SPECIALIZED CONCRETE WORK. PROC.: EDISON ELECTRIC INSTITUTE, NATIONAL COAL ASSOC., BUREAU OF MINES SYMPOS. PITTSBURGH 1967, S. 235-249, COMPILED BY FABER, J.H. ; CAPP, P. ; SPENCER, J.D.

/B32/ BUTTLER, F.G. ; MORGAN, S.R. ; RICHARDS, P.W.: THE REACTION OF CALCIUM HYDROXIDE LIBERATED ON HYDRATION OF PORTLAND CEMENT WITH FLY ASH IN MORTARS. IN: SILICATES INDUSTRIELS 47 (1982), NR. 4, S. 103-107

/B33/ BANFILL, P.F.G.: AN EXPERIMENTAL STUDY OF THE EFFECT OF PFA ON THE RHEOLOGY OF FRESH CONCRETE AND CEMENT PASTE. PROC.: INT. SYMPOS.: THE USE OF PFA IN CONCRETE DEPT. CIV. ENG. LEEDS UNIV. 1982, VOL. 1, S. 161-171

/B34/ BUTTLER, F.G.: EFFECT OF REACTION BETWEEN PULVERISED FUEL ASH AND CALCIUM HYDROXIDE ON CONCRETE PROPERTIES. BOSTON : MATERIAL RESEARCH SOCIETY IN: EFFECTS OF FLY-ASH INCORPORATION IN CEMENT AND CONCRETE : PROCEEDINGS SYMPOSIUM N ANNUAL MEETING (DIAMOND, S.(ED.)) BOSTON, NOVEMBER 16-18, 1981, S. 155-164

/B35/ BARKER, W.R.: THE DISTINCTION BETWEEN WATER REDUCTION AND REACTIVITY AS THE FACTORS INFLUENCING FLY-ASH PERFORMANCE IN CONCRETE. BOSTON : MATERIAL RESEARCH SOCIETY IN: EFFECTS OF FLY-ASH INCORPORATION IN CEMENT AND CONCRETE : PROCEEDINGS SYMPOSIUM N ANNUAL MEETING (DIAMOND, S., (ED.)) BOSTON, NOVEMBER 16-18, 1981, S. 289-295

/B36/ BESLAGIC, S. ; JAKIC, M.: POSSIBILITY OF USING FLY ASH. IN: BEOGRAD (1979), S. 93-103

/B37/ BIJEN, J.M.J.M.: VLIEGAS IN DE BOUW (NETHERL.). IN: MATERIAL TESTING AND CONSULTING RAPPORT NR. 80137

/B38/ BIJEN, J.M.J.M.: CEMENTS FROM SLAGS AND FLY ASH. INTRON BV MAASTRICHT

/B39/ BUILDING RESEARCH ADVISORY SERVICE: "POZZOLAN" A CLASSIFIED PULVERISED-FUEL ASH FOR USE IN CONCRETE - FINAL REPORT. BUILDING RESEARCH STATION JOB NUMBER: WJ 2516, GARSTON, 1979

/B40/ BAMFORTH, P.B.: IN SITU MEASUREMENT OF THE EFFECT OF PARTIAL PORTLAND CEMENT REPLACEMENT USING EITHER FLY ASH OR GROUND GRANULATED BLAST-FURNACE SLAG ON THE PERFORMANCE OF MASS CONCRETE. IN: PROC. INSTN. CIV. ENG., PART 2, (1981) 69, S. 777-801, 71, S. 601-613

/B41/ BAMFORTH, P.B.: AN INVESTIGATION INTO THE INFLUENCE OF PARTIAL PORTLAND CEMENT REPLACEMENT USING EITHER FLY ASH OR GROUND GRANULATED BLAST-FURNACE SLAG ON THE EARLY AGE AND LONGTERM BEHAVIOUR OF CONCRETE. TAYLOR WOODROW RESEARCH REPORT NO. 014J/78/2067, SOUTHALL, 1978

/B42/ BERETKA, J. ; BROWN, T.: PROPERTIES AND POZZOLANIC BEHAVIOUR OF AUSTRALIAN FLY ASH. IN: JOURNAL OF THE AUSTRALIAN CERAMIC SOCIETY 12 (1976), NR. 1, S. 13-19

/B43/ BLENKINSOP, J.C. ; BRADBURY, H.W.: THE USE OF POZZOLANIC AND OTHER ADMIXTURES IN PREBLENDED CEMENT AND GROUTS. PROC. 1ST INT. CONF. ON READY-MIXED CONCRETE, DUNDEE, OCT. 1975 EDITED BY DHIR, R.K.

/B44/ BARKER, W.R.: AN INVESTIGATION OF FLY ASH CONCRETE. IN: CONCRETE, BETON NR. 29, 1983.3

/B45/ BACH, T.: PERFORMANCE OF AIR-ENTRAINING AGENTS IN CONCRETES CONTAINING FLY ASH. IN: CBL INTERN. REPORT NO. 24, 1980

/B46/ BRAUN, H. ; GEBAUER, J.: MOEGLICHKEITEN UND GRENZEN DER VERWENDUNG VON FLUGASCHEN IM ZEMENT. IN: ZEMENT-KALK-GIPS (ZKG) 36 (1983), NR. 5, S. 254-258

/B47/ BERG, VOM, W.: FLUGASCHE ALS BETONZUSATZSTOFF NACH DIN 1045. IN: VGB KRAFTWERKSTECHNIK 62 (1982), NR. 11, S. 967-977

/B48/ BUTTLER, F.G. ; DECTER, M.H. ; SMITH, G.R.: STUDIES ON THE DESICCATION AND CARBONATION OF SYSTEMS CONTAINING PORTLAND CEMENT AND FLY ASH. DETROIT : AMERICAN CONCRETE INSTITUTE ACI SP-79, 1983 - IN: PROCEEDINGS OF THE CANMET/ACI FIRST INTERNATIONAL CONFERENCE ON THE USE OF FLY ASH, SILICA FUME, SLAG AND OTHER MINERAL BY-PRODUCTS IN CONCRETE VOL. 1 (1983), S. 367-381

/B49/ BORDONADO, G. ; NISSOUX, J.L.: ROAD -BUILDING CONCRETES INCORPORATING FLY ASH OR SLAG. DETROIT : AMERICAN CONCRETE INSTITUTE ACI SP-79, 1983 - IN: PROCEEDINGS OF THE CANMET/ACI FIRST INTERNATIONAL CONFERENCE ON THE USE OF FLY ASH, SILICA FUME, SLAG AND OTHER MINERAL BY-PRODUCTS IN CONCRETE VOL. 1 (1983), S. 471-493

/B50/ BIJEN, J.M.J.M.: FLY ASH AGGREGATES. DETROIT : AMERICAN CONCRETE INSTITUTE ACI SP-79, 1983 - IN: PROCEEDINGS OF THE CANMET/ACI FIRST INTERNATIONAL CONFERENCE ON THE USE OF FLY ASH, SILICA FUME, SLAG AND OTHER MINERAL BY-PRODUCTS IN CONCRETE VOL. 1 (1983), S. 495-517

/B51/ BAKKER, R.F.M.: PERMEABILITY OF BLENDED CEMENT CONCRETES. DETROIT : AMERICAN CONCRETE INSTITUTE, ACI SP-79, 1983 -IN: PROCEEDINGS OF THE CANMET/; ACI FIRST INTERNATIONAL CONFERENCE ON THE USE OF FLY ASH, SILICA FUME, SLAG AND OTHER MINERAL BY-PRODUCTS IN CONCRETE VOL. 1, (1983), S. 589-605

/B52/ BOEHM, E. ; KUNZ, P.: VERWENDUNGSMOEGLICHKEITEN VON STEINKOHLE- FLUGASCHEN IM BAUWESEN. IN: BAUWIRTSCHAFT (1983), NR. 16, S. 612-616

/B53/ BROWN, G.E. ; OATES, D.B.: AIR ENTRAINMENT IN SULFATE-RESISTANT CONCRETE. IN: CONCRETE INTERNATIONAL 5 (1983), NR. 1, S. 36-39

/B54/ BARBER, E.G. ; JONES, G.T. ; KNIGHT, P.G.K. ; MILES, M.M.: PFA UTILISATION. CENTRAL ELECTRICITY GENERATING BOARD, 1972

/B55/ BOTTKE, R.: MINDESTZEMENTGEHALT UND VERWENDUNG VON FLUGASCHE. IN: BETONWERK- UND FERTIGTEILTECHNIK 48 (1982), NR. 5, S. 311-314

/B56/ BOTTKE, R.: BETONPOROSITAET ALS QUALITAETSMASSSTAB. IN: BAUWIRTSCHAFT (1981) NR. 14, S. 463-465, NR. 21, S. 749-750

/B57/ BLUEMEL, O.W. ; LUKAS, W.: KRITISCHE BEMERKUNGEN ZUM SULFATANGRIFF AUF BETON. SOURCE UNKNOWN

/B58/ BLUEMEL, O.W. ; LUKAS, W.: DAS KALKBINDEVERMOEGEN (CA(OH)2) VON FLUGASCHE IM BEISPIEL DER ASCHE AUS ST. ANDRAE/LAVANTTAL-KAERNTEN. IN: MITTEILUNGEN AUS DEM INSTITUT FUER BAUSTOFFLEHRE UND MATERIALPRUEFUNG AN DER UNIVERSITAET INNSBRUCK, JAHRESBERICHT 1975/76, S.33-34

/B59/ BRACETT, C.E.: PRODUCTION AND UTILISATION OF ASH IN THE UNITED STATES. ASH UTILISATION, 1973, BUREAU OF MINES INFORMATION CIRCULAR 8640/1974, S.12-18

/B60/ BARENBERG, E.J.: UTILISATION OF ASH IN STABILISED BASE CONSTRUCTION. ASH UTILISATION, 1973, BUREAU OF MINES INFORMATION CIRCULAR 8640/1974, S. 180-196

/B61/ BRINK, R.H.: USE OF WASTE SULFATE ON TRANSPO '72 PARKING LOT. ASH UTILISATION, 1973, BUREAU OF MINES INFORMATION CIRCULAR 8640/1974, S. 197-207

/B62/ BLOCKER, W.V. ; MORRISON, R.E. ; MORTON, W.E. ; BABCOCK, A.W.: MARKETING POWERPLANT AGGREGATES AS A ROAD BASE MATERIAL. ASH UTILIZATION, 1973, BUREAU OF MINES INFORMATION CIRCULAR 8640/1974, S. 208-223

/B63/ BARBER, G.: LAND RECLAMATION AND ENVIROMENTAL BENEFITS OF ASH UTILIZATION. ASH UTILIZATION, 1973, BUREAU OF MINES INFORMATION CIRCULAR 8640/1974, S. 246-257

/B64/ BOEHM, E. ; KUNZ, P.: CHEMISCHE UND PHYSIKALISCHE EIGENSCHAFTEN VON KRAFTWERKSRUECKSTAENDEN. ESSEN : VGB-KRAFTWERKSTECHNIK GMBH. - IN: MOEGLICHKEITEN DER VERSTAERKTEN VERWERTUNG VON RUECKSTAENDEN AUS KRAFTWERKSFEUERUNGEN = REIHE VGB TECHNISCHWISSENSCHAFTLICHE BERICHTE "WAERMEKRAFTWERKE" (1982), NR. VGB-TW 205, S. 12-17

/B65/ BOEHM, E. ; KUNZ, P.: VERWERTUNGSMENGEN UND EINSATZGEBIETE VON RUECKSTAENDEN AUS STEINKOHLEKRAFTWERKEN. ESSEN : VGB-KRAFTWERKSTECHNIK GMBH. - IN: MOEGLICHKEITEN DER VERSTAERKTEN VERWERTUNG VON RUECKSTAENDEN AUS KRAFTWERKFEUERUNGEN = REIHE VGB TECHNISCH- WISSENSCHAFTLICHE BERICHTE "WAERMEKRAFTWERKE" (1982), NR. VGB-TW 205, S. 22-25

/B66/ BOEHM, E. ; KUNZ, P.: PRODUKTION VON BAUSTOFFEN; INTERNATIONALE SITUATION DER ASCHEVERWERTUNG. ESSEN : VGB-KRAFTWERKSTECHNIK GMBH. - IN: MOEGLICHKEITEN DER VERSTAERKTEN VERWERTUNG VON RUECKSTAENDEN AUS KRAFTWERKSFEUERUNGEN = REIHE VGB TECHNISCHWISSENSCHAFTLICHE BERICHTE "WAERMEKRAFTWERKE" (1982), NR. VGB-TW 205, S. 56-67

/B67/ BOEHM, E. ; KUNZ, P.: NEUERE VERWERTUNGSMOEGLICHKEITEN FUER FLUGASCHE. ESSEN : VGB-KRAFTWERKSTECHNIK GMBH. - IN: MOEGLICHKEITEN DER VERSTAERKTEN VERWERTUNG VON RUECKSTAENDEN AUS KRAFTWERKSFEUERUNGEN = REIHE VGB TECHNISCHWISSENSCHAFTLICHE BERICHTE "WAERMEKRAFTWERKE" (1982), NR. VGB-TW 205, S.130-143

/B68/ BERRY, E.E.: FLY ASH FOR USE IN CONCRETE. PART I: A CRITICAL REVIEW OF THE CHEMICAL, PHYSICAL AND PUZZOLANIC PROPERTIES OF FLY ASH. CANMET REPORT (1976), 76-25

/B69/ BRADBURY, H.W.: THE USE OF FLY ASH IN PRE-BLENDED CEMENT. PROC.: INT. SYMPOS.: THE USE OF PFA IN CONCRETE DEPT. CIV. ENG. LEEDS UNIV. 1982, VOL. 1, S. 41-49 ALSO IN: SILICATES INDUSTRIELS (1982), NR. 12, S. 283-288

/B70/ BROWN, P.W. ; BERGER, R.L. ; CLIFTON, J.R. ; FROHNSDORFF, G.: LIMITATIONS TO FLY ASH USE IN BLENDED CEMENTS. U.S. DEPARTMENT OF COMMERCE, NATIONAL BUREAU OF STANDARDS

/B71/ BOGUE, R.H.: PORTLAND CEMENT ASSOCIATION FELLOWSHIP, PAPER NO. 55, 1939

/B72/ BUILDING RESEARCH ADVISORY SERVICE: "POZZOLAN" A CLASSIFIED PULVERISED FUEL ASH FOR USE IN CONCRETE - FINAL REPORT. BUILDING RESEARCH STATION JOB NUMBER: WJ 2516, GARSTON (1979)

/B73/ BROTSCHI, J. ; MEHTA, P.K.: TEST METHODS FOR DETERMINING POTENTIAL ALKALI-SILICA REACTIVITY IN CEMENTS. IN: CEMENT AND CONCRETE RESEARCH 8 (1978), NR. 2, S. 191-200

/B74/ BAKKER, R.F.M.: ABOUT THE CAUSE OF THE RESISTANCE OF BLASTFURNACE CEMENT CONCRETE TO THE ALKALI-SILICA REACTION, PROCEEDINGS OF THE FIFTH INTERN. CONFERENCE ON ALKALI-AGGREGATE REACTION IN CONCRETE, 1981, S252/29

/B75/ BUTLER, F.G. ; MORGAN, S.R. ; WALKER, E.J.: STUDIES ON THE RATE AND EXTENT OF REACTION BETWEEN CALCIUM HYDROXIDE AND PULVERIZED FUEL ASH AT 38 °C. PROCEEDINGS OF THE FITH INTERNATIONAL CONFERENCE ON ALKALI-AGGREGATE REACTION IN CONCRETE, 1981, S252/58

/B76/ BERTACCHI, P.: DETERIORATION OF CONCRETE CAUSED BY CARBONIC ACID. RILEM INTERNATIONAL SYMPOSIUM ON DURABILITY OF CONCRETE, TECHNICAL UNIVERSITY, PRAGUE, 1969, ACADEMIA (PRAGUE), PP C159-C168

/B77/ BIJEN, J. ; WALTJE, H.: ALTERNATIVE RAW MATERIALS AND BONDING MATERIALS FOR PORTLAND CEMENT. PART IV.: SLAG-FLY ASH CEMENT. MINISTRY OF HOUSING, PHYSICAL PLANNING AND ENVIRONMENT, LEIDSCHENDAM, NETHERLANDS, 1985

/B78/ BIJEN, J.: USE OF POWDER COAL FLY ASH AND RESIDUAL HEAT OF COAL FIRED UTILITIES FOR THE PRODUCTION OF AUTOCLAVED AERATED CONCRETE. INTRON REPORT NO. 84074 ON BEHALF OF THE PROJECT OFFICE FOR ENERGY RESEARCH OF THE NETHERLANDS ENERGY FOUNDATION, MAASTRICHT, MAY 1984

/B79/ BRAMA, S.P.: USE OF LIGNITE FLY ASH AS A MINERAL FILLER IN BITUMINOUS CONCRETE. NORTH DAKOTA ST. UNIV. OF FARGO, PP. 26, 1968

/C1/ CANNON, R.W.: PROPORTIONING FLY ASH CONCRETE MIXES FOR STRENGTH AND ECONOMY. IN: JOURNAL OF THE AMERICAN CONCRETE INSTITUTE 65 (1968), S.969-979

/C2/ CAPP, J.B. ; SPENCER, J.D.: FLY ASH UTILIZATION, A SUMMARY OF APPLICATION AND TECHNOLOGY. U.S. DEPT. OF INTERIOR, BUREAU OF MINES 1970

/C3/ CARLES-GIBERGUES, A.: CONTRIBUTION MINERALOGIQUE A L' ETUDE DES PROPRIETES POUZZOLANIQUES DES CENDRES VOLANTES DE PRINCIPALES CENTRALES THERMIQUES FRANCAISES. DISSERTATION AN DER UNIVERSITAET TOULOUSE, 1961

/C4/ CARLES-GIBERGUES, A. ; STAMBOLIEVA, R. ; VAQUIER, A.: ROLE INITIAL DES SULFATES D'UNE CENDRE VOLANTE DANS SON CARACTERE POUZZOLANIQUE. IN: MATERIAUX ET CONSTRUCTIONS 6 (1973), S. 29-38

/C5/ CHOPRA, S.K. ; KISHAN LAL: SINTERED LIGHTWEIGHT AGGREGATE FROM INDIAN FLY ASHES. IN: INDIAN CONCRETE JOURNAL 35 (1961), S.150-153

/C6/ CHOPRA, S.K. ; NARAIN, S.N.: MEASUREMENT OF THE SPECIFIC SURFACE OF THE FLY ASH. IN: MATERIALS RESEARCH AND STANDARDS 4 (1964), S. 487-490

/C7/ CLENDENNING, T.G. ; DURIE, N.D.: PROPERTIES AND USE OF FLY ASH FROM A STEAM PLANT OPERATING UNDER VARIABLE LOAD. PROC.: ASTM 62 (1962), S. 1019-1037

/C8/ CABRERA, J.G. ; PLOWMAN, C.: HYDRATION MECHANISMS, MINERALOGY AND MORPHOLOGY OF THE C3A-PFA SYSTEM. PROC.: INT. SYMPOS.: THE USE OF PFA IN CONCRETE, DEPT. CIV. ENG. LEEDS UNIV. 1982, VOL.1, S. 111-120

/C9/ CABRERA, J.G. ; HOPKINS, C.J.: THE EFEECTS OF PFA ON THE RHEOLOGY OF CEMENT PASTES. LEEDS : DEPARTMENT OF CIVIL ENGINEERING, 1982 - IN: PROCEEDINGS OF THE INTERNATIONAL SYMPOSIUM ON THE USE OF PFA IN CONCRETE, S. 141-149

/C10/ COPELAND, B.G.T.: PFA CONCRETE FOR HYDRAULIC TUNNELS AND SHAFTS, DINORWIC PUMPED STORAGE SCHEME-CASE HISTORY. PROC.: INT. SYMPOS.: THE USE OF PFA IN CONCRETE, DEPT. CIV. ENG. LEEDS UNIV. 1982, VOL. 1, S. 323-342

/C11/ CAPP, P.: FLY ASH IN AGRICULTURE. PROC.: EDISON ELECTRIC INSTITUTE, NATIONAL COAL ASSOC., BUREAU OF MINES SYMPOS., PITTSBURGH 1967, S.210-220, COMPILED BY FABER, J.H. ; CAPP, P. ; SPENCER, J.D.

/C12/ CORYELL, G.L.: FUTURE OF FLY ASH USE. PROC.: EDISON ELECTRIC INSTITUTE, NATIONAL COAL ASSOC., BUREAU OF MINES SYMPOS. PITTSBURGH 1967, S. 250-258, COMPILED BY FABER, J.H. ; CAPP, P. ; SPENCER, J.D.

/C13/ COSTA, U. ; MASAZZA, F.: NATURAL POZZOLANAS AND FLY ASHES. BOSTON : MATERIAL RESEARCH SOCIETY - IN: EFFECTS OF FLY-ASH INCORPORATION IN CEMENT AND CONCRETE : PROCEEDINGS SYMPOSIUM N ANNUAL MEETING (DIAMOND, S. (ED.)) BOSTON, NOVEMBER 16-18, 1981, S. 134-144

/C14/ CROW, R.D. ; DUNSTAN, E.R.: PROPERTIES OF FLY ASH CONCRETE. BOSTON : MATERIALS RESEARCH SOCIETY - IN: EFFECTS OF FLY-ASH INCORPORATION IN CEMENT AND CONCRETE : PROCEEDINGS SYMPOSIUM N ANNUAL MEETING (DIAMOND, S. (ED)) BOSTON, NOVEMBER 16-18, S. 214-224

/C15/ CAIN, C.J.: EFFECTS OF VARIOUS TYPES OF FLY ASH ON BEHAVIOR AND PROPERTIES OF CONCRETE. BOSTON : MATERIAL RESEARCH SOCIETY - IN: EFFECTS OF FLY-ASH INCORPORATION IN CEMENT AND CONCRETE : PROCEEDINGS SYMPOSIUM N ANNUAL MEETING (DIAMOND,S.(ED.)) BOSTON, NOVEMBER 16-18, 1981, S. 260-268

/C16/ CASTRO, M.J.S.: CEMENTOS QUE CONTIENEN 20 Y 30 POR CIENTO DE CENIZAS VOLANTES DE PUENTES DE GARCIA RODRIGUEZ Y DE PONFERRADA. IN: MATERIALES DE CONSTRUCCION NUM. 179 (1980), S. 9-37

/C17/ CLENDENNING, T.G. ; LOUGHBOROUGH, M.T. ; CIVIL RESEARCH DEPARTMENT: A VERSATILE MIX-PROPORTIONING CONCEPT FOR FLY ASH CONCRETE. TORONTO, ONTARIO : CIVIL RESEARCH DEPARTMENT - REPORT, 1985 -

/C18/ CHRISTENSEN, J.K. ; KRYGER, L. ; PIND, N.: THE DETERMINATION OF TRACES OF CADMIUM, LEAD AND THALLIUM IN FLY ASH BY POTENTRIOMETRIC STRIPPING ANALYSIS. IN: ANALYTICA CHIMICA ACTA, 141 (1982), S. 131-146

/C19/ CHRISTENSEN, J.K. ; KRYGER, L. ; PIND, N.: EFFECT OF ELECTROACTIVE SUBSTANCES FORMING WATER - SOLUBLE REDUCED SPECIES ON THE POTENTIOMETRIC STRIPPING DETERMINATION OF MERCURY SOLUBLE METALS. IN: ANALYTICA CHIMICA ACTA, 136 (1982), S. 39-50

/C20/ CAIN, C.J.: FLY ASH - A NEW RESOURCE MATERIAL. IN: DES PLAINES/1L, 1983 PRESENTED AT NRMCA CONVENTION, 18.01.1983, PHOENIX/ARIZONA ALSO IN: CONCRETE (1983), NR. 11, S. 28-32

/C21/ CHATTERJI, S. ; COLLEPARDI, M. ; MORICONI, G.: POZZOLANIC PROPERTY OF NATURAL AND SYNTHETIC POZZOLANS : A COMPARATIVE STUDY. DETROIT : AMERICAN CONCRETE INSTITUTE SP-79, 1983 IN: PROCEEDINGS OF THE CANMET/ACI FIRST INTERNATIONAL CONFERENCE ON THE USE OF FLY ASH, SILICA FUME, SLAG AND OTHER MINERAL BY-PRODUCTS IN CONCRETE VOL. 1 (1983), S. 221-233

/C22/ COSTA, U. ; MASSAZZA, F.: SOME PROPERTIES OF POZZOLANIC CEMENTS CONTAINING FLY ASHES. DETROIT : AMERICAN CONCRETE INSTITUTE ACI SP-79, 1983 - IN: PROCEEDINGS OF THE CANMET/ACI FIRST INTERNATIONAL CONFERENCE ON THE USE OF FLY ASH, SILICA FUME, SLAG AND OTHER MINERAL BY-PRODUCTS IN CONCRETE VOL. 1 (1983), S. 235-254

/C23/ CARETTE, G. ; MALHOTRA, V.M.: EARLY-AGE STRENGTH DEVELOPMENT OF CONCRETE INCORPORATING FLY ASH AND CONDENSED SILICA FUME. PROC.: THE USE OF FLY ASH, SILICA FUME, SLAG AND OTHER MINERAL BY-PRODUCTS IN CONCRETE; MONTEBELLO, 1983 ACI PUBLICATION SP-79, VOL. II, PP. 765-784

/C24/ COOK, J.E.: FLY ASH IN CONCRETE - TECHNICAL CONSIDERATIONS. IN: CONCRETE INTERNATIONAL (1983), NR. 9,

/C25/ CAPP, J.P. ; GILLMORE, D.W.: SOIL-MAKING POTENTIAL OF POWERPLANT FLY ASH IN MINED-LAND RECLAMATION. ASH UTILISATION, 1973, BUREAU OF MINES INFORMATION CIRCULAR 8640/1974, S. 258-268

/C26/ CEMBUREAU: REPORT TO CEMBUREAU TECHNICAL COMMITTEE ABOUT SPECIFICATIONS FOR FLY ASH AS A CONCRETE ADDITIVE. 1982

/C27/ CABRERA, J.G. ; PLOWMAN, C.: THE HYDRATION OF TRICALCIUM ALUMINATE AND TETRACALCIUM ALUMINOFERRITE EXTRACTED FROM CEMENT. PROC. INTERN. SEMINAR ON CALCIUM ALUMINATES, SEPT. 1982, TURIN (ITALY). (MURAT, M. ; BACHIORRINI, A. ; GUILHOT, B. ; NEGRO, A. ; REGOURD, M. ; SOUSTELLE, M. (ED.)), PP. 182-198

/C28/ CARLES-GIBERGUES, B. ; THENOZ, B. ; VAQUIER, A.: HYDRATION MECHANISM OF A FLY ASH PRODUCED FROM CALCAREOUS LIGNITE. PARIS ; EDITIONS SEPTIMA, 1980 - IN: 7TH INTERNATIONAL CONGRESS ON THE CHEMISTRY OF CEMENT (1980), VOL. III, S. IV.53-IV.59

/C29/ COUTINHO, A.: ASPECTS OF SULFATE ATTACK ON CONCRETE. IN: ASTM CEMENT CONCRETE AND AGGREG. 1 (1979), NR. 1, S. 10-12

/C30/ CALLEJA, J.: SOME CONSIDERATIONS ON THE PRESENT STATE AND FUTURE TRENDS OF KNOWLEDGE ON CALCIUM ALUMINATES, AS A POSSIBLE BASIS FOR DISCUSSING THE FUTURE WORK IN THIS FIELD. PARIS : EDITIONS SEPTIMA, 1980 - IN: 7TH INTERNATIONAL CONGRESS ON THE CHEMISTRY OF CEMENT, VOL. IV, S. 613

/C31/ CHEN, B. ; HONG, D. GUO, H. ; ZHUANG, Y.: TEN-YEAR FIELD EXPOSURE TESTS ON THE ENDURANCE OF REINFORCED CONCRETE IN HARBOR WORKS. IN: CEMENT AND CONCRETE RESEARCH 13 (1983), NR. 5, S. 603-610

/C32/ CHATTERJI, S. ; JEFFERY, J.W.: A NEW HYPOTHESIS OF SULFATE EXPANSION. IN: MAGAZINE OF CONCRETE RESEARCH 15 (1963), S. 83-86

/C33/ CHU, T.Y. ET AL: SOIL STABILIZATION WITH LIME-FLY ASH MIXTURES : PRELIMINARY STUDIES WITH SILT AND CLAYED SOILS. HIGHWAY RESEARCH BOARD BULLETIN 108, 1955

/D1/ DATTA, R.K. ; GARG, S.K. ; REHSI, S.S.: STUDIES ON THE INCIDENCE OF CORROSION OF STEEL REINFORCEMENT IN CONCRETE CONTAINING FLY ASH. PROC.: RILEM SYMPOS. ON CONCR. AND REINF. CONCR. IN HOT COUNTRIES HAIFA 1971, BUILDING RES. STATION, TECHNION HAIFA VOL. II, S. 363-380

/D2/ DAVIS, R.E. ; DAVIS, H.E. ; KELLY, J.W.: WEATHERING RESISTANCE OF CONCRETES CONTAINING FLY-ASH CEMENTS. IN: JOURNAL OF THE AMERICAN CONCRETE INSTITUTE 37 (1941), S.281-293

/D3/ DERDECKA-GRYZMEK, A. ; ROSZCZYNIALSKY, W. ; GUSTAW, K.: DER EINFLUSS VON ZUSAETZEN AUF DIE EIGENSCHAFTEN VON ZEMENTEN. IN: BAUSTOFFINDUSTRIE 19 (1976) AUSG. A, S.14-16

/D4/ DEV, R.S. ; KAUTZ, K. ; KIRSCH, H.: MORPHOLOGISCHE UND PHASENANALYTISCHE UNTERSUCHUNGEN VON KRAFTWERKSTAEUBEN. ESSEN : HAUS DER TECHNIK, 1976 VORTRAGSVEROEFFENTLICHUNG, NR.368, S.46-48, S. 94-95

/D5/ DHIR, R.K. ; DARFOUR, E.S. ; MUNDAY, J.G.L.: STRENGTH CHARACTERISTICS OF CONCRETE CONTAINING PFA ADDITIVE. IN: SILICATES INDUSTRIELS 44 (1979), S. 23-29

/D6/ DIAMOND, S.: EFFECTS OF TWO DANISH FLY ASHES ON ALKALI CONTENTS OF PORE SOLUTIONS OF CEMENT-FLY-ASH PASTES. IN: CEMENT AND CONCRETE RESEARCH 11 (1981), NR. 3, S. 383-394

/D7/ DIAMOND, S. ; RAVINA, D. ; LOVELL, J.: THE OCCURRENCE OF DUPLEX FILMS ON FLY-ASH SURFACES. IN: CEMENT AND CONCRETE RESEARCH 10 (1980), S. 297-300

/D8/ DOERICH, J. ; JUNKERS, G.: ERFAHRUNGEN MIT EINER MISCHANLAGE FUER HYDRAULISCH GEBUNDENES MATERIAL AUF FLUGSTAUBBASIS ZUR HERSTELLUNG VON

TRAGSCHICHTEN IM STRASSENBAU. IN: VGB KRAFTWERKSTECHNIK 61 (1981), S. 218-220

/D9/ DRON, R.: LES PUOZZOLANES ET LA PUOZZOLANICITE. IN: REVUE DES MATERIAUX DE CONSTRUCTION (1975), NR. 692, S. 27-29

/D10/ DUCREUX, R.: EMPLOI DES CENDRES VOLANTES COMME AGREGAT OU COMME LIANT DANS LES BETONS ETUVES. IN: REVUE DES MATERIAUX DE CONSTRUCTION (1960), NR. 533, S. 43-48

/D11/ DUCREUX, R. ; JARRIGE, A.: QUELQUES EXPERIENCES RELATIVES A L' INFLUENCE DE LA FINESSE SUR LE COMPORTEMENT DES CENDRES VOLANTES POUZZOLANIQUES EN ADDITION AU CIMENT. IN: SILICATES INDUSTRIELS 27 (1962), S. 517-527

/D12/ DALZIEL, J.A.: THE BEHAVIOUR OF PULVERIZED FUEL ASH WITH DIFFERENT PORTLAND CEMENTS. PROC.: INT. SYMPOS.: THE USE OF PFA IN CONCRETE, DEPT. CIV. ENG. LEEDS UNIV. 1982, VOL. 1, S. 191-200

/D13/ DUNSTAN, M.R.H.: THE USE OF HIGH FLY ASH CONTENT CONCRETE IN ROADS. PROC.: INT. SYMPOS.: THE USE OF PFA IN CONCRETE, DEPT. CIV. ENG. LEEDS UNIV. 1982, VOL. 1, S. 277-288

/D14/ DEDMAN, H.W.G.: THE COMMERCIAL UTILIZATION OF PULVERIZED FUEL ASH FROM POWER STATIONS OF THE CENTRAL ELECTRICITY GENERATING BOARD. PROC.: EDISON ELECTRIC INSTITUTE, NATIONAL COAL ASSOC., BUREAU OF MINES SYMPOS., PITTSBURGH 1967, S. 156-164, COMPILED BY FABER, J.H. ; CAPP, P. ; SPENCER, J.D.

/D15/ DEDMAN, H.W.G.: ECONOMIC COMMISSION FOR EUROPE MEETING SUMMARY. PROC.: EDISON ELECTRIC INSTITUTE, NATIONAL COAL ASSOC., BUREAU OF MINES SYMPOS. PITTSBURGH 1967, S. 259-261, COMPILED BY FABER, J.H. ; CAPP, P. ; SPENCER, J.D.

/D16/ DIAMOND, S.: INTIMATE ASSOCIATION OF COAL PARTICLES AND INORGANIC SPHERES IN FLY ASH. IN: CEMENT AND CONCRETE RESEARCH 12 (1982), S. 405-407

/D17/ DHIR, R.K. ; APTE, A.G. ; MUNDAY, G.L.: EFFECT OF IN-SOURCE VARIABILITY OF PULVERIZED-FUEL ASH UPON THE STRENGTH OF OPC/PFA CONCRETE. IN: MAGAZINE OF CONCRETE RESEARCH 33 (1981), S.199-207

/D18/ DODSON, V.H.: THE STRENGTH CONTRIBUTION OF FLY ASH CONCRETE - A NEW APPROACH TO ITS ESTIMATION. BOSTON : MATERIAL RESEARCH SOCIETY - IN: EFFECTS OF FLY-ASH INCORPORATION IN CEMENT AND CONCRETE : PROCEEDINGS SYMPOSIUM N ANNUAL MEETING (DIAMOND,S.(ED.)) BOSTON, NOVEMBER 16-18, 1981, S. 184-193

/D19/ DIAMOND, S.: THE CHARACTERIZATION OF FLY ASHES. BOSTON : MATERIAL RESEARCH SOCIETY - IN: EFFECTS OF FLY-ASH INCORPORATION IN CEMENT AND CONCRETE : PROCEEDINGS SYMPOSIUM N ANNUAL MEETING (DIAMOND,S.(ED.)) BOSTON, NOVEMBER 16-18, 1981, S. 12-23

/D20/ DIAMOND, S. ; LOPEZ-FLORES, F.: ON THE DISTINCTION IN PHYSICAL AND CHEMICAL CHARACTERISTICS BETWEEN LIGNITIC AND BITUMINOUS FLY ASHES. BOSTON : MATERIAL RESEARCH SOCIETY - IN: EFFECTS OF FLY-ASH INCORPORATION IN CEMENT AND CONCRETE : PROCEEDINGS SYMPOSIUM N ANNUAL MEETING (DIAMOND, S. (ED.)) BOSTON, NOVEMBER 16-18, 1981, S. 34-44

/D21/ DIAMOND, S. ; LOPEZ-FLORES, F.: COMPARATIVE STUDIES OF THE EFFECTS OF LIGNITIC AND BITUMINOUS FLY ASHES IN HYDRATED CEMENT SYSTEMS. BOSTON : MATERIAL RESEARCH SOCIETY - IN: EFFECTS OF FLY-ASH INCORPORATION IN CEMENT AND CONCRETE : PROCEEDINGS SYMPOSIUM N ANNUAL MEETING (DIAMOND, S. (ED.)) BOSTON, NOVEMBER 16-18, 1981, S. 112-123

/D22/ DODSON, V.H.: THE EFFECT OF FLY ASH ON THE SETTING TIME OF CONCRETE CHEMICAL OR PHYSICAL. BOSTON : MATERIAL RESEARCH SOCIETY - IN: EFFECTS OF FLY-ASH INCORPORATION IN CEMENT AND CONCRETE : PROCEEDINGS SYMPOSIUM N ANNUAL MEETING (DIAMOND,S.(ED.)) BOSTON, NOVEMBER 16-18, 1981, S. 166-171

/D23/ DROLJC, S.: MECHANISM OF BONDING OF WASTE MINERAL MATERIALS. IN: SJL BEOGRAD REF.1/6 (1972)

/D24/ DROLJC, S. ; DIMIC, D.: FLY ASH - USE IN CIVIL ENGINEERING. IN: SAVETOVANJE, BEOGRAD (1979), S. 65-68

/D25/ DRAGISEVIC, L. ; MILETIC, S.: RESEARCH WORK ON POSSIBILITY OF USING FLY ASH "KOLUBARA" IN CEMENT INDUSTRY. IN: BEOGRAD (1979), S. 105-115

/D26/ DHIR, R.K. ; MUNDAY, J.G.L.: STRENGTH VARIABILITY OF OPC/PFA CONCRETE. IN: CONCRETE (1981), S. 33-37

/D27/ DOERICH, J.: ANFORDERUNGEN AN FLUGASCHEN AUS SICHT DER VERWENDER. VGB-KONFERENZ 1982: VERWERTUNG VON VERBRENNUNGSRUECKSTAENDEN. IN: VGB-KRAFTWERKSTECHNIK GMBH, VGB-TB 202, S. 280-288

/D28/ DIAMOND, S.: ON THE GLASS PRESENT IN LOW-CALCIUM AND IN HIGH-CALCIUM FLY-ASHES. IN: CEMENT AND CONCRETE RESEARCH 13 (1983), S. 459-464

/D29/ DOERICH, J. ; JUNKERS, G.: STRASSENBAU MIT HYDRAULISCH GEBUNDENER TRAG-SCHICHT AUF FLUGSTAUBBASIS. IN: VGB KRAFTWERKSTECHNIK 60 (1980), NR. 5, S. 384-389

/D30/ DEUSTER, D.: NEUE BAUSTOFFE, EIN BEITRAG ZUM UMWELTSCHUTZ. TEILPROJEKT: FLUGASCHEZEMENT. BUNDESMINISTERIUM FUER FORSCHUNG UND TECHNOLOGIE, FORSCHUNGSBERICHT T 80-176

/D31/ DMITRIEV, A.M. ; ENTINE, Z.B. ; GOLDSTEIN, L.J. ; CHATOKHINA, L.P.: HYDRATATION, MORPHOLOGIE ET PROPRIETES DES CIMENTS DE CENDRES. PARIS : EDITIONS SEPTIMA, 1980 - IN: 7TH INTERNATIONAL CONGRESS ON THE CHEMISTRY OF CEMENT, VOL. III (1980), S. IV.98-IV.103

/D32/ DIAMOND, S.: HYDRATION REACTIONS OF C2A CONTAINED IN AN UNUSUAL FLY ASH. PARIS : EDITIONS SEPTIMA, 1980 - IN: 7TH INTERNATIONAL CONGRESS ON THE CHEMISTRY OF CEMENT, VOL. III (1980), S. IV.19-IV.23

/D33/ DALZIEL, J.A.: THE EFFECT OF CURING TEMPERATURE ON THE DEVELOPMENT OF STRENGTH OF MORTAR CONTAINING FLY ASH. PARIS : EDITIONS SEPTIMA, 1980 - IN: 7TH INTERNATIONAL CONGRESS ON THE CHEMISTRY OF CEMENT, VOL. III (1980), S. IV.93-IV.97

/D34/ DUNSTAN, M.R.H.: CIRIA TECHN. NOTE 105, MAY 1981, PP. 59

/D35/ DAVIS, R.E. ; CARLSON, R.W. ; KELLY, J.W. ; DAVIS, H.E.: PROPERTIES OF CEMENT AND CONCRETE CONTAINING FLY ASH. IN: JOURNAL OF THE AMERICAN CONCRETE INSTITUTE 33 (1937), S. 577-612

/D36/ DIKEOU, J.T.: FLY ASH INCREASES RESISTANCE OF CONCRETE TO SULFATE ATTACK. IN: RESEARCH REPORT, A WATER RESOURCES TECHNICAL PUBLICATION, BUREAU OF RECLAMATION (1970), NR. 23, S. 1-17 DENVER, COL.

/D37/ DUNSTAN, E.R.: PERFORMANCE OF LIGNITE AND SUB-BITUMINOUS FLY ASH IN CONCRETE - A PROGRESS REPORT, 23 PP, REPT. REC-ERC-76-1, BUREAU OF RE-CLAMATION, 1976 DENVER, COL.

/D38/ DUNSTAN, E.R. JR.: A SPEC ODYSSEY - SULFATE RESISTANT CONCRETE FOR THE 1980'S. DETROIT : AMERICAN CONCRETE INSTITUTE (ACI), 1982 - IN: SYMPOSIUM ON

SULFATE RESISTANCE OF CONCRETE, ACI SPECIAL PUBLICATION SP-77, (VERBECK, G. (ED)), S. 41-62

/D39/ DIAMOND, S.: CHAPTER 40: CHEMICAL REACTIONS OTHER THAN CARBONATES, SIGNIFICANCE OF TESTS, AND PROPERTIES OF CONCRETE AND MAKING MATERIALS, ASTM - STP 169B (1978), PP. 708-721

/D40/ DIAMOND, S.: A REVIEW OF ALKALI-SILICA REACTION AND EXPANSION MECHANISMS. IN: CEMENT AND CONCRETE RESEARCH 6 (1976), NR. 4, S. 549-560

/D41/ DUNCAN, M.A.G. ; SWENSON, E.G. ; GILLOT, J.E. ; FORAN, M.R.: ALKALI- AGGREGATE REACTION IN NOVA SCOTIA I: SUMMARY OF A FIVE-YEAR STUDY. IN: CEMENT AND CONCRETE RESEARCH 3 (1973), NR. 1, S. 55-69

/D42/ DOLAR-MANTUANI, L.: HANDBOOK OF CONCRETE AGGREGATES - A PETROGRAPHIC AND TECHNOLOGICAL EVALUATION, NOYES PUBLICATIONS, 1983, 345 PP

/D43/ DONELLY, J.R. ; JONS, E.: SYNTHETIC GRAVEL FROM DRY FLUE GLASS DESULPHURIZATION END-PRUDUCTS. PROCEEDINGS OF THE 6TH INTERNATIONAL ASH UTILIZATION SYMPOSIUM, RENO, NEVADA, MARCH 1982

/D44/ DI GIOIA, A.M. ; NUZZO, W.L.: FLY ASH AS A STRUCTURAL FILL. IN: JOURNAL POWER DIVISION, ASCE 98 (1972), PO1, S. 77-92

/E1/ EFES, Y.: ANWENDUNGSBEISPIELE FUER RASTERELEKTRONENMIKROSKOPISCHE UNTERSUCHUNGEN IM BEREICH DER NICHTMETALLISCH-ANORGANISCHEN BAUSTOFFORSCHUNG. ESSEN : HAUS DER TECHNIK, 1976. - VORTRAGS- VEROEFFENTLICHUNG NR.368 (1976), S. 50-51, S. 95-96

/E2/ EFES, Y.: UNTERSUCHUNGEN UEBER EINFLUESSE AUF DIE SPEZIFISCHE OBERFLAECHE NACH BLAINE VON STEINKOHLENFLUGASCHEN UND UEBER DIE AUSWIRKUNGEN DES BLAINE-WERTES AUF ANDERE EIGENSCHAFTEN. IN: TIZ- FACHBERICHTE 104 (1980), NR. 1, S. 20-29

/E3/ EFES, Y. ; LUEHR, H.-P.: ZUM EINFLUSS DER KORNGROESSENVERTEILUNG UND DER GLASPHASE AUF DIE PUZZOLANISCHE WIRKSAMKEIT GLUEHVERLUSTARMER STEINKOHLENFLUGASCHEN. OPLADEN : WESTDEUTSCHER VERLAG. - IN: FORSCHUNGSBERICHTE DES LANDES NORDRHEIN-WESTFALEN, (1978), NR. 2761

/E4/ ENDELL, J.: BRAUNKOHLENFILTERASCHE ALS BAUSTOFFBINDEMITTEL. SOURCE UNKNOWN

/E5/ ENTIN, Z.B. ; YASHINA, E.T. ; LEPESKENKOVA, G.G. ; RYAZANTSEVA, N.Z.: ON THE HYDRATION AND HARDENING OF CEMENTS WITH FLY-ASH ADDITION. IN: VI. INT. CONGR. ON CHEM. OF CEM., SUPPL. PAP. III-3, MOSCOW 1974, S. 95-99

/E6/ ERGIN, S.: THE EFFECTS OF FLY ASH ON SULFATE CORROSION OF CONCRETE. IN: FLY-ASH SYMPOS., ANKARA 1970

/E7/ ERYTHROPEL, H.: EIN NEUER HOCHWERTIGER INDUSTRIEMAUERSTEIN AUS STEINKOHLENFLUGASCHE. IN: BETONSTEIN-ZEITUNG 18 (1952), S.41-45

/E8/ ERYTHROPEL, H.: DERZEITIGER STAND DER ASCHEVERWERTUNGSMOEGLICHKEITEN IN DEUTSCHLAND. IN: MITTEILUNGEN VGB 43 (1963), S.310-316

/E9/ ERYTHROPEL, H.: ANFALL UND VERWERTUNG VON KRAFTWERKSASCHEN. IN: TECHNISCHE MITTEILUNGEN 59 (1966), S.427-431

/E10/ ERYTHROPEL, H.: INTERNATIONALE ERFAHRUNGEN AUF DEM GEBIET DER ASCHEVERWERTUNG. IN: MITTEILUNGEN VGB 50 (1970), S.191-197

/E11/ ERYTHROPEL, H.: ASCHE UND SCHLACKE. IN: BRENNSTSTOFF-WAERME-KRAFT 24 (1972), S.159-160

/E13/ EWERS, J. ; GUESFELDT, K.H.: RINGUNTERSUCHUNG AN FUELLERN. IN: STRASSE UND AUTOBAHN 15 (1965), S.236-240

/E14/ ENTIN, Z.B. ; KUZNETSOVA, T.V. ; DMITRIEV, A.M. ; LEPESHENKOVA, G.G.: PROPERTIES OF FLY ASH AND HYDRATION OF ASH PARTICLES IN HARDENING CONCRETE. PROC.: INT. SYMPOS.: THE USE OF PFA IN CONCRETE, DEPT. CIV. ENG. LEEDS UNIV. 1982, VOL.1, S. 95-100

/E15/ ELLIS, C.E.: THE APPLICATION OF THE TWO-POINT WORKABILITY TEST AND BRITISH STANDARD TESTS TO OPC/PFA CONCRETES. PROC.: INT. SYMPOS.: THE USE OF PFA IN CONCRETE, DEPT. CIV. ENG. LEEDS UNIV. 1982, VOL.1, S. 121-131

/E16/ ERYTHROPEL, H.: ASH PRODUCTION AND UTILIZATION IN THE GERMAN FEDERAL REPUBLIC. PROC.: EDISON ELECTRIC INSTITUTE, NATIONAL COAL ASSOC., BUREAU OF MINES SYMPOS. PITTSBURGH 1967, S. 165-175, COMPILED BY FABER, J.H. ; CAPP, P. ; SPENCER, J.D.

/E17/ EEROLA, M. ; KEPPO, M. ; KORKMAN, J. ; KUKKO, H. ; VANHANEN, A.: POWER PLANT ASHES AND THEIR UTILIZATION PART 10: SUMMARY. TECHNICAL RESEARCH CENTRE OF FINLAND CONCRETE AND SILICATE LABORATORY, REPORT 68

/E18/ ERIKSEN, K.: COMBINED USE OF FLY ASH AND SUPERPLASTICIZER IN THE PRODUCTION OF CONCRETE WITH HIGH STRENGTH (IN DAN.). IN: CBL SPECIAL PUBLICATION (1980) NR. 7, S. 113-115

/E19/ ERIKSEN, K. ; NEPPER-CHRISTENSEN, P.: EXPERIENCE IN THE USE OF SUPERPLASTICIZERS IN SOME SPECIAL FLY ASH CONCRETES. IN: ACI SP 68-1, PP. 1-20

/E20/ EUW, VON, M.: THE USE OF FLY-ASH IN THE CEMENT INDUSTRY. IN: SOCIETE DES CIMENTS FRANCAIS - FRANCE, PP. II C.1 - II C.14

/E21/ EFES, Y.: FRUEHFESTIGKEITEN EINIGER FLUGASCHEZEMENTE (FAZ). IN: BETONWERK UND FERTIGTEIL-TECHNIK (1983), NR. 2, S. 101-104

/E22/ EEROLA, M. ; ALKIO, R.: UTILIZATION OF POWER PLANT ASHES; PART 7; USE IN ROAD CONSTRUCTION. TECHNICAL RESEARCH CENTRE OF FINLAND, ROAD AND TRAFFIC LABORATORY, REPORT 54

/E23/ ELFERT, R.J. JR.: BUREAU OF RECLAMATION EXPERIENCES WITH FLY ASH AND OTHER POZZOLANS IN CONCRETE. ASH UTILISATION, 1973, BUREAU OF MINES INFORMATION CIRCULAR 8640/1974, S. 80-93

/E24/ EFES, Y.: INFLUENCE OF BLAST-FURNACE SLAG ON THE DURABILITY OF CEMENT MOR TAR BY CARBONIC ACID ATTACK - PROBLEMS CONNECTED WITH TESTS ON CORRODED SPECIMENS. PHILADELPHIA : AMERICAN SOCIETY FOR TESTING AND MATERIALS, 1980 IN: DURABILITY OF BUILDING MATERIALS AND COMPONENTS - PROCEEDINGS OF THE FIRST INTERNATIONAL CONFERENCE, ASTM STP 691, (SEREDA, P.J. ; LITVAN, G.G. (ED)), S. 364-376

/F1/ FELLMAN, - ; FUEHRER, D.: STEINKOHLENFILTERASCHE FUER BITUMENGEBUNDENE TRAGSCHICHTEN IM STRASSENBAU. IN: BAUSTOFFINDUSTRIE 14 (1971), S.266-269

/F2/ FLIERT VAN DER, C.: VLIEGAS IN BETON. IN: BETON DE ING. 68 (1956), S.9-17

/F3/ FOREST, J. ; DEMOULIAN, E.: RECHERCHE D' UNE METHODE D' APPRECIATION RAPIDE DE LA REACTIVITE DES CENDRES VOLANTES ET DES POUZZOLANES ADDITIONNEES AU CIMENT. IN: SILICATES INDUSTRIELS 29 (1964), S.265-278

/F4/ FOREST, J. ; DEMOULIAN, E.: APPRECIATION DE L' ACTIVITE DES CENDRES VOLANTES ET DES POUZZOLANES. IN: REVUE DES MATERIAUX DE CONSTRUCTION ET DE TRAVAUX PUBLICS NO 577 (1951), S. 312-317

/F5/ FROHNSDORFF, G. ; CLIFTON, J.R.: FLY ASHES IN CEMENTS AND CONCRETE: TECHNICAL NEEDS AND OPPORTUNITIES. WASHINGTON : U.S.DEPARTMENT OF COMMERCE, NATIONAL BUREAU OF STANDARDS, 1981

/F6/ FULTON, A.A. ; MARSHALL, W.T.: THE USE OF FLY ASH AND SIMILAR MATERIALS IN CONCRETE. IN: PROCEEDINGS OF THE INSTITUTION OF CIVIL ENGINEERS - PART I: GENERAL ORDINARY MEETINGS AND OTHER SELECTED PAPERS (NOV. 1956) 5 (1956), NR. 6, S. 714-730, PAPER-NR. 6154

/F7/ FUNGK, E. ; GREULING, P. ; ILGNER, R.: BRAUNKOHLENFILTERASCHE ALS FEINSTZUSCHLAGSSTOFF FUER BETON. IN: BAUSTOFFINDUSTRIE 17 (1974), AUSG.B, S. 17-18

/F8/ FUNGK, E. ; ILGNER, R. ; LANG, E.: BRAUNKOHLENFILTERASCHEN DER DDR ALS ZUMAHLSTOFF IN DER ZEMENTINDUSTRIE. IN: SILIKATTECHNIK 20 (1969), S. 302-307

/F9/ FUNGK, E. ; ILGNER, R. ; LANG, E.: BRAUNKOHLENFILTERASCHEN DER DDR ALS ZUMAHLSTOFF IN DER ZEMENTINDUSTRIE. IN: SILIKATTECHNIK 21 (1970), S. 386-388

/F10/ FURLAN, V. ; HOUST, Y.: LES MATERIAUX POUZZOLANIQUES ET LEUR UTILISATION. IN: CHANTIERS 11 (1980), S. 29-32

/F11/ FABER, J.H. ; STIRLING, H.T.: COAL AND ASH CHARACTERISTICS THAT AFFECT THE USE OF POZZOLANS IN THE UNITED STATES. PROC.: INT. SYMPOS.: THE USE OF PFA IN CONCRETE, DEPT. CIV ENG. LEEDS UNIV. 1982, VOL.1, S. 23-33

/F12/ FALECKI, Z.: THE UNITED NATIONS ECONOMIC COMMISION FOR EUROPE AND ITS WORK IN THE FIELD OF THE UTILIZATION OF ASH PRODUCED BY THERMAL POWER PLANTS. PROC.: EDISON ELECTRIC INSTITUTE, NATIONAL COAL ASSOC., BUREAU OF MINES SYMPOS., PITTSBURGH 1967, S. 135-142, COMPILED BY FABER, J.H. ; CAPP, P. ; SPENCER, J.D.

/F13/ FABER, J.H.: ABSTRACT COAL ASH UTILIZATION IN MASONRY CONSTRUCTION. BOULDER, COLORADO : UNIVERSITY OF COLORADO; THE MASONRY SOCIETY - IN: PROCEEDINGS OF THE NORTH AMERICAN MASONRY CONFERENCE, S.60.1-60.14

/F14/ FELDMAN, R.F.: PORE STRUCTURE FORMATION DURING HYDRATION OF FLY-ASH AND SLAG CEMENT BLENDS. BOSTON : MATERIAL RESEARCH SOCIETY - IN: EFFECTS OF FLY-ASH INCORPORATION IN CEMENT AND CONCRETE : PROCEEDINGS SYMPOSIUM N ANNUAL MEETING (DIAMOND,S.(ED.)) BOSTON, NOVEMBER 16-18, 1981, S. 124-133

/F15/ FOMBONNE, J.: LES CENDRES VOLANTES DANS LA FABRICATION DES CIMENTS HYPOCALCIQUES. (1974) SOURCE UNKNOWN

/F16/ FERJAN, M.: MORTARS PRODUCED OF LIGNITE FLY ASH TE SOSTANJ. IN: INFORMACIJE ZRMK 12 (1961), S.1-4

/F17/ FERJAN, M.: CEILING CONSTRUCTION OF FLY ASH ELEMENTS. IN: INFORMACIJE ZRMK 5 (1964), S. 1-4

/F18/ FERJAN, M. ; DROLJC, S. ; JENCEK, L.A. ; HRIBERNIK, E.: FORMSTEINE AUS ELEKTROFILTERASCHE. IN: INT. SYMP. FUER DAMPFGEHAERTETE KALZIUMSILIKATBAUSTOFFE 2 HANNOVER (1969) VORTRAG 35

/F19/ FERJAN, M.: BUILDING WITH FLY ASH ELEMENTS. IN: INFORMACIJE 99 (1969)

/F20/ FERJAN, M. ; NAMORS, V.: LIGHTWEIGHT AGGREGATES OF FLY ASH. IN: DOKUMENTACIJA ZA GRADEVINARSTVO I ARHITEKTURU DGA-1065 (1970)

/F21/ FERJAN, M. ; NAMORS, V.: USE OF FLY ASH. IN: SIMPOZIJUM SJL BEOGRAD (1970), S. 1272-1276

/F22/ FERJAN, M.: POSSIBILITY OF USING FLY ASH IN BUILDING CONSTRUCTIONS. IN: INFORMACIJE ZRMK 119 (1971)

/F23/ FRANKLIN, R.E.: DER EINFLUSS VON FLUGASCHE AUF DIE FESTIGKEIT VON DECKENBETON FUER DEN STRASSENBAU = THE EFFECT OF PULVERISED FUEL ASH ON THE STRENGTH OF PAVEMENT-QUALITY CONCRETE. CROWTHORNE, BERKSHIRE : TRANSPORT AND ROAD RESEARCH LABORATORY, 1981 - IN: TRRL LABORATORY REPORT (1981), NR. 982

/F24/ FUNDAL, E.: GESINTERTER ZUSCHLAGSTOFF AUS FLUGASCHE UND KOHLENSCHIEFER. IN: ZEMENT-KALK-GIPS (ZKG) 36 (1983), NR. 5, S. 259-265

/F25/ FORSS, B.: SLAG AND FLY ASH AS BINDING MATERIAL IN CONCRETE. IN: CBL SPECIAL PUBLICATION (1980) NR. 7, S. 12-18

/F26/ FELDMAN, R.F.: SIGNIFICANCE OF POROSITY MEASUREMENTS ON BLENDED CEMENT PERFORMANCE. DETROIT : AMERICAN CONCRETE INSTITUTE ACI SP-79, 1983 - IN: PROCEEDINGS OF THE CANMET/ACI FIRST INTERNATIONAL CONFERENCE ON THE USE OF FLY ASH, SILICA FUME, SLAG AND OTHER MINERAL BY-PRODUCTS IN CONCRETE VOL. 1 (1983), S.415-433

/F27/ FRANKLIN, R.E. ; GIBBS, W.E.; SHERWOOD, P.T.: THE USE OF PULVERISED FUEL ASH IN LEAN CONCRETE ROADBASES. PART 1, LABORATORY STUDIES. TRANSPORT AND ROAD RESEARCH LABORATORY CROWTHORNE, 1982, 31S., SUPPLEMENTARY REPORT 736

/F28/ FUEHRES, M. ; LENZ, W.: BRAUNKOHLENASCHE ALS ZUSCHLAGSTOFF BEIM STRASSENBAU. IN: TIEFBAU 25 (1983), NR.7, S. 424-426

/F29/ FINGER, W.: DIE ENTWICKLUNG DER FLUGASCHE ST. ANDRAE ZUM HOCHWERTIGEN BINDEMITTEL. IN: MITTEILUNGEN AUS DEM INSTITUT FUER BAUSTOFFLEHRE UND MATERIALPRUEFUNG AN DER UNIVERSITAET INNSBRUCK, JAHRESBERICHT 1975/76, S.7-11

/F30/ FRATINI, N.: CONTROLLO CHIMICA DEI CEMENTI POZZOLANICI. ANNALI DI CHIMICA APPLICATE 44 (1954), PP 709-718

/F31/ FELDMAN, R.F.: PORE STRUCTRUE DAMAGE IN BLENDED CEMENTS CAUSED BY MERCURY INTRUSION. IN: JOURNAL OF THE AMERICAN CERAMIC SOCIETY 67 (1984), NR. 1, S. 30-33

/F32/ FELDMAN, R.F. ; SEREDA, P.J.: CHARACTERISTICS OF SORPTION AND EXPANSION ISOTHERMS OF REACTIVE LIMESTONE AGGREGATE. IN: JOURNAL OF THE AMERICAN CONCRETE INSTITUTE 58 (1961), NR. 2, PP 204-214

/F33/ FRIEDE, H.: DEPTH OF THE CORRODED ZONE IN CONCRETE EXPOSED TO CARBONIC ACID. PHILADELPHIA : AMERICAN SOCIETY FOR TESTING AND MATERIALS, 1980 - IN: DURABILITY OF BUILDING MATERIALS AND COMPONENTS PROCEEDINGS OF THE FIRST INTERNATIONAL CONFERENCE, ASTM STP 691, (SEREDA, P.J. ; LITVAN, G.G. (ED)), S. 355-363

/F34/ FTIKOS, CH. ; PARISSAKIS, G.: THE COMBINED ACTION OF MG2+ AND CL- IONS IN CEMENT PASTES. IN: CEMENT AND CONCRETE RESEARCH 15 (1985), NR. 4, S. 593-599

/F35/ FATTUHI, N.I. ; HUGHES, B.P.: EFFECT OF ACID ATTACK ON CONCRETE WITH DIFFERENT ADMIXTURES OR PROTECTIVE COATINGS. IN: CEMENT AND CONCRETE RESEARCH 13 (1983), NR. 5, S. 655-665

/F36/ FELDMAN, R.F. ; HUANG, C.-Y.: RESISTANCE OF MORTARS CONTAINING SILICA FUME TO ATTACK BY A SOLUTION CONTAINING CHLORIDES. IN: CEMENT AND CONCRETE RESEARCH 15 (1985), NR. 3, S. 411-420

/F37/ FELDMAN, R.F. ; RAMACHANDRAN, V.S.: NEW ACCELERATED METHODS FOR PREDICTING DURABILITY OF CEMENTITIOUS MATERIALS. PHILADELPHIA : AMERICAN SOCIETY FOR TESTING AND MATERIALS, 1980 - IN: DURABILITY OF BUILDING MATERIALS AND COMPONENTS PROCEEDINGS OF THE FIRST INTERNATIONAL CONFERENCE, ASTM STP 691, (SEREDA, P.J. ; LITVAN, G.G.(ED)), S. 313-325

/F38/ FELLER, M.: PRUEFUNG UND BEWERTUNG VON FUELLSTOFFEN. IN: DIE STRASSE 12 (1972)

/F39/ FINNECY, E.E.: THE TREATMENT AND DISPOSAL OF HAZARDOUS WASTES IN GREAT BRITAIN. PAPER PRESENTED AT "AVFALL '79" JOENKOEPING, SWEDEN 1979

/G1/ GAMBS, G.C.: POWER PLANT ASH - FOR USE IN CEMENT. IN: TRANSACTIONS SOCIETY OF MINING ENGINEERS 235 (1966), S. 422-428

/G2/ GASPAR, L.: LES CENDRES VOLANTES ET LE LAITIER GRANULE EN CONSTRUCTION ROUTIERE. IN: BULL. LIAISON LABOR. PONTS ET CHAUSSEES NO 86 (1976), S. 135-143

/G3/ GHOSH, R.S.: PROPORTIONING OF CONCRETE INCORPORATING FLY ASH. TORONTO, ONTARIO : CIVIL RESERACH DEPARTMENT, ONTARIO HYDRO - IN: CANADIAN JOURNAL OF CIVIL ENGINEERING, 3 (1976), S. 68-82

/G4/ GHOSH, R.S. ; TIMUSK, J.: CREEP OF FLY ASH CONCRETE. IN: JOURNAL OF THE AMERICAN CONCRETE INSTITUTE 52 (1981), S. 351-357

/G5/ GOLDSCHMIDT, K. ; DEUSTER, D. ; BARHO, W.: FLUGASCHEZEMENT. IN: VGB KRAFTWERKSTECHNIK, ESSEN, 59 (1979), NR. 1, S. 33-36

/G6/ GRAHAM, D.E.: FLY ASH AND ITS USE IN CONCRETE. MARYLAND : SILVER SRING, 1972 - IN: NATIONAL READY MIXED CONCRETE ASSOC.PUPLICATION (1972), NR. 138

/G7/ GRAMATKE, J. ; REICHERT, G.: ERFAHRUNGEN MIT DEN STEINKOHLENFEUERUNGEN IM KRAFTWERK SCHOLVEN. IN: VGB KRAFTWERKSTECHNIK 55 (1975), S. 277-286

/G8/ GRIEB, W.E. ; WOLF, D.O.: CONCRETE CONTAINING FLY ASH AS A REPLACEMENT FOR PORTLAND BLAST-FURNACE SLAG CEMENT. IN: PROCEEDINGS ASTM 61 (1961), S. 1143-1153

/G9/ GROSS, G. ; GRIGOLEIT, G.: FILTERASCHE ALS FEINSTZUSCHLAGSTOFF IN ZEMENTGEBUNDENEM BETON. BAUINFORMATION DDR, BERLIN 1979

/G10/ GUIDA, K.: THE USES OF FLY ASH IN A FERRO CEMENT MIX DESIGN. THIRD INT. ASH UTIL. SYMPOS., PAP. NR.A-6 PITTSBURGH, PENNSYLVANIA 1973

/G11/ GUILLAUME, L.: L' ACTIVITE POUZZOLANIQUE DES CENDRES VOLANTES DANS LES CIMENTS AU LAITIER. IN: SILICATES INDUSTRIELS 28 (1963), S. 297-300

/G12/ GUTT, W.H. ; NIXON, P.J.: USE OF WASTE MATERIALS IN THE CONSTRUCTION INDUSTRY - ANALYSIS OF THE RILEM SYMPOSIUM BY CORRESPONDENCE. IN: MATERIAUX ET CONSTRUCTIONS 12 (1979), S. 255-306

/G13/ GIFFORD, P.M. ; WARD, M.A.: RESULTS OF LABORATORY TESTS ON LEAN MASS CONCRETE UTILIZING TO A HIGH LEVEL OF CEMENT REPLACEMENT. PROC.: INT.SYMPOS.: THE USE OF PFA IN CONCRETE, DEPT. CIV. ENG. LEEDS UNIV. 1982, VOL.1, S. 221-230

/G14/ GHOSE, A. ; PRATT, P.L.: STUDIES OF THE HYDRATION REACTIONS AND MICROSTRUCTURE OF CEMENT-FLY-ASH PASTES. BOSTON : MATERIAL RESEARCH SOCIETY - IN: EFFECTS OF FLY-ASH INCORPORATION IN CEMENT AND CONCRETE : PROCEEDINGS SYMPOSIUM N ANNUAL MEETING (DIAMOND,S.(ED.)) BOSTON, NOVEMBER 16-18, 1981, S. 82-91

/G15/ GRUTZECK, M.W. ; ROY, D.M. ; SCHEETZ, B.E.: HYDRATION MECHANISMS OF HIGH-LIME FLY ASH IN PORTLAND-CEMENT COMPOSITIONS. BOSTON : MATERIAL RESEARCH SOCIETY - IN: EFFECTS OF FLY-ASH INCORPORATION IN CEMENT AND CONCRETE : PROCEEDINGS SYMPOSIUM N ANNUAL MEETING (DIAMOND,S.(ED.)) BOSTON, NOVEMBER 16-18, 1981, S. 92-101

/G16/ GUILLAUME, L.: UNE EXPERIENCE DES PLUS DE 10 ANS DANS L'UTILISATION EN CIMENTERIE DES CENDRES VOLANTES POUZZOLANIQUES. IN: REV. IND. MIN. (1963), S. 232-237

/G17/ GRIFFITH, J.S. ; DUESK, J.T. ; BAILY, E.C.: A NEW USE FOR FLY ASH - A LIGHTWEIGHT CERAMIC BUILDING MATERIAL. IN: AMER. SOC.OF MECH. ENG. PAPER 61-WA-291

/G18/ GUTT, W.H. ; NIXON, P.J. ; SMITH, M.A. ; HARRISON, W.H. ; RUSSELL, A.D.: A SURVEY OF THE LOCATIONS, DISPOSAL AND PROSPECTIVE USES OF THE MAJOR INDUSTRIAL BY-PRODUCTS AND WASTE MATERIALS. IN: BUILDING RESEARCH ESTABLISHMENT CURRENT PAPER 19/74, GARSTON (1974)

/G19/ GIORDANI, M.: LIGHT MORTARS WITH FLY ASH (ITAL.). IN: IL CEMENTO 79 (1982), 3/1

/G20/ GAZE, M.E. ; NIXON, P.J.: THE EFFECT OF PFA UPON ALKALI - AGGREGATE REACTION. IN: MAGAZINE OF CONCRETE RESEARCH 35 (1983) NR. 123, S. 107-110

/G21/ GEBLER, S. ; KLIEGER, P.: EFFECT OF FLY ASH ON THE AIR-VOID STABILITY OF CONCRETE. DETROIT : AMERICAN CONCRETE INSTITUTE, ACI SP-79, 1983 - IN: PROCEEDINGS OF THE CANMET/ACI FIRST INTERNAITONAL CONFERENCE ON THE USE OF FLY ASH, SILICA FUME, SLAG AND OTHER MINERAL BY-PRODUCTS IN CONCRETE VOL. 1, (1983), S. 103-142

/G22/ GUYOT, R. ; RANC, R. ; VARIZAT, A.: COMPARISON OF THE RESISTANCE TO SULFATE SOLUTIONS AND TO SEA WATER OF DIFFERENT PORTLAND CEMENTS WITH OR WITHOUT SECONDARY CONSTITUENTS. DETROIT : AMERICAN CONCRETE INSTITUTE ACI SP-79, 1983 - IN: PROCEEDINGS OF THE CANMET/ACI FIRST INTERNATIONAL CONFERENCE ON THE USE OF FLY ASH, SILICA FUME, SLAG AND OTHER MINERAL BY-PRODUCTS IN CONCRETE VOL. 1 (1983), S. 453-469

/G23/ GOTTWALD, S. ; TOUSSAINT, A.: VERWERTUNG VON RUECKSTAENDEN AUS BRAUN- UND STEINKOHLEKRAFTWERKEN IM STRASSENBAU. VGB-KONFERENZ 1982: VERWERTUNG VON VERBRENNUNGSRUECKSTAENDEN IN: VGB-KRAFTWERKS- TECHNIK GMBH, VGB-TB 202, S. 161-170

/G24/ GUILLAUME, L.: CONTROLE SIMPLE DES CENDRES VOLANTES POUZZOLANIQUES AUX AJOUTEES AUX CIMENTS PORTLAND ARTIFICIELS. IN: REVUE DES MATERIAUX NR. 517 (1958), PP 272-273

/G25/ GILLOTT, J.E.: ALKALI-AGGREGATE REACTIONS IN CONCRETE. IN: ENGINEERING GEOLOGY 9 (1975), PP 309-326

/G26/ GRATTAN-BELLEW, P.E.: PREVENTIVE MEASURES TO COUNTERACT EXPANSION OF CONCRETE CONTAINING ALKALI-REACTIVE AGGREGATES. IN: DURABILITY OF BUILDING MATERIALS 1 (1983), S. 363-376

/G27/ GRATTON-BELLEW, P.E.: PREVENTIVE MEASURES TO COUNTERACT EXPANSIVITY OF CONCRETE AGGREGATES, PROCEEDINGS OF THE FITH INTERNATIONAL CONFERENCE ON ALKALI-AGGREGATE REACTION IN CONCRETE, 1981, S252/9, 13PP

/G28/ GUTT, W.H. ; EVERETT, L.H.: DURABILITY OF SOME COMMON BUILDING MATERIALS. PHILADELPHIA : AMERICAN SOCIETY FOR TESTING AND MATERIALS, 1980 - IN: DURABILITY OF BUILDING MATERIALS AND COMPONENTS - PROCEEDINGS OF THE FIRST INTERNATIONAL CONFERENCE, ASTM STP 691, (SEREDA, P.J. ; LITVAN, G.G.(ED)), S. 131-144

/G29/ GJORV, O.E. ; VENNESLAND, O.: DIFFUSION OF CHLORIDE IONS FROM SEAWATER INTO CONCRETE. IN: CEMENT AND CONCRETE RESEARCH 9 (1979), NR. 2, S. 229-238

/G30/ GAFFNEY, J.A. ; HUTCHINSON, A.T.: USE OF PULVERIZED FUEL ASH IN A LARGE EMBANKMENT. INTERNATIONAL CONFERENCE ON THE USE OF BY PROD. AND WASTE IN CIVIL ENG., VOL II, S. 349-355, PARIS, 1978

/H1/ HANQUEZ, E.: UTILISATION DES CENDRES VOLANTES ET DES SCHISTES HOUILLERS: PROCEDES SURCHISTE ET SUREX. IN: REVUE DES MATERIAUX DE CONSTRUCTION ET DE TRAVAUX PUBLICS NO 703 (1976), S. 336-339 ALSO IN: L'INDUSTRIE CERAMIQUE NO 696 (1976) S.443-446

/H2/ HARRISON, W.H. ; MUNDAY, R.S.: AN INVESTIGATION INTO THE PRODUCTION OF SINTERED PFA AGGREGATE. GARSTON : BUILDING RESEARCH ESTABLISHMENT, NR. CP2/75

/H3/ HENK, B.: UEBERLEGUNGEN EINES BETONINGENIEURS ZUM THEMA FLUGASCHE. IN: BAUGEWERBE 56 (1976), S. 31, 35-36

/H4/ HENNIG, K. ; SOPORA, H.: TECHNOLOGIE UND STOFFLICHE VORAUSSETZUNGEN FUER DIE ERZEUGUNG VON PUZZOLANZEMENT UNTER VERWENDUNG VON FLUGASCHE. IN: BAUSTOFFINDUSTRIE 12 (1969), S. 271-275

/H5/ HENNIG, K. ; SOPORA, H.: TECHNOLOGIE DER PUZZOLANZEMENTHERSTELLUNG UND ERGEBNISSE DER MOERTELTECHNISCHEN VERSUCHE. IN: BAUSTOFF- INDUSTRIE 12 (1969), S. 306-307

/H6/ HERMANN, A.: UEBER KALKSANDSTEINMASSEN MIT ZUSAETZEN VON KALKMEHL, KRAFTWERKSFLUGASCHEN ODER ZERKLEINERTEN KRAFTWERKSGRANULATEN. 2. INT. SYMPOS. FUER DAMPFGEHAERTETE KALZIUMSILIKAT-BAUSTOFFE, HANNOVER 1969

/H7/ HESTER, J.A. ; SMITH, O.F.: USE OF FLY ASH IN CONCRETE BY THE ALABAMA HIGHWAY DEPARTMENT. SYMPOSIUM ON FLY ASH IN CONCRETE, 43RD ANNUAL MEETING 1964, IN: HIGHWAY RESEARCH RECORD, WASHINGTON (1965), NR.73, S.32-40

/H8/ HRIBERNIK, E. ; FERSAN, M. ; DROLJC, S. ; JENEK, A.L.: FORMSTEINE AUS ELEKTROFILTERASCHE. 2. INT. SYMPOS. FUER DAMPFGEHAERTETE KALZIUMSILIKATBAUSTOFFE, HANNOVER 1969

/H9/ HUANG, SHI YUAN: HYDRATATION OF LIME-FLY ASH PASTE AT HIGH TEMPERATUR. IN: CBI FORSKNING FO RESEARCH 10:79 (1979)

/H10/ HUBER, H.: DIE VERWENDUNG VON FLUGASCHE BEI DER BETONHERSTELLUNG IM KRAFTWERKS- UND TUNNELBAU. IN: ZEMENT UND BETON 23 (1978), S. 162-170

/H11/ HUGHES, R.D.: EXPERIMENTAL CONCRETE PAVEMENT CONTAINING FLY ASH ADMIXTURES. SYMPOS. ON FLY ASH IN CONCR. 43RD ANN. MEET. 1964, HIGHWAY RES. REC. NR.73 (1965), S. 41-44

/H12/ HYLAND, E.J.: FLY ASH QUALITY CONTROL: IMPORTANT TO USERS. IN: MOD. CONCRETE 43 (1979), S. 60-62, 64

/H13/ HYLAND, J.: FLY ASH IN READY-MIX CONCRETE. PROC.: EDISON ELECTRIC INSTITUTE, NATIONAL COAL ASSOC., BUREAU OF MINES SYMPOS., PITTSBURGH 1967, S. 80-86, COMPILED BY FABER, J.H. ; CAPP, P. ; SPENCER, J.D.

/H14/ HESTER, J.A.: FLY ASH IN ROADWAY CONSTRUCTION. PROC.: EDISON ELECTRIC INSTITUTE, NATIONAL COAL ASSOC., BUREAU OF MINES SYMPOS., PITTSBURGH 1967, S. 87-100, COMPILED BY FABER, J.H. ; CAPP, P. ; SPENCER, J.D.

/H15/ HOLUM, K.: THE PUBLIC CONCERN FOR ENVIRONMENTAL IMPROVEMENT. PROC.: EDISON ELECTRIC INSTITUTE, NATIONAL COAL ASSOC., BUREAU OF MINES SYMPOS., PITTSBURGH 1967, S. 321-325, COMPILED BY FABER, J.H. ; CAPP, P. ; SPENCER, J.D.

/H16/ HOY, L.: CONSUMER ECONOMICS: USE OF FLY ASH IN CONCRETE. PROC.: EDISON ELECTRIC INSTITUTE, NATIONAL COAL ASSOC., BUREAU OF MINES SYMPOS., PITTSBURGH 1967, S. 204-209, COMPILED BY FABER, J.H. ; CAPP, P. ; SPENCER, J.D.

/H17/ HOBBS, D.W.: INFLUENCE OF PULVERIZED-FUEL ASH AND GRANULATED BLAST-FURNACE SLAG UPON EXPANSION CAUSED BY THE ALKALI-SILICA REACTION. IN: MAGAZINE OF CONCRETE RESEARCH 34 (1982), S. 83-94

/H18/ HOBBS, D.W.: EXPANSION DUE TO ALKALI-SILICA REACTION AND THE INFLUENCE OF PULVERIZED FUEL ASH. IN: PROCEEDINGS OF THE 5TH INTERNATIONAL CONFERENCE ON ALKALI-AGGREGATE REACTION IN CONCRETE, CAPE TOWN - SOUTH AFRICA (1981), S. 252/30

/H19/ HOBBS, D.W.: THE EFFECT OF PULVERIZED-FUEL ASH UPON THE WORKABILITY OF CEMENT PASTE AND CONCRETE. IN: MAGAZINE OF CONCRETE RESEARCH 32 (1980), NR. 113, S. 219-226

/H20/ HUET, J. ; CHOQUET, F. ; VERHASSELT, A.: LES CENDRES VOLANTES BELGES. BRUXELLES : CENTRE DE RECHERCHES ROUTIERES, 1981

/H21/ HUBER, H.: DER ARLBERG STRASSENTUNNEL. SONDERDRUCK DER ARLBERG STRASSENTUNNEL AG, INNSBRUCK

/H22/ HENRIKSEN, C. ET AL: FLY ASH IN DANISH ROAD CONSTRUCTION (DAN.). IN: DANSK VEJTIDSSKRIFT 5 (1980), S. 3-7

/H23/ HOBBS, D.W.: POSSIBLE INFLUENCE OF SMALL ADDITIONS OF PFA, GBFS AND LIMESTONE FLUOR UPON EXPANSION CAUSED BY THE ALKALI - SILIKA REACTION. IN: MAGAZINE OF CONCRETE RESEARCH 35 (1983), NO. 122, S. 55-58

/H24/ HUBER, H. ; CICHOCKI, B. ; FEISCHL, M. ; LANDL, K. ; ROTH, H.: DER BETON DES KUEHLTURMS FUER DAS KRAFTWERK VOIGTSBERG 3. IN: BETON- UND STAHLBETONBAU 78 (1983), NR. 5, S. 121-124

/H25/ HOBBS, D.W.: INFLUENCE OF FLY ASH UPON THE WORKABILITY AND EARLY STRENGTH OF CONCRETE. DETROIT : AMERICAN CONCRETE INSTITUTE, ACI SP-79, 1983 - IN: PROCEEDINGS OF THE CANMET/ACI FIRST INTERNATIONAL CONFERENCE ON THE USE OF FLY ASH, SILICA FUME, SLAG AND OTHER MINERAL BY-PRODUCTS IN CONCRETE VOL. 1 (1983), S. 289-306

/H26/ HO, D.W.S. ; LEWIS, R.K.: CARBONATION OF CONCRETE INCORPORATING FLY ASH OR A CHEMICAL ADMIXTURE. DETROIT : AMERICAN CONCRETE INSTITUTE ACI SP-79, 1983 - IN: PROCEEDINGS OF THE CANMET/ACI FIRST INTERNATIONAL CONFERENCE ON THE USE OF FLY ASH, SILICA FUME, SLAG AND OTHER MINERAL BY-PRODUCTS IN CONCRETE VOL. 1 (1983), S. 333-346

/H27/ HARTLEN, J.: VERWERTUNG VON VERBRENNUNGSRUECKSTAENDEN VON KOHLEFEUERUNGEN IN SCHWEDEN. VGB-KONFERENZ 1982: VERWERTUNG VON VERBRENNUNGSRUECKSTAENDEN IN: VGB-KRAFTWERKSTECHNIK GMBH, VGB-TB 202, S. 23-34

/H28/ HILDEBRANDT, H.: ERFAHRUNGEN MIT DEM EINSATZ VON FLUGSTAUB IM BETONBAU. IN: VGB-KRAFTWERKSTECHNIK GMBH, VGB-TB 202, S. 130-138 (VGB-KONFERENZ 1982: VERWERTUNG VON VERBRENNUNGSRUECKSTAENDEN)

/H29/ HYCNAR, J.: THE INDUSTRIAL UTILIZATION OF FLY-ASH PRODUCED IN POWER STATIONS. SOURCE UNKNOWN

/H30/ HAQUE, M.N. ; LANGAN, B.W. ; WARD, M.A.: HIGH FLY ASH CONCRETES. IN: JOURNAL OF THE AMERICAN CONCRETE INSTITUTE 81 (1984), NR. 1, S. 54-60

/H31/ HENNING, O.: UEBER EINIGE KRAFTWERKSSPEZIFISCHE EINFLUESSE AUF DIE EIGENSCHAFTEN VON BRAUNKOHLEFLUGASCHEN. IN: BAUSTOFFINDUSTRIE (1984), NR. 3,

/H32/ HEUFERS, H. ; MAGEL, PH.: FLUGASCHEZEMENT - HERSTELLUNGSVERFAHREN, QUALITAET UND WIRTSCHAFTLICHKEIT. IN: ZEMENT-KALK-GIPS 37 (1984), NR. 2, S. 55-61

/H33/ HEYDEN, VAN DER, L. ; BIERVLIET, VAN, L. ; MORTELMANS, F.: A DESIGN METHOD OF MIX PROPORTIONING FOR CONCRETE USING FLY ASH. 1ST INTERNATIONAL CONFERENCE ON THE USE OF FLY ASH, SILICA FUME, SLAG AND OTHER MINERAL BY-PRODUCTS IN CONCRETE, MONTEBELLO, 31. JULY - 5. AUG. 1983

/H34/ HALSE, Y. ; PRATT, P.L.: THE DEVELOPMENT OF MICROSTRUCTURE DURING THE HY DRATION OF CEMENT BLENDS I. OPC BLENDED WITH 30% OF DRAX P.F.A.. 1ST INTERNATIONAL CONFERENCE ON THE USE OF FLY ASH, SILICA FUME, SLAG AND OTHER MINERAL BY-PRODUCTS IN CONCRETE, MONTEBELLO, 31. JULY - 5. AUG. 1983

/H35/ HYLAND, E.J.: FACTORS AFFECTING POZZOLAN MARKETING. ASH UTILISATION, 1973, BUREAU OF MINES INFORMATION CIRCULAR 8640/1974, S.114-122

/H36/ HUBBARD, F.H. ; DHIR, R.K.: A COMPOSITIONAL INDEX OF THE POZZOLANIC POTENTIAL OF PULVERIZED-FUEL ASH. IN: JOURNAL OF MATERIALS SCIENCE LETTERS 3 (1984), S. 958-960

/H37/ HUANG, S.: HYDRATION OF FLY ASH CEMENT AND MICROSTRUCTURES OF FLY ASH CEMENT PASTES. RESEARCH REPORT (B1-2-81) 79 PP., SWEDISH CEMENT AND CONCRETE RESEARCH INSTITUTE, STOCKHOLM, 1981

/H38/ HUAKUN, L. ; ZHONGYA, L. ; SHENGJIE, L.: COMPOSITION AND HYDRATION OF HIGH-CALCIUM FLY ASH. PARIS : EDITIONS SEPTIMA, 1980 - IN: 7TH INTERNATIONAL CONGRESS ON THE CHEMISTRY OF CEMENT, VOL. III, S. IV.7-IV.12

/H39/ HANSEN, W.C.: THE CHEMISTRY OF SULPHATE-RESISTING PORTLAND CEMENTS. TORONTO : UNIVERSITY OF TORONTO PRESS, 1968 - IN: PERFORMANCE OF CONCRETE (SWENSON, E.G. (ED)), S. 18-55

/H40/ HARBOE, E.M.: LONGTIME STUDIES AND FIELD EXPERIENCE WITH SULFATE ATTACK. DETROIT : AMERICAN CONCRETE INSTITUTE (ACI), 1982 - IN: SYMPOSIUM ON SULFATE RESISTANCE OF CONCRETE, ACI SPECIAL PUBLICATION SP-77, (VERBECK, G. (ED)), S. 1-20

/H41/ HURST, W.D.: EXPERIENCE IN THE WINNIPEG AREA WITH SULPHATE-RESISTING CEMENT CONCRETE. TORONTO : UNIVERSITY OF TORONTO PRESS, 1968 - IN: PERFORMANCE OF CONCRETE, (SWENSON, E.G. (ED)), S. 125-134

/H42/ HANSEN, W.C.: STUDIES RELATING TO THE MECHANISM BY WHICH THE ALKALI-AGGREGATE REACTION PRODUCES EXPANSION IN CONCRETE. IN: JOURNAL OF THE ACI 40 (1944), NR. 3, S. 213-227

/H43/ HADLEY, D.W.: ALKALI-REACTIVITY OF CARBONATE ROCKS EXPANSION AND DE-DOLOMITIZATION. IN: PROCEEDINGS, HIGHWAY RESEARCH BOARD, (1961), VOL. 40, S. 462-474

/H44/ HOBBS, D.W.: THE ALKALI-SILICA REACTION - A MODEL FOR PREDICTING EXPANSION IN MORTAR. IN: MAGAZIN OF CONCRETE RESEARCH 33 (1981), NR. 117, S. 208-219

/H45/ HJORTH, L.: CEMENT SPECIFICATIONS FOR CONCRETE EXPOSED TO CHLORIDES AND SULPHATES. LYNGBY : DEPARTEMENT OF STRUCTURAL ENGINEERING -IN:

CEB-RILEM WORKSHOP DURABILITY OF CONCRETE STRUCTURES, COPENHAGEN, 1983, S. 229-235

/H46/ HOLDEN, W.R. ; PAGE, C.L. ; SHORT, N.R.: THE INFLUENCE OF CHLORIDES AND SULPHATES ON DURABILITY. CHICHESTER, ENGLAND : ELLIS HORWOOD LIMITED, 1983 IN: CORROSION OF REINFORCEMENT IN CONCRETE CONSTRUCTION : MEETING HELD IN LONDON, ENGLAND ON 13-15 JUNE 1983, (CRANE, A.P.(ED)), S. 143-150

/H47/ HANSSON, I.L.H. ; HANSSON, C.M.: ELECTRICAL RESISTIVITY MEASUREMENTS OF PORTLAND CEMENT BASED MATERIALS. IN: CEMENT AND CONCRETE RESEARCH 13 (1983), NR. 5, S. 675-683

/H48/ HUANG, SHI YUAN: HYDRATATION OF LIME-FLY ASH PASTE AT HIGH TEMPERATURE. IN: CBI FORSKNING FO RESEARCH 10:79 (1979) ISSN 0346-6906

/I1/ ILGNER, R.: EINFLUSS DES ZUSATZES VON BRAUNKOHLEFILTERASCHEN AUF DIE HYDRATATION DES BINDEMITTELS UND DIE ZEMENTEIGENSCHAFTEN. 5. INT. BAUSTOFF- UND SILIK.-TAG., WEIMAR 1973, S. 463-472

/I2/ IVANOV, Y. ; ZACHARIEVA, S.: INFLUENCE OF FLY ASH ON THE RHEOLOGY OF FRESH CONCRETE. PROC.: INT. SYMPOS.: THE USE OF PFA IN CONCRETE. DEPT. CIV. ENG. LEEDS UNIV. 1982, VOL.1, S. 133-139

/I3/ IDORN, G.M.: INTERNATIONAL ASPECTS OF DEVELOPMENT OF THE USES OF FLY ASH WITH CEMENT. BOSTON : MATERIAL RESEARCH SOCIETY IN: EFFECTS IN FLY-ASH INCORPORATION IN CEMENT AND CONCRETE : PROCEEDINGS SYMPOSIUM N ANNUAL MEETING (DIAMOND,S.(ED.)) BOSTON, NOVEMVER 16-18, 1981, S. 244-259

/I4/ ISAKOVIC, S. ; GEDER, J.: THE USE OF FLY ASH FOR CONSTRUCTING AN EMBANKMENT ACROSS A MARSH. IN: C.R.CONF.INT.SOUS-PRODUITS ET DECHETS DANS LE GENIE CIVIL, PARIS (1978), S. 355-358

/J1/ JACOBS, J.: VERWERTUNG VON VERBRENNUNGSRUECKSTAENDEN AUS KRAFTWERKSFEUERUNGEN. IN: VGB KRAFTWERKSTECHNIK 58 (1978), S. 342-353

/J2/ JAMBOR, J.: HYDRATATIONSPRODUKTE DER KALK-PUZZOLAN-BINDEMITTEL. IN: ZEMENT-KALK-GIPS 16 (1963), S. 177-186

/J3/ JAMBOR, J.: RELATION BETWEEN PHASE COMPOSITION, OVER-ALL POROSITY AND STRENGTH OF HARDENED LIME-POZZOLANA PASTES. IN: MAGAZINE OF CONCRETE RESEARCH 15 (1963), S. 131-142

/J4/ JARRIGE, A.: DIE VERWENDUNG VON FLUGASCHE BEI DER HERSTELLUNG VON BETONWAREN. IN: BETONSTEIN-ZEITUNG 29 (1963), S. 570-572

/J5/ JARRIGE, A.: DIE VERWENDUNG VON FLUGASCHE IN FRANKREICH. IN: MITTEILUNGEN VGB 39 (1959), S. 345-355

/J6/ JARRIGE, A.: LES UTILISATIONS DES CENDRES VOLANTES DANS LE CONSTRUCTION. IN: ANNALES DE L' INST. TECH. DU BATIMENT ET DE TRAVAUX PUBL. NR. 138 (1959), S. 521-544

/J7/ JARRIGE, A.: QUELQUES ASPECTS DE L' EFFET POUZZOLANIQUE EN RAPPORT AVEC LES UTILISATION DES CENDRES VOLANTES. IN: SILICATES INDUSTRIELS 28 (1963), S. 123-134

/J8/ JARRIGE, A.: LES CENDRES VOLANTES DANS LES BETONS. IN: REVUE DES MATERIAUX DE CONSTRUCTION ET DE TRAVAUX PUBLICS NO 655 (1970), S. 109-116

/J9/ JARRIGE, A.: LES CENDRES VOLANTES. PARIS : EYROLLES, 1971

/J10/ JARRIGE, A.: UEBER DIE PUZZOLANEIGENSCHAFTEN VON FLUGASCHEN. IN: ZEMENT- KALK-GIPS 16 (1963), S. 390-392

/J11/ JEFFERS, P.E.: HOW LIGHTWEIGHT AGGREGATE IS MADE FROM FLY ASH. IN: BRICK AND CLAY RECORD 160 (1972), S. 28-29

/J12/ JENSEN, S.A.: VERWENDUNGSMOEGLICHKEITEN FUER KRAFTWERKSASCHE. IN: VGB KRAFTWERKSTECHNIK 57 (1977), S. 845-848

/J13/ JINDAL, B.K.: PROPERTIES OF STRUCTURAL LIGHTWEIGHT CONCRETE USING SINTERED FLY ASH AGGREGATE. IN: INDIAN CONCRETE JOURNAL 38 (1964), S. 413-418

/J14/ JOSHI, R.C. ; ROSAUER, E.A.: POZZOLANIC ACTIVITY IN SYNTHETIC FLY ASHES ; PART I: EXPERIMENTAL PRODUCTION OF SYNTHETIC FLY ASHES ; PART II: CHARACTERIZATION OF SYNTHETIC FLY ASHES ; PART III: POZZOLANIC BAHAVIOR. AMES ; IOWA : ENGINEERING RESEARCH INSTITUTE, 1971 -(ERI -71029 PREPRINT PRO- JECT 703) (PRESENTD IN PART AT THE 73RD. ANNUAL MEETING, THE AMERICAN CERAMIC SOCIETY, CHICAGO, 1971 (SECOND SYMPOSIUM ON CEMENTS))

/J15/ JARRIGE, A.: AN ATTEMPT TO EXPLAIN FRENCH SUCCESS IN THE UTILIZATION OF FLY ASH. PROC.: EDISON ELECTRIC INSTITUTE, NATIONAL COAL ASSOC., BUREAU OF MINES SYMPOS., PITTSBURGH 1967, S. 143-155, COMPILED BY FABER, J.H. ; CAPP, P. ; SPENCER, J.D.

/J16/ JAWED, I. ; SKALNY, J.: HYDRATION OF TRICALCIUM SILICATE IN THE PRESENCE OF FLY ASH. BOSTON : MATERIAL RESEARCH SOCIETY - IN: EFFECTS OF FLY-ASH INCORPORATION IN CEMENT AND CONCRETE : PROCEEDINGS SYMPOSIUM N ANNUAL MEETING (DIAMONT,S.(ED.)) BOSTON, NOVEMBER 16-18, 1981, S. 60-70

/J17/ JARRIGE, A.: LA POSITION DES PRODUCTEURS DE CENDRES. IN: REVUE DE L'INDUSTRIE MINERALE (1963), S. 238-239

/J18/ JENCEK, L.A.: X-RAY ANALYSIS OF FINE GRADED BROWN COAL AND LIGNITE FLY ASH FROM SOME SOURCES IN SLOVENIA. IN: CEMENT, ZAGREB 15 (1971), S. 55-62

/J19/ JONGE, DE, L.S.: VLIEGAS VOOR DE BAKSTEENINDUSTRIE. IN: KLEI/GLAS/KERAMIEK (1981), S.110-112

/J20/ JELLEY, B.: THE USE OF PULVERISED FUEL ASH IN CONCRETE IN THE WESTERN CAPE. IN: CONCRETE, BETON NR. 29, 1983.3

/J21/ JOSHI, R.C. ; WARD, M.A.: COAL ASH PRODUCTION, UTILIZATION, AND PROPERTIES AND ASH MANAGEMENT PRACTICE. ORIGINAL PRESENTATION AT A SHORT COURSE ON ASH UTILIZATION, UNIVERSITY OF CALGARY, NOVEMBER 1982

/J22/ JOSHI, R.C. ; NATT, G.S.: ROLLER COMPACTED HIGH FLY ASH CONCRETE (GEOCRETE). DETROIT : AMERICAN CONCRETE INSTITUTE ACI SP-79, 1983 - IN: PROCEEDINGS OF THE CANMET/ACI FIRST INTERNATIONAL CONFERENCE ON THE USE OF FLY ASH, SILICA FUME, SLAG AND OTHER MINERAL BY-PRODUCTS IN CONCRETE VOL. 1 (1983), S. 347-366

/J23/ JACOBS, J.: EINSATZ VON FLUGSTAUB AUS STEINKOHLENFEUERUNGEN IM BETONBAU. VGB-KONFERENZ 1982: VERWERTUNG VON VERBRENNUNGSRUECKSTAENDEN IN: VGB- KRAFTWERKSTECHNIK GMBH, VGB-TB 202, S. 72-85

/J24/ JAMBOR, J.: UNE NOUVELLE METHODE DE DETERMINATION DE L'ACTIVITE POUZZOLANIQUE. IN: REVUE DES MATERIAUX NR. 564 (1962), S. 240-256

/J25/ JOSHI, R.C. ; CARETTE, G.G. ; MALHOTRA, V.M.: PAPER PRESENTED AT FIRST INTERNATIONAL CONFERENCE USE OF FLY ASH, SILICA FUME, SLAG AND OTHER MINERAL BY-PRODUCTS IN CONCRETE, JULY 1983, MONTEBELLO (CANADA)

/J26/ JOSHI, R.C. ; WARD, M.A.: CEMENTITIOUS FLY ASHES - STRUCTURAL AND HYDRATION MECHANISM. PARIS : EDITIONS SEPTIMA, 1980 - IN: 7TH INTERNATIONAL CONGRESS ON THE CHEMISTRY OF CEMENT, VOL. III (1980), S. IV.78-IV.83

/J27/ JENSEN, A.D. ; CHATTERJI, S. ; CHRISTENSEN, P. ; THAULOW, N.: STUDIES OF ALKALI-SILICA REACTION - PART II: EFFECT OF AIR-ENTRAINMENT ON EXPANSION. IN: CEMENT AND CONCRETE RESEARCH 14 (1984), NR. 3, S. 311-314

/J28/ JOHNSTON, C.D.: WASTE GLASS AS COARSE AGGREGATE FOR CONCRETE. IN: ASTM JOURNAL OF TESTING AND EVALUATION 2 (1974), NR. 5, S. 311-350

/J29/ JOSHI, R.: USE OF FLY ASH FOR SOIL STABILIZATION. PROC. INT. CONF. ON THE USE OF BY PRODUCTS AND WASTE IN CIV. ENG., VOL. II, PP 359-364, PARIS, 1978

/J30/ JOSHI, R.C. ET AL: NEW AND CONVENTIONAL USES OF FLY ASH. ASCE ANNUAL AND NAT. ENG. CONV., PP 1-30, KANSAS CITY, 1974

/K1/ KALOUSEK, G.L. ; PORTER, L.C. ; BENTON, E.J.: CONCRETE FOR LONG-TIME SERVICE IN SULFATE ENVIRONMENT. IN: CEMENT AND CONCRETE RESEARCH 2 (1972), S. 79-89

/K2/ KANTOR, A.J.: FLY ASH SINTERING. IN: PIT AND QUARRY 49 (1957), S. 88-90

/K3/ KARPINSKI, J.Y.: LA DETERMINATION DE L' ACTIVITE POUZZOLANIQUE DES CENDRES VOLANTES PAR L' ESSAI A LA CORROSION ACCELEREE. IN: REVUE DES MATERIAUX DE CONSTRUCTION ET DE TRAVAUX PUBLICS, NO 510 (1958), S. 63-74

/K4/ KASSAI, G. ; SASAKI, J. ; TANAKA, H.: POZZOLAN CEMENT PRODUCED BY MIXING FLY ASH AND MIXING GRINDING FLY ASH (IN JAPANESE). PROC.: 10TH GEN. MEET. JAPAN. CEM. ASSOC. 1956, S. 227-232

/K5/ KELLER, H.: BEEINFLUSSUNG BESTIMMTER BETONEIGENSCHAFTEN DURCH EFA-FUELLER. IN: BAU UND DIE BAUINDUSTRIE 22 (1969), S. 112-115

/K6/ KIKAS, V.K. ; PIKSARV, E.I. ; KHAIN, A.A. ; LAUL, I.A.: EFFECTIVE ASH CEMENTS ON THE BASIS OF THE SOLID FUEL FLY ASHES. VI. INT. CONGR. ON CHEM. OF CEM., SUPPL. PAP. III-3, MOSCOW, 1974

/K7/ KIRSCH, H.: DIE VERWERTUNG VON FLUGSTAEUBEN UND SCHLACKEN AUS STEIN-KOHLENGEFEUERTEN KRAFTWERKEN. IN: MASCHINENMARKT 67 (1961) S. 9-12

/K8/ KISHITANI, K. ; TOMOSAWA, F. ; OKA, S.: RHEOLOGICAL STUDY OF SUPERPLASTICIZED CONCRETE DEVELOPMENTS IN THE USE OF SUPERPLASTICIZERS, PUBL: SP 68-24. IN: AMERICAN CONCRETE INSTITUTE, DETROIT (1981) S. 423-439

/K9/ KOBAYASHI, M. ; MIYAKE, N. ; KOKUBU, M.: FLY ASH FOR LONG-TERM HIGH-ALUMINA CEMENT STRENGTH. VI. INT. CONGR. ON CHEM. OF CEM., SUPPL. PAP. III-3, MOSCOW, 1974

/K10/ KOEHLING, R. ; STOENNER, A.: MOEGLICHKEITEN ZUR ASCHEVERWERTUNG IN DER BUNDESREPUBLIK DEUTSCHLAND. IN: BETONWERK UND FERTIGTEIL-TECHNIK 42 (1976) S. 340-344

/K11/ KOENIG, P.: EINSATZ VON BRAUKOHLENFILTERASCHE IN DER BETONFERTIGTEIL-INDUSTRIE DER DDR. IN: BAUSTOFFINDUSTRIE 20 (1977) AUSG. B, S. 15-17

/K12/ KOERTH, W.: UNTERSUCHUNG EINIGER GRUNDFRAGEN DES FLUGASCHEZEMENTS. IN: BAUSTOFFINDUSTRIE 11 (1968), S. 343-346, 380, 382-383

/K13/ KOKUBU, M.: FLY ASH AND FLY ASH CEMENT. PROC.: 5TH INT. SYMPOS. ON CHEM., PART IV, S. 75-105, TOKYO, 1968

/K14/ KOKUBU, M. ; YAMADA, J.: FLY ASH CEMENTS. IN: PROCEEDINGS OF THE VI. INTERNATIONAL CONGRESS ON THE CHEMISTRY OF CEMENT, SEC. III, III-3, S. 1-51, MOSCOW, 1974

/K15/ KOVACS, R.: HYDRATATION PROCESS AND DURABILITY OF FLY-ASH CEMENTS. VI. INT. CONGR. ON CHEM. OF CEM., SUPPL. PAP. III-3, MOSCOW, 1974

/K16/ KOVACS, R.: EFFECT OF THE HYDRATATION PRODUCTS ON THE PROPERTIES OF FLY ASH CEMENTS. IN: CEMENT AND CONCRETE RESEARCH 5 (1975) S. 73-82

/K17/ KRAUSE, M.: BRAUNKOHLENFILTERASCHEN DER DDR ALS ZUMAHLSTOFF IN DER ZEMENTINDUSTRIE. IN: BAUSTOFFINDUSTRIE 14 (1971), S. 19-23

/K18/ KRONSBEIN, W.: DIE HYDRAULISCHEN EIGENSCHAFTEN VON STEINKOHLENFLUGASCHE UND IHR EINFLUSS AUF DIE SULFATBESTAENDIGKEIT VON PORTLANDZEMENT. IN: ZEMENT-KALK-GIPS 4 (1951), S. 123-127

/K19/ KRONSBEIN, W.: FLUGASCHEN ALS HYDRAULISCHE ZUSCHLAEGE FUER DIE HERSTELLUNG VON BINDEMITTELN FUER MOERTEL UND BETON. IN: ZEMENT 30 (1941), S. 503-506

/K20/ KRUEGER, J.E.: FLY ASH IN SOUTH AFRICA. IN: COAL GOLD AND BASE MINERALS 24 (1976), NR. 6, S. 73, 75, 77, 79

/K21/ KUCERA, K.: EIN NEUER KUENSTLICHER ZUSCHLAGSTOFF FUER BITUMINOESE FAHRBAHNBEFESTIGUNGEN IN DER TSCHECHOSLOWAKEI. IN: BITUMEN 38 (1976), S. 177- 178

/K22/ KUNZE, W.: GESINTERTE FLUGASCHEPELLETS ALS ZUSCHLAG FUER KONSTRUKTIONSLEICHTBETON. IN: BETONWERK UND FERTIGTEIL-TECHNIK 40 (1974), S. 50-55

/K23/ KURDOWSKI, W. ; POLESZAK, M.: VERWENDUNG VON FLUGASCHEN ZUR ZEMENTPRODUKTION. IN: TONINDUSTRIE ZEITUNG (TIZ) 102 (1978), S. 696-700

/K24/ KOVACS, R. ; VAMOS, G.: INVESTIGATION OF THE PHYSICAL AND CHEMICAL PROPERTIES OF HUNGARIAN FLY ASHES. PROC.: INT. SYMPOS.: THE USE OF PFA IN CONCRETE, DEPT. CIV. ENG. LEEDS UNIV. 1982, VOL.1, S. 17-21

/K25/ KORDINA, K. ; SCHWICK, W.: UNTERSUCHUNGEN VON BETONZUSATZSTOFFEN ZUR VERMEIDUNG DER ALKALI - ZUSCHLAG - REAKTION. IN: BETONWERK UND FERTIGTEILTECHNIK 47 (1981), S. 328-331

/K26/ KOCUVAN, I.: CLASSIFICATION OF FLY ASH. IN: BEOGRAD (1979), S. 69-75

/K27/ KORAC, V. ; UKRAINCIK, U. ; HLAVANJA, I.: FLY ASH FROM THE THERMOELECTRIC PLANT KAKANJ AS ADDITIVE TO THE PORTLANDCEMENT CLINKER BY PRODUCTION OF CEMENT FOR DAMS. IN: SAVETOVANJE TROGIR (1980), S. 281-303

/K28/ KUKKO, H. ; KEPPO, M.: USE OF FLY ASH IN CONCRETE (FIN.). IN: RAKENNUSTEKNIIKA 36 (1980), S. 95-98

/K29/ KEPPO, M. ; YLINEN, P.: POWER PLANT ASHES AND THEIR UTILIZATION (FIN.) : PART 1: THE QUANTITIES OF ASHES PRODUCED IN FINLAND AND THEIR QUALITY. TECH. RES. CENTER OF FINLAND, CONCRETE AND SILICATE LABORATORY, REPORT 61

/K30/ KARJALAINEN, T.: UTILIZATION OF ASH AS FERTILIZER AND SOIL AMENDMENT (FIN.). KEMIRA OY, AGRICULTURAL RESEARCH HELSINKI (1979)

/K31/ KUKKO, H. ; LINDBAECK, G.: POWER PLANT ASHES AND THEIR UTILIZATION PART 2: USE OF COAL ASHES AS CONSTITUENT OF CEMENT AND CONCRETE (FIN.). TECHNICAL RESEARCH CENTRE OF FINLAND, CONCRETE AND SILICATE LABORATORY, REPORT 62

/K32/ KEPPO, M. ; PERANDER, T.: POWER PLANT ASHES AND THEIR UTILIZATION PART 5: USE IN LIGHTWEIGHT AGGREGATE PRODUCTION (FIN.). TECHNICAL RESEARCH CENTRE OF FINLAND, CONCRETE AND SILICATE LABORATORY, REPORT 65

/K33/ KEPPO, M.: POWER PLANT ASHES AND THEIR UTILIZATION PART 6: USE IN THE PRODUCTION OF BRICKS (FIN.). TECHNICAL RESEARCH CENTRE OF FINLAND, CONCRETE AND SILICATE LABORATORY, REPORT 66

/K34/ KRUEGER, J.E.: THE USE OF INDUSTRIAL RESIDUES IN CEMENT AND CONCRETE. IN: CONCRETE, BETON NR. 29, 1983.3

/K35/ KAITILA, H.: WORK ON CONCRETE NORM AND USE OF MATERIALS SUBSTITUTING CEMENT IN CONCRETE, SOME VIEWPOINTS (DAN.). IN: CBL SPECIAL PUBLICATION (1980), NR. 7. S. 26-29

/K36/ KUKKO, H.: FLYGAKSA SOM DELMATERIAL I BETONG (DAN.). IN: CBL SPECIAL PUBLICATION (1980) NR. 7, S. 88-89

/K37/ KELLER, H.: DAUERHAFTER BETON MIT GEEIGNETEM FLUGSTAUB. IN: BAUWIRTSCHAFT (1983), NR. 22

/K38/ KELLER, H.: FLUGSTAUB ALS BETONZUSATZSTOFF. IN: BETON- UND STAHLBETONBAU (1983), NR. 3, S. 78-80

/K39/ KORAC, V. ; UKRAINCIK, V.: STUDIES INTO THE USE OF FLY ASH IN CONCRETE FOR WATER DAM STRUCTURES. DETROIT : AMERICAN CONCRETE INSTITUTE ACI SP-79, 1983 - IN: PROCEEDINGS OF THE CANMET/ACI FIRST INTERNATIONAL CONFERENCE ON THE USE OF FLY ASH, SILICA FUME, SLAG AND OTHER MINERAL BY-PRODUCTS IN CONCRETE VOL. 1 (1983), S. 173-185

/K40/ KASAI, Y. ET AL: AIR PERMEABILITY AND CARBONATION OF BLENDED CEMENT MORTARS. DETROIT : AMERICAN CONCRETE INSTITUTE ACI SP-79, 1983 - IN: PROCEEDINGS OF THE CANMET/ACI FIRST INTERNATIONAL CONFERENCE ON THE USE OF FLY ASH, SILICA FUME, SLAG AND OTHER MINERAL BY-PRODUCTS IN CONCRETE VOL. 1 (1983), S. 435-451

/K41/ KOHNO, K. ; HARII, K. ; FUKUSHIMA, H.: USE OF FLY ASH, BLAST-FURNACE SLAG AND CONDENSED SILICA FUME FOR CONCRETE BLOCK STRIPPED IMMEDIATELY AFTER MOLDING. PROC.: THE USE OF FLY ASH, SILICA FUME, SLAG AND OTHER MINERAL BY-PRODUCTS IN CONCRETE; MONTEBELLO, 1983, ACI PUBLICATION SP-79, VOL. II, PP. 1165-1176

/K42/ KELLER, H. ; SCHOLZ, H.: DER WEG ZUM ERFOLG BEI DER VERWERTUNG VON VERBRENNUNGSRUECKSTAENDEN IM BAUWESEN. VGB-KONFERENZ 1982: VERWERTUNG VON VERBRENNUNGSRUECKSTAENDEN IN: VGB-KRAFTWERKS- TECHNIK GMBH, VGB-TB 202, S. 138-147

/K43/ KOKUBU, M. ; NAGATAKI, S.: CARBONATION OF CONCRETE CORRELATING WITH THE CORROSION OF REINFORCEMENT IN FLY ASH CONCRETE. IN: RILEM, INT. SYMPOSIUM : DURABILITY OF CONCRETE - 1969, FINAL REPORT, PART II, S.D71-D79

/K44/ KIKKERT, J. ; BURMEISTER, J.: PROBENVORBEREITUNG UND ROENTGENFLUORESZENZANALYSE VON KOHLE UND FLUGASCHE IN DER ZEMENTINDUSTRIE. IN: ZEMENT-KALK- GIPS (ZKG) 36 (1983), NR. 5, S. 268-270

/K45/ KOVACS, R.: THE INFLUENCE OF THE PRODUCTION TECHNOLOGY OF FLY ASH CEMENTS UPON THEIR HYDRATION AND HARDENING. IN: 7TH INT. CONGRESS ON THE CHEMISTRY OF CEMENT, PARIS 1980, VOL. III, PP. IV104 - IV109

/K46/ KRUEGER, J.E. ; ROSSOUW, A.F.G. ; DIJK, VAN, J.: THE USE OF PFA IN CEMENT AND CONCRETE : RESEARCH AND UTILISATION IN THE REPUBLIC OF SOUTH AFRICA. PRETORIA : CSIR, 1982 - IN: CSIR PUBLICATION NR. R/BOU 982 - ALSO AFRICA PROC. WORKSHOP, RESEARCH AND DEVELOPMENT NEEDS FOR USE OF FLY ASH IN CEMENT AND CONCRETE, MARCH 1981, PALO ALTO, (BAKKER, W.T. (ED.)), PP. 3.48-3.58, EPRI (1981)

/K47/ KONDO, R.: CHEMICAL RESISTIVITIES OF VARIOUS TYPES OF CEMENTS. 4TH INTNL. SYMP. ON CHEM. CEM., PP 881-886, WASHINGTON, 1960

/K48/ KALOUSEK, G.L. ; BENTON, E.J.: MECHANISM OF SEAWATER ATTACK ON CEMENT PASTES. IN: JOURNAL OF THE ACI 67 (1970), NR. 2, S. 187-192

/K49/ KEENE, P.W.: SOME TESTS ON THE DURABILITY OF CONCRETE MIXES OF SIMILAR COMPRESSIVE STRENGTH. IN: MAGAZINE OF CONCRETE RESEARCH 37 (1960), NR.13, S. 3-12

/K50/ KATAWAKI, K.: CORROSION MECHANISMS OF REINFORCING STEEL IN CONCRETE IN CHLORIDE ENVIRONMENT, TRANSACTIONS OF THE JAPAN CONCRETE INSTITUTE 4 (1982), PP 65-72

/L1/ LAFUMA, H.: L' ADDITION DES CENDRES AU CIMENT PORTLAND. SOURCE UNKNOWN

/L2/ LANG, E.: UEBER DEN PUZZOLANCHARAKTER VON BRAUNKOHLENFILTERASCHEN. IN: BAUSTOFFINDUSTRIE 13 (1970), S. 197-201

/L3/ LARSON, C.H.: FIELD SUBSTITUTION OF FLY ASH FOR A PORTION OF CEMENT IN AIR-ENTRAINED CONCRETE. PROC.: OF THE HIGHWAY RES. BOARD (1954), S. 258-264

/L4/ LEGG, F.E.: EXPERIMENTAL FLY-ASH CONCRETE PAVEMENT IN MICHIGAN. IN: HIGHWAY RESEARCH RECORD NR. 73 (1965), S. 1-12

/L5/ LEONARD, G.K. ; ASCE, M. ; SCHWAB, P.A.: TVA USES NON-SPECIFICATION FLY ASH. IN: CIVIL ENGINEERING 28 (1958), S. 188-192

/L6/ LEONARD, R.J. ; DAVIDSON, D.T.: POZZOLANIC REACTIVITY STUDY OF FLY ASH. IN: HIGHWAY RESEARCH BOARD BULLETIN 231 (1958), S. 1-13

/L7/ LEWANDOWSKI, R.: VERWENDUNG VON RUECKSTAENDEN INDUSTRIELLER PROZESSE ZUR HERSTELLUNG VON BETON. BERLIN : ERNST & SOHN, 1979. - IN: FORSCHUNGSBEITRAEGE FUER DIE BAUPRAXIS (EIBL, J.(ED.)), S. 91-99

/L8/ LEWANDOWSKI, R. ; JACOBS, J.: ERWEITERUNG DER VERWERTUNGSMOEGLICHKEITEN VON VERBRENNUNGSRUECKSTAENDEN. IN: VGB KRAFTWERKSTECHNIK 60 (1980), S. 972-985

/L9/ LOHTIA, R.P. ; NAUTIYAL, B.D. ; JAIN, O.P.: CREEP OF FLY ASH CONCRETE. IN: JOURNAL OF THE AMERICAN CONCRETE INSTITUTE 47 (1976), S. 469-472

/L10/ LOVEWELL, C.E.: PORTLAND-POZZOLAN CEMENT. THIRD INT. ASH UTILIZATION SYMP., PAP. B-2, PITTSBURGH, PENNSYLVANIA 1973

/L11/ LOVEWELL, C.E. ; WASHA, G.A.: PROPORTIONING CONCRETE MIXTURES USING FLY ASH. IN: JOURNAL OF THE ACI 29 (1958), S. 1093-1102

/L12/ LUEHR, H.-P.: DERZEITIGER STAND DER ZULASSUNGSVERFAHREN FUER STEINKOHLENFLUGASCHE ALS BINDEMITTEL-KOMPONENTE IM BETON UND STAHLBETON IN DER BRD. IN: BETONSTEIN-ZEITUNG 37 (1971), S. 16-21

/L13/ LUEHR, H.-P.: SYMPOSIUM ON THE USE OF ASH IN PARTICULAR IN PRODUCTION AND PREFABRICATED CONSTRUCTION ELEMENTS IN ANKARA (IN GERMAN). IN: BETON-STEIN-ZEITUNG 37 (1971), S. 789-792

/L14/ LUEHR, H.-P.: ZUR VERWENDUNG VON STEINKOHLENFLUGASCHE (ELEKTROFILTERSTAUB) ALS BETONZUSATZSTOFF. IN: BETONWERK UND FERTIGTEIL-TECHNIK 38 (1972), NR. 7, S. 511-517

/L15/ LUEHR, H.-P.: ANFORDERUNGEN AN KRAFTWERKSNEBENPRODUKTE BEI DER VERWENDUNG IM BAUWESEN. IN: VGB KRAFTWERKSTECHNIK 58 (1978), S. 354-358

/L16/ LUEHR, H.-P. ; EFES, Y.: INFLUENCE OF THE GRANULOMETRY OF FLY ASHES WITH LOW IGNITION LOSSES ON THE STRENGTH DEVELOPMENT OF MORTAR PRISMS.

MOSCOW, 1974 - IN: PROCEEDINGS OF THE VI. INTERNATIONAL CONGRESS ON THE CHEMISTRY OF CEMENT, SEC. III, III-3, S. 1-27

/L17/ LUEHR, H.-P. ; THOLEN, J.: BEITRAG ZUM EINFLUSS DER DEN GLUEHVERLUST AUSMACHENDEN BESTANDTEILE VON STEINKOHLENFLUGASCHE AUF EIGENSCHAFTEN FLUGASCHEHALTIGER ZEMENTMOERTELPRISMEN. OPLADEN : WESTDEUTSCHER VERLAG. - IN: FORSCHUNGSBERICHTE DES LANDES NRW (1972), NR. 2276

/L18/ LUKAS, W.: THE INFLUENCE OF AN AUSTRIAN FLY ASH ON THE REACTION PROCESSES IN THE CLINKER PHASES OF PORTLAND-CEMENTS. IN: MATERIAUX ET CONSTRUCTIONS 9 (1976), S. 331-337

/L19/ LUKAS, W.: DER EINFLUSS VON FLUGASCHE AUF DIE HYDRATATIONSWAERME-ENTWICKLUNG IN PORTLANDZEMENT-FLUGASCHE-MISCHUNGEN. IN: BETONWERK UND FERTIGTEIL-TECHNIK 45 (1979), S. 747-752

/L20/ LUKAS, W.: REAKTIONSABLAUF BEI EINER OESTERREICHISCHEN FLUGASCHE MIT KALKHYDRAT. IN: TONINDUSTRIEZEITUNG 100 (1976), NR. 10, S. 358-362

/L21/ LUKAS, W.: REAKTIONSVORGAENGE IM GEMISCH FLUGASCHE - GIPS - KALKHYDRAT. IN: TIZ-FACHBERICHTE 104 (1980), NR. 9, S. 630-633

/L22/ LUKAS, W. ; BLUEMEL, O.W.: DER EINFLUSS VON FLUGASCHE IN ZEMENTFLUGASCHEMISCHUNGEN AUF DIE HYDRATATIONSWAERMEENTWICKLUNG. INSTITUT FUER BAUSTOFFLEHRE UND MATERIALPRUEFUNG UNIV. INNSBRUCK, BM-MITT. 3 (1977), S. 37-41

/L23/ LEWANDOWSKI, R.: VERSUCHSREIHE MIT FLUGASCHE - EINFLUSS VON FLUGASCHE-STAEUBEN UNTERSCHIEDLICHER QUALITAET AUF DIE BETONEIGENSCHAFTEN. IN: BAUGEWERBE 61 (1982), NR. 21, S. 44-49, NR.22, S. 23-25

/L24/ LANGER, A.: ELEKTROFILTERASCHE FUER DEN BAU DER NEUEN B 236 N. IN: TIEFBAU, INGENIEURBAU, STRASSENBAU 24 (1982), S. 646-647

/L25/ LAWSON, E.M. ; NIXON, P.J.: A SURVEY OF THE LOCATIONS, DISPOSAL AND PROSPECTIVE USES OF THE MAJOR INDUSTRIAL BY-PRODUCTS AND WASTE MATERIALS IN SCOTLAND. IN: BUILDING RESEARCH ESTABLISHMENT CURRENT PAPER 50/78 GARSTON (1978)

/L26/ LAHTI, -: CALCULATION OF THE VALUE OF POWER PLANT ASHES IN SOIL AMENDMENT (FIN.) ESPOO 1979. KEMIRA OY, ESPOO RESEARCH CENTRE. 12S +LIITT. 5S. (12 P. +APP.5.)

/L27/ LEWANDOWSKI, R.: EINFLUSS UNTERSCHIEDLICHER FLUGASCHEQUALITAETEN UND -ZUGABEMENGEN AUF DIE BETONEIGENSCHAFTEN. IN: BETONWERK- UND FERTIGTEILTECHNIK 49 (1983), NR. 1, S. 11-15, NR. 2, S. 105-110, NR. 3, S. 152-158

/L28/ LEKSO, S.: TEST ROAD ON FARO, STABILIZING SAND BY USING FLY ASH (DAN.). IN: STATENS VEJLABORATORIUM, INTERNAL NOTES NO. 115, APRIL 1981

/L29/ LEWANDOWSKI, R.: ZUR WITTERUNGSBESTAENDIGKEIT VON BETONEN MIT ZUSAETZEN VON STEINKOHLEFLUGASCHEN - ERFAHRUNGEN UND UNTERSUCHUNGEN. IN: BETON- UND STAHLBETONBAU 78 (1983), NR. 7, S. 184-190

/L30/ LANE, R.O.: EFFECT OF FLY ASH ON FRESHLY MIXED CONCRETE. IN: CONCRETE INTERNATIONAL 5 (1983), NR. 10, S. 50-52

/L31/ LAMOND, J.F.: TWENTY-FIVE YEARS EXPERIENCE USING FLY ASH IN CONCRETE. DETROIT : AMERICAN CONCRETE INSTITUTE SP-79, 1983 - IN: PROCEEDINGS OF THE CANMET/ACI FIRST INTERNATIONAL CONFERENCE ON THE USE OF FLY ASH, SILICA FUME, SLAG AND OTHER MINERAL BY-PRODUCTS IN CONCRETE VOL. 1 (1983), S. 47-69

/L32/ LUKAS, W.: DIE VERWERTUNG VON FLUGASCHE IN DER OESTERREICHISCHEN ZEMENTINDUSTRIE - PROBLEME, ANFORDERUNGEN, TECHNOLOGISCHE EFFEKTE. VGB-KONFERENZ 1982: VERWERTUNG VON VERBRENNUNGSRUECKSTAENDEN IN: VGB-KRAFTWERKSTECHNIK GMBH, VGB-TB 202, S. 18-23

/L33/ LEWANDOWSKI, R.: ZUR WITTERUNGSBESTAENDIGKEIT VON BETONEN MIT FLUGASCHE ZUSAETZEN. VGB-KONFERENZ 1982: VERWERTUNG VON VERBRENNUNGSRUECKSTAENDEN VGB-KRAFTWERKSTECHNIK GBMH, VGB-TB 202, S. 109-129

/L34/ LUDWIG, U.: ZUR VERWERTUNG VON VERBRENNUNGSRUECKSTAENDEN MIT HOEHEREN KALKGEHALTEN. VGB-KONFERENZ 1982: VERWERTUNG VON VERBRENNUNGSRUECKSTAENDEN IN: VGB-KRAFTWERKSTECHNIK GMBH, VGB-TB 202, S. 246-256

/L35/ LEWANDOWSKI, R.: VERWENDUNG MINERALISCHER NEBENPRODUKTE ZUR BETONHERSTELLUNG. IN: BETONWERK UND FERTIGTEILTECHNIK 50 (1984), NR. 1, S. 36-40, NR. 2, S. 101-104

/L36/ LAMB, D.W.: ASH DISPOSAL IN DAMS, MOUNDS, STRUCTURAL FILLS AND RETAINING WALLS. ASH UTILISATION, 1973, BUREAU OF MINES INFORMATION CIRCULAR 8640/1974, S.170-179

/L37/ LUKAS, W.: REAKTIONSVERSUCHE VON FLUGASCHE MIT KALKHYDRAT. INNSBRUCK : INSTITUT FUER BAUSTOFFLEHRE UND MATERIALPRUEFUNG AN DER UNIVERSITAET INNSBRUCK, 1974 - IN: JAHRESBERICHT DES INSTITUTS FUER BAUSTOFFLEHRE UND MATERIALPRUEFUNG 1974, S. 23-32

/L38/ LANE, R.O. ; BEST, J.F.: PROPERTIES AND USE OF FLY ASH IN PORTLAND CEMENT CONCRETE. IN: CONCRETE INTERNATIONAL 4 (1982), NR. 7, S. 81-92

/L39/ LEA, F.M.: THE CHEMISTRY OF CEMENT AND CONCRETE. THIRD EDITION LONDON : EDWARD ARNOLD PUBL. LTD., 1970

/L40/ LANE, R.O. ; BEST, J.F.: LABORATORAY STUDIES ON THE EFFECTS OF SUPERPLASTICIZERS ON THE ENGINEERING PROPERTIES OF PLAIN AND FLY ASH CONCRETE. DETROIT : AMERICAN CONCRETE INSTITUTE, ACI-SP-62, 1979 - IN: SUPERPLASTICIZERS IN CONCRETE, (MALHOTRA, V.M. (ED)), S. 193-207

/L41/ LARSEN, J.T. ; PAGE, G.C.: FLY ASH FOR STRUCTURAL CONCRETE IN AGGRESSIVE ENVIRONMENTS, 4TH ASH UTILIZ. SYMP., PP 572-588, ST. LOUIS, 1976

/L42/ LUDWIG, U.: THEORETICAL AND PRACTICALL RESEARCH ON THE ALKALI-SILIKA REACTION IN CONCRETE, 1981, S252/44, 6PP

/L43/ LIMA, J.V. ; RODNEY, P.B.: HAZARDOUS WASTE MANAGEMENT : THE ESI CONCEPT. PAPER OF ASTM SYMPOSIUM ON TESTING OF HAZARDOUS SOLID WASTES, 1981

/M1/ MACKOWSKY, M..: MIKROSKOPISCHE BEOBACHTUNGEN AN FLUGSTAEUBEN. HANDBUCH DER MIKROSKOPIE IN DER TECHNIK BD. 2, T.1, 1952, S. 739-759

/M2/ MADAAN, D.S. ; KUKREJA, C.B. ; SASTRY, V.V.: STRENGTH OF FLY ASH CONCRETE FOR STRUCTURAL PURPOSES. IN: INDIAN CONCRETE JOURNAL 53 (1979), S. 294-300

/M3/ MAENGEL, S. ; WEBER, J.W.: VERAENDERUNGEN VON BETONEIGENSCHAFTEN DURCH ZUSATZ VON FILTERASCHEN. IN: BETONSTEIN ZEITUNG 37 (1971), S. 146-150

/M4/ MANZ, O.E.: ASH FROM LIGNITE. SECOND ASH UTILIZATION SYMPOS. U.S. DEPT. OF THE INTERIOR, BUREAU OF MINES, PITTSBURGH 1970

/M5/ MANZ, O.E.: CONCRETE UTILIZING AMERICAN LIGNITE FLY ASH. SYMPOS. ON THE USE OF ASH, IN PARTICULAR, IN PRODUCTION OF CONCRETE AND PREFABRICATED CONSTRUCTION ELEMENTS, ANKARA 1970

/M6/ MARCINKOWSKI, S.A.: DIE VERWENDUNG DER FLUGASCHE ALS BAUSTOFF IN POLEN. IN: BETONWERK UND FERTIGTEIL-TECHNIK 41 (1975), S. 604-606

/M7/ MARCINKOWSKI, S.A.: DER EINFLUSS VON KRAFTWERKSEINRICHTUNGEN AUF DIE QUALITAET VON FLUGASCHEN. IN: BETON 28 (1978), S. 55-56

/M8/ MARCINKOWSKI, S.A.: RADIOAKTIVITAETSMESSUNGEN VON FLUGASCHEN AUS DER BUNDESREPUBLIK DEUTSCHLAND. IN: TONINDUSTRIE ZEITUNG 103 (1979), S. 341

/M9/ MARCINKOWSKI, S.A. ; PENSKO, J.: RADIOAKTIVITAETSMESSUNGEN AN IN POLEN ZUR HERSTELLUNG VON BAUSTOFFEN UND BETONEN GENUTZTEN RANDPRODUKTEN DER KRAFTWERKSINDUSTRIE. IN: TONINDUSTRIE ZEITUNG 103 (1979), S. 272,274,277

/M10/ MASSAZZA, M.F.: STRUCTURE OF POZZOLANA AND FLY-ASH AND THE HYDRATION OF POZZOLANIC AND FLY ASH CEMENTS. 7TH INT. CONGR. ON CHEM. OF CEM. PARIS, 1980 PROC.: VOL. IV, S. 85-91, PARIS 1981

/M11/ MASSAZZA, M.F. ; COSTA, U.: ASPECTS OF THE POZZOLANIC ACTIVITY AND PROPERTIES OF POZZOLANIC CEMENTS. IN: IL CEMENTO 76 (1979), S. 3-18

/M12/ MCINTOSH, J.D. ; JORDAN, J.P.R. ; CALLAGHAN, W.O.: BEEINFLUSSUNG DER EIGENSCHAFTEN DES BETONS BEI TEILWEISEM ERSATZ DES PORTLANDZEMENTES DURCH FLUGASCHE. IN: MITTEILUNGEN DER VGB 39 (1959), S. 432-439

/M13/ MEIXNER, A.: STEINKOHLEASCHE - EIN WERTVOLLES HYDRAULISCHES BINDEMITTEL. IN: TONINDUSTRIE ZEITUNG 86 (1962), S. 30-35

/M14/ MIELENZ, R.C.: ASTM SPECIFICATIONS ON FLY ASH FOR USE IN CONCRETE. PROC. EDISON ELECTRIC INSTITUTE, NATIONAL COAL ASSOC. BUREAU OF MINES SYMPOS., PITTSSBURGH 1967, S. 271-286 COMPILED BY FABER, J.H.; CAPP, P.; SPENCER, J.D.

/M15/ MIELENZ, R.C.: SPECIFICATIONS AND METHODS OF USING FLY-ASH IN PORTLAND CEMENT CONCRETE. THIRD INT. ASH UTILIZATION SYMPOS., PAP.B-1 PITTSBURGH, PENNSYLVANIA, 1973

/M16/ MILLER, R.H. ; COLLINS, R.J.: WASTE MATERIALS AS POTENTIAL REPLACEMENT FOR HIGHWAY AGGREGATES. PROC.: OF THE WORLD CONFERENCE ON TRANSPORT RES. WASHINGTON 1976, TRANSP. RES. BOARD NCHRP REP. NO 166

/M17/ MINNICK, L.J.: INVESTIGATIONS RELATING TO THE USE OF FLY ASH AS A POZZOLANIC MATERIAL AND AS AN ADMIXTURE IN PORTLAND CEMENT CONCRETE. IN: PROCEEDINGS OF THE AMERICAN SOCIETY FOR TESTING AND MATERIALS ASTM 59 (1954), S. 1129-1177

/M18/ MINNICK, L.J.: FUNDAMENTAL CHARACTERISTICS OF PULVERIZED COAL FLY ASH. PROC.: ASTM 59 (1959), S. 1155-1177

/M19/ MINNICK, L.J.: REACTIONS OF HYDRATED LIME WITH PULVERIZED COAL FLY ASH. PROC.: EDISON ELECTRIC INSTITUTE, NATIONAL COAL ASSOC., BUREAU OF MINES SYMPOS. 1967, S. 287-315, COMPILED BY FABER, J.H.; CAPP, J.P.; JOHN, D.S.

/M20/ MINNICK, L.J.: REACTIONS OF CALCIUM AND MAGNESIUM COMPOUNDS WITH PULVERIZED COAL FLY ASH. IN: THE ANNUAL MEETING OF THE AMERICAN INST. OF MINING, METALLURGICAL AND PETROLEUM ENGRS., NEW YORK 1968 -

/M21/ MINNICK, L.J. ; MEYERS, W.F.: PROPERTIES OF LIME-FLY ASH-SOIL COMPOSITIONS EMPLOYED IN ROAD-CONSTRUCTIONS. IN: HIGHWAY RESEARCH BOARD BULLETIN NR. 69 (1953), S. 1-28

/M22/ MINNICK, L.J. ; WEBSTER, W.C. ; PURDY, E.J.: PREDICTIONS OF THE EFFECTS OF FLY ASH IN PORTLAND CEMENT MORTAR AND CONCRETE. IN: JOURNAL OF MATERIALS 6 (1971), S. 163-187

/M23/ MONTGOMERY, D.G. ; HUGHES, D.C. ; WILLIAMS, R.I.T.: FLY ASH IN CONCRETE - A MICROSTRUCTURE STUDY. IN: CEMENT AND CONCRETE RESEARCH 11 (1981), S. 591-603

/M24/ MUNDAY, J.H.L. ; ONG, L.T. ; WONG, L.B. ; DHIR, R.K.: LOAD-INDEPENDENT MOVEMENTS IN OPC/PFA CONCRETE. PROC.: INT. SYMPOS.: THE USE OF PFA IN CONCRETE, DEPT. CIV. ENG. LEEDS UNIV. 1982, VOL.1, S. 243-254

/M25/ SEE/M14/

/M26/ MOHAN, K. ; TAYLOR, F.W.: PASTES OF TRICALCIUM SILICATE WITH FLY ASH ANALYTICAL ELECTRON MICOSCOPY, TRIAMTHYL SILICATION AND OTHER STUDIES. BOSTON : MATERIAL RESEARCH SOCIETY - IN: EFFECTS OF FLY-ASH INCORPORATION IN CEMENT AND CONCRETE : PROCEEDINGS SYMPOSIUM N ANNUAL MEETING (DIAMOND,S.(ED.)) BOSTON, NOVEMBER 16-18, 1981, S. 54-59

/M27/ MONTGOMERY, D.G.: FLY ASH IN CEMENT PASTES : SOME EFFECTS OF PARTICLE SIZE AND DISTRIBUTION. BOSTON : MATERIAL RESEARCH SOCIETY - IN: EFFECTS OF FLY-ASH INCORPORATION IN CEMENT AND CONCRETE : PROCEEDINGS SYMPOSIUM N ANNUAL MEETING (DIAMOND,S.(ED.)) BOSTON, NOVEMBER 16-18, 1981, S. 194-203

/M28/ MANZ, O.E.: AMERICAN AND FOREIGN CHARACTERIZATION OF FLY ASH FOR USE IN CONCRETE. BOSTON : MATERIAL RESEARCH SOCIETY - IN: EFFECTS OF FLY-ASH INCORPORATION IN CEMENT AND CONCRETE : PROCEEDINGS SYMPOSIUM N ANNUAL MEETING (DIAMOND,S.(ED.)) BOSTON, NOVEMBER 16-18, 1981, S. 269-279

/M29/ MARTIN, R. ; HILTON, R.G.: A SUCCESSFUL APPLICATION OF THE USE OF FLY ASH IN SPECIFICATION AIR-ENTRAINED CONCRETE. BOSTON : MATERIAL RESEARCH SOCIETY - IN: EFFECTS OF FLY-ASH INCORPORATION IN CEMENT AND CONCRETE : PROCEEDINGS SYMPOSIUM N ANNUAL MEETING (DIAMOND,S.(ED.)) BOSTON, NOVEMBER 16-18, 1981, S. 306-314

/M30/ MARKOVIC, M.: USE OF FLY ASH IN CIVIL ENGINEERING. IN: BEOGRAD (1979), S. 77-91

/M31/ MUSTONEN, R. ; SINKKO, K.: RADIOACTIVITY OF PEAT, COAL AND THEIR PRODUCTS OF COMBUSTION (FIN.). IN: SAETEILYTURVALLISUUSLAITOS, RAPPORT I STL-B-36 HELSINKI (1981)

/M32/ MEYER, J.S.F.: PULVERISED FUEL ASH IN SOUTH AFRICA. IN: CONCRETE, BETON NR. 29, 1983.3

/M33/ MARTIN, K.: WAERMEDAEMMENDE KALK-SILIKAT-LEICHTBAUSTEINE AUS STEINKOHLEASCHEN. IN: BAUWIRTSCHAFT (1982) NR. 39, S. 1447-1450

/M34/ MOURITZEN, J.: FLY ASH FOR THE CEMENT AND CONCRETE INDUSTRY (DAN.). IN: CBL SPECIAL PUBLICATION (1980), NR. 7, S. 96-105 ; NORDISK BETONG, NO. 2, 1980, PP. 27-36

/M35/ MOURITZEN, J.: 400 - 500.000 TONS OF FLY ASH FROM THE POWER PLANTS WILL BE USABLE IN THE CEMENT AND CONCRETE INDUSTRY (DAN.). IN: ELEKTRO- TEKNIKEREN (1979), NR. 18, S. 450-452

/M36/ MEHTA, P.K. ; GJORV, O.E.: PROPERTIES OF PORTLAND CEMENT CONCRETE CONTAINING FLY-ASH AND CONDENSED SILICA-FUME. IN: CEMENT AND CONCRETE RESEARCH 12 (1982), NR. 5, S. 587-595

/M37/ MAGEL, P.: FLUGASCHE - CHANCE UND HERAUSFORDERUNG FUER DIE DEUTSCHE ZEMENTINDUSTRIE. IN: ZEMENT-KALK-GIPS (ZKG) 36 (1983), NR. 5, S. 249-253

/M38/ MEHTA, P.K.: POZZOLANIC AND CEMENTITIOUS BYPRODUCTS AS MINERAL ADMIXTURES FOR CONCRETE - A CRITICAL REVIEW. DETROIT : AMERICAN CONCRETE INSTITUTE SP-79, 1983 - IN: PROCEEDINGS OF THE CANMET/ACI FIRST INTERNAIONAL CONFERENCE ON THE USE OF FLY ASH, SILICA FUME, SLAG AND OTHER MINERAL BY-PRODUCTS IN CONCRETE VOL. 1 (1983), S. 1.46

/M39/ MANZ, O.E.: REVIEW OF INTERNATIONAL SPECIFICATIONS FOR USE OF FLY ASH IN PORTLAND CEMENT CONCRETE. DETROIT : AMERICAN CONCRETE INSTITUTE, ACI SP-79, 1983 - IN: PROCEEDINGS OF THE CANMET/ACI FIRST INTERNATIONAL CONFERENCE ON THE USE OF FLY ASH, SILICA FUME, SLAG AND OTHER MINERAL BY-PRODUCTS IN CONCRETE VOL. 1 (1983), S. 187-200

/M40/ MATHER, B.: CEMENTS USERS' EXPECTATIONS WITH REGARD TO BLENDED CEMENTS. DETROIT : AMERICAN CONCRETE INSTITUTE ACI SP-79, 1983 - IN: PROCEEDINGS OF THE CANMET/ACI FIRST INTERNATIONAL CONFERENCE ON THE USE OF FLY ASH, SILICA FUME, SLAG AND OTHER MINERAL BY-PRODUCTS IN CONCRETE VOL. 1 (1983) S. 255-265

/M41/ MUNDAY, J.G.L. ; ONG, L.T. ; DHIR, R.K.: MIX PROPORTIONING OF CONCRETE WITH PFA - A CRITICAL REVIEW. DETROIT : AMERICAN CONCRETE INSTITUTE ACI SP-79, 1983 - IN: PROCEEDINGS OF THE CANMET/ACI FIRST INTERNATIONAL CONFERENCE ON THE USE OF FLY ASH, SILICA FUME, SLAG AND OTHER MINERAL BY- PRODUCTS IN CONCRETE VOL. 1 (1983), S. 267-288

/M42/ MAILVAGANAM, N.P. ; BHAGRATH, R.S. ; SHAW, K.L.: EFFECTS OF ADMIXTURES ON PORTLAND CEMENT CONCRETES INCORPORATING BLAST FURNACE SLAG AND FLY ASH. DETROIT : AMERICAN CONCRETE INSTITUTE SP-79, 1983 - IN: PROCEEDINGS OF THE CANMET/ACI FIRST INTERNATIONAL CONFERENCE ON THE USE OF FLY ASH, SILICA FUME, SLAG AND OTHER MINERAL BY-PRODUCTS IN CONCRETE VOL. 1, (1983), S. 519-537

/M43/ MELAND, I.: INFLUENCE OF CONDENSED SILICA FUME AND FLY ASH ON THE HEAT EVOLUTION IN CEMENT PASTES. PROC.: THE USE OF FLY ASH, SILICA FUME, SLAG AND OTHER MINERAL BY-PRODUCTS IN CONCRETE; MONTEBELLO, 1983 ACI PUBLICATION SP-79, VOL. II, PP. 665-676

/M44/ MAGEL, P.: ENTSORGUNGSMOEGLICHKEITEN VON FLUGASCHE DURCH DIE ZEMENTINDUSTRIE. VGB-KONFERENZ 1982: VERWERTUNG VON VERBRENNUNGSRUECKSTAENDEN IN: VGB-KRAFTWERKSTECHNIK GMBH, VGB-TB 202, S. 101-109

/M45/ MONK, M.: PORTLAND-PFA CEMENT: A COMPARISON BETWEEN INTERGRINDING AND BLENDING. IN: MAGAZINE OF CONCRETE RESEARCH 35 (1983), NR. 124, S.131-141

/M46/ MARCIALIS, A. ; MASSIDDA, L. ; SANNA, V.: LOW TEMPERATURE STEAM CURING HYDRATION OF LIME-FLY-ASH COMPACTS. IN: CEMENT AND CONCRETE RESEARCH 13 (1983), S. 887-894

/M47/ MANMOHAN, D. ; MEHTA, P.K.: INFLUENCE OF PUZZOLANIC, SLAG AND CHEMICAL ADMIXTURES ON PORE SIZE DISTRIBUTION AND PERMEABILITY OF HARDENEND CEMENT PASTES. IN: CEMENT CONCRETE AGGREGATES 3 (1981), NR. 1, S. 63-67

/M48/ MEHTA, P.K.: SULFATE RESISTANCE OF BLENDED PORTLAND CEMENTS CONTAINING POZZOLANS AND GRANULATED BLASTFURNACE SLAG. PROCEEDINGS: 5TH. INT. SYMPOSIUM ON CONCRETE TECHNOLOGY, MEXICO CITY 1981, PP. 35-50

/M49/ MUNN, R.L. ; SAMARIN, A.: THE INFLUENCE OF CONCRETE TECHNOLOGY ON DESIGN FOR SERVICEABILITY. 11TH BIENNAL CONFERENCE PAPERS, SYDNEY 1983

/M50/ MINNICK, L.J.: MULTIPLE BYPRODUCT UTILISATION. ASH UTILISATION, 1973, BUREAU OF MINES INFORMATION CIRCULAR 8640/1974, S.44-55

/M51/ MOULTON, L.K.: BOTTOM ASH AND BOILER SLAG. ASH UTILISATION, 1973, BUREAU OF MINES INFORMATION CIRCULAR 8640/1974, S.148-169

/M52/ MARTENS, D.C. ; PLANK, C.O.: BASIC SOIL BENEFITS FROM ASH UTILISATION. ASH UTILISATION, 1973, BUREAU OF MINES INFORMATION CIRCULAR 8640/1974, S. 269-279

/M53/ MANZ, O.E. ; FABER, J.H. ; TAKAGI, H.: WORLDWIDE PRODUCTION OF FLY ASH AND UTILIZATION IN CONCRETE. TRONDHEIM : CANMET/ACI INTERNATIONAL CONFERENCE, 1989 - IN: FLY ASH, SILICA FUME, SLAG AND NATURAL POZZOLANS IN CONCRETE - SUPPLEMENTARY PAPERS, (MALHOTRA, V.M.(ED)), S. 1-15

/M54/ MURPHY, CH.H.: HANDBOOK OF PARTICLE SAMPLING AND ANALYSIS METHODS. VERLAG CHEMIE INTERNATIONAL, 1984

/M55/ MUKHERJEE, W.C. ; LOUGHBOROUGH, M.T. ; MALHOTRA, V.M.: DEVELOPMENT OF HIGH- STRENGTH CONCRETE INCORPORATING A LARGE PERCENTAGE OF FLY ASH AND SUPERPLASTICIZERS. IN: ASTM CEMENT CONCRETE AND AGGREGATES 4 (1982), NR. 2, S. 81-86

/M56/ MINDESS, S. ; YOUNG, J.F.: CONCRETE. NEW JERSEY : PRENTICE-HALL, 1981.- VOL. XVI

/M57/ MATTHEWS, J.D. ; GUTT, W.H.: STUDIES OF FLY ASH AS A CEMENTITIOUS MATERIAL. PAPER 4.10, CONF. ASH TECH. AND MARKETING, OCT. 1978, LONDON, CENT. ELECT. GEN. BOARD, LONDON, 1978

/M58/ MICHIHIRO, S.: STUDIES OF IMPROVEMENT OF THE INITIAL STRENGTH OF FLY ASH CEMENT, KASAI DENRYOKU KABUSHIKI KAISHA SOGO GIDUTSU KENKYUSHO SOKEN HOKOKU 29 (1982), PP 185-189

/M59/ MALHOTRA, V.M.: MECHANICAL PROPERTIES AND DURABILITY OF SUPER- PLASTICIZED SEMI-LIGHTWEIGHT CONCRETE. DETROIT : AMERICAN CONCRETE INSTITUTE, ACI SP-68 1981 - IN: DEVELOPMENTS IN THE USE OF SUPERPLASTICIZERS, (MALHOTRA, V.M. (ED)), S. 283-305

/M60/ MALHOTRA, V.M. ; CARETTE, G.G. ; BREMNER, T.W.: DURABILITY OF CONCRETE IN MARINE ENVIRONMENT CONTAINING GRANULATED BLAST FURNACE SLAG, FLY ASH OR BOTH. DETROIT, MICHIGAN : AMERICAN CONCRETE INSTITUTE, 1980 (ACI SP-65) IN: PERFORMANCE OF CONCRETE IN MARINE ENVIRONMENT, S. 157-168

/M61/ MEHTA, P.K.: PERFORMANCE TESTS FOR SULFATE RESISTANCE AND ALKALI-SILICA REACTIVITY OF HYDRAULIC CEMENTS. PHILADELPHIA : AMERICAN SOCIETY FOR TESTING AND MATERIALS, 1980 - IN: DURABILITY OF BUILDING MATERIALS AND COMPONENTS PROCEEDINGS OF THE FIRST INTERNATIONAL CONFERENCE, ASTM STP 691, (SEREDA, P.J. ; LITVAN, G.G.(ED)), S. 336-345

/M62/ MATHER, K.: FACTORS AFFECTING SULFATE RESISTANCE OF MORTARS. PARIS : EDITIONS SEPTIMA, 1980 - IN: 7TH INTERNATIONAL CONGRESS ON THE CHEMISTRY OF CEMENT, VOL. IV (1980), S. 580-585

/M63/ MARSH, B.K.: RELATATIONSHIPS BETWEEN ENGINEERING PROPERTIES AND MICRO-STRUCTURAL CHARACTERISTICS OF HARDENED CEMENT PASTE CONTAINING PULVERIZED FUEL ASH AS A PARTIAL CEMENT REPLACEMENT. PHD THESIS, THE HATFIELD POLYTECHNIC ; CEMENT AND CONCRETE ASSOCIATION, 1984

/M64/ MEHTA, P.K.: MECHANISM OF SULFATE ATTACK ON PORTLAND CEMENT CONCRETE - ANOTHER LOOK. IN: CEMENT AND CONCRETE RESEARCH 13 (1983), NR. 3, S. 401-406

/M65/ MEHTA, P.K. ; GJORV, O.E.: A NEW METHOD FOR TESTING SULFATE RESISTANCE OF CEMENTS. IN: JOURNAL OF TESTING AND EVALUATION 6 (1974), NR. 2, S. 510-515

/M66/ MEHTA, P.K.: EVALUATION OF SULFATE-RESISTING CEMENTS BY A NEW TEST METHOD. IN: JOURNAL OF THE ACI 72 (1975), NR. 10, S. 573-575

/M67/ MATHER, B.: A DISCUSSION OF THE PAPER "A REVIEW OF ALKALI SILICA REACTION AND EXPANSION MECHANISMS". IN: CEMENT AND CONCRETE RESEARCH 6 (1976), NR. 6, S. 813-813

/M68/ TANG, M.-S. ; YE, Y.-F. ; YUAN, M.-Q. ; ZHEN, S.-H.: THE PREVENTIVE EFFECT OF MINERAL ADMIXTURES ON ALKALI-SILICA REACTION AND ITS MECHANISMS. IN: CEMENT AND CONCRETE RESEARCH 13 (1983), NR. 2, S. 171-176

/M69/ MLODECKI, J.: TESTING THE RESISTANCE OF MORTARS AND CONCRETES TO ACID AND CARBONIC ACID ATTACK BY STATIONARY ACCELERATED METHOD AND BY FLOW METHOD. PRAGUE : ACADEMIA, 1969 - IN: RILEM SYMPOSTIUM ON DURABILITY OF CONCRETE 1969, S. C221-C240

/M70/ MEHTA, P.K.: DURABILITY OF CONCRETE IN MARINE ENVIRONMENT : A REVIEW. DETROIT, MICHIGAN : AMERICAN CONCRETE INSTITUTE, 1980 (ACI SP-65) IN: PERFORMANCE OF CONCRETE IN MARINE ENVIRONMENT, S. 1-20

/M71/ MIDGLEY, H.G. ; ILLSTON, J.M.: THE PENETRATION OF CHLORIDES INTO HARDENED CEMENT PASTES. IN: CEMENT AND CONCRETE RESEARCH 14 (1984), NR. 4, S. 546-558

/M72/ MARSH, B.K. ; DAY, R.L. ; BONNER, D.G.: PORE STRUCTURE CHARACTERISTICS AFFECTING THE PERMEABILITY OF CEMENT PASTES CONTAINING FLY-ASH. IN: CEMENT AND CONCRETE RESEARCH 15 (1985), NR. 6, S. 1027-1038

/M73/ MEHTA, P.K.: EFFECT OF LIME ON HYDRATION OF PASTES CONTAINING GYPSUM AND CALCIUM ALUMINATES OR CALCIUM SULPHOALUMINATES. IN: JOURNAL OF THE CERAMIC SOCIETY 56 (1973), PP 315-319

/M74/ MCCOY, W.J. ; CALDWELL, A.G.: NEW APPROACH TO INHIBITING ALKALI-AGGREGATE REACTION. IN: JOURNAL OF THE ACI 47 (1951) PP 693-706

/M75/ MEHTA, P.K. ; GERWICK, B.C. JR.: CRACKING-CORROSION INTERACTION IN CONCRETE EXPOSED TO MARINE ENVIRONMENT. IN: CONCRETE INTERNATIONAL 4 (1982), NR. 10, S. 45-51

/M76/ MEDELSKY, V.: EMPLOI DES DECHETS EN COUCHES D'ISOLATION THERMIQUE DES ROUTES. INT. CONF. ON THE USE OF BY PROD. AND WASTE IN CIVIL ENG., VOL II, PP 375-379, PARIS, 1978

/M77/ MATEOS, M.: STABILIZATION OF SOILS WITH FLY-ASH ALONE. IOWA STATE UNIVERSITY SOILS RESEARCH LAB., CONTRIB. 63:21, 1961

/N1/ NASSER, K.W. ; MARZOUK, H.M.: PROPERTIES OF MASS CONCRETE CONTAINING FLY ASH AT HIGH TEMPERATURES. IN: JOURNAL OF THE ACI 50 (1979), S. 537-550

/N2/ NICKOL, D.: FLUGASCHE SPART ZEMENT UND LEICHTZUSCHLAGSTOFFE EIN. IN: BAUSTOFFINDUSTRIE 9 (1966), S. 271-273

/N3/ NISHIBAYASHI, S. ; YAMURA, K. ; INOUE, S.: RHEOLOGICAL PROPERTIES OF SUPERPLASTICIZED CONCRETE. DEVELOPMENTS IN THE USE OF SUPERPLASTICIZERS, PUBL. SP 68-25 , S. 441-466 AMER. CONCR. INST., DETROIT 1981

/N4/ NURSE, R.W.: THE UTILIZATION OF FLY ASH FOR BUILDING MATERIAL. IN: JOURNAL OF THE INSTITUTE OF FUEL 29 (1956), S. 85-88

/N5/ NATIONAL ASH ASSOCIATION; N.N.: HOW FLY ASH IMPROVES CONCRETE BLOCK, READY MIX CONCRETE, CONCRETE PIPE. CONCR. IND. YEAR BOOK 1973-74 NATIONAL ASH ASSOC., WASHINGTON

/N6/ N.N.: MAKE LIGHTWEIGHT AGGREGATE FROM FLY ASH. IN: ROCK PRODUCTS 62 (1959), S. 87, 149

/N7/ N.N.: NOCHMALS ARMIERTER GASBETON AUF FLUGASCHEBASIS. IN: DIE BAUWIRTSCHAFT 24 (1970), S. 1677

/N8/ N.N.: UTILIZATION OF INDIAN FLY ASHES, A TECHNICAL APPRAISAL. CEMENT RES. INST. INDIA (ED.), SP-3 NEW DEHLI 1974

/N9/ NICOLESCU, L.D.: SOME PROPERTIES OF ROMANIAN FLY ASHES. PROC.: INT. SYMPOS. THE USE OF PFA IN CONCRETE, DEPT. CIV. ENG. LEEDS UNIV. 1982, VOL.1, S. 35-40

/N10/ NAGATAKI, S. ; SAKAI, E. ; MAEDA, M.: EFFECTIVE USE OF FLY ASH IN PRECAST CONCRETE PRODUCED BY AUTOCLAVE CURING. PROC.: INT. SYMPOS.: THE USE OF PFA IN CONCRETE DEPT. CIV. ENG. LEEDS UNIV. 1982, VOL.1, S. 201-208

/N11/ N.N.: REPORT OF SPECIAL INVESTIGATION NO:3083 ON POZZOLAN - A CLASSIFIED PULVERISED-FUEL ASH FOR USE IN CONCRETE. SOURCE UNKNOWN

/N12/ SEE/N13/

/N13/ NICOLESCU, L.: CEMENT AND CONCRETE WITH ADDED FLY ASH. EOROPEAN CONFERENCE ON BUILDING MATERIALS, LANCASTER 1980

/N14/ NIXON, P.J. ; GAZE, M.E.: THE USE OF FLY ASH AND GRANULATED BLASTFURNACE SLAG TO REDUCE EXPANSION DUE TO ALKALI-AGGREGATE REACTION. IN: PROCEEDINGS OF THE 5TH INTERNATIONAL CONFERENCE ON ALKALI-AGGREGATE REACTION IN CONCRETE, CAPE TOWN - SOUTH AFRICA (1981), S. 252/32

/N15/ N.N.: VERWANGENDE GRONDSTOFFEN EN ALTERNATIEVE BINDMIDDELEN VOOR PORTLAND-CEMENT. VAR 32/80, 1980 MINISTERIE VAN VOLKSGEZONDHEID EN MILIEUHYGENIE, NETHERLANDS

/N16/ N.N.: VERWANGENDE GRONDSTOFFEN EN ALTERNATIEVE BINDMIDDELEN VOOR PORTLAND-CEMENT. DEEL II: DE MARKT EN MARKETINGMOGELIJKHEDEN. VAR 9/82, 1982 MINISTERIE VAN VOLKSGEZONDHEID EN MILIEUHYGIENE, NETHERLANDS

/N17/ NORHOLM, A. ; OSBAECK, B.: FLY ASH USAGE AND RELATED R&D ACTIVITIES IN DENMARK. WORKSHOP PROCEEDINGS: R&D NEEDS FOR USE OF FLY ASH IN CEMENT AND CONCRETE. SPECIAL REPORT, EPRI CS - 2616 - SR 1982

/N18/ NIELSEN, H.C.A.: PREPARATION OF FLY ASH CEMENTS. IN: PROC. OF THE 7TH INT. CONGRESS ON THE CHEMISTRY OF CEMENT. PARIS 1980, VOL. III, IV -72. (FLS-REWIEW NO. 32, F.L. SMIDTH & CO. A/S, DENMARK)

/N19/ NASSER, K.W. ; MARZOUK, H.M.: PROPERTIES OF CONCRETE MADE WITH SULFATE RESISTING CEMENT AND FLY ASH. DETROIT : AMERICAN CONCRETE INSTITUTE ACI SP-79, 1983 - IN: PROCEEDINGS OF THE CANMET/ACI FIRST INTERNATIONAL CONFERENCE ON THE USE OF FLY ASH, SILICA FUME, SLAG AND OTHER MINERAL BY- PRODUCTS IN CONCRETE VOL. 1 (1983), S. 383-395

/N20/ NEWMAN, J.B. ; SULLIVAN, P.J.E. ; BELL, A.M.: IN-SERVICE PERFORMANCE OF REINFORCED CONCRETE STRUCTURES CONTAINING PULVERISED FUEL ASH. IN: CONCRETE (1983) NR. DECEMBER, S. 9-12

/N21/ NOWAK, Z.: IRON AND ALUMINA EXTRACTION FROM POWERPLANT FLY ASH IN POLAND. ASH UTILISATION, 1973, BUREAU OF MINES INFORMATION CIRCULAR 8640/1974, S.224-230

/N22/ NOWACKI, P.: LIGNITE TECHNOLOGY. - NOYES DATA CORP. USA, 1980

/N23/ NAGATAKI, S. ; SAKAI, E. ; TAKEUCHI, T.: THE FLUIDITY OF FLY-ASH CEMENT PASTE WITH SUPERPLASTICIZER. IN: CEMENT AND CONCRETE RESEARCH 14 (1984), NR. 5, S. 631-638

/N24/ NIXON, P.J. ; COLLINS, R.J. ; RAYMENT, P.L.: THE CONCENTRATION OF ALKALIES BY MOISTURE MIGRATION IN CONCRETE - A FACTOR INFLUENCING ALKALI AGGREGATE REACTION. IN: CEMENT AND CONCRETE RESEARCH 9 (1979), NR. 4, S. 417-423

/N25/ NICOLESCU, L.D.: CONCRETE WITH FLY ASH AND BLAST FURNACE CEMENT SHOWING HIGH RESISTANCE TOWARDS CHEMICAL AGENTS, RESOURCE CONSERVATION 9 (1982), PP 149-153

/N26/ NAGATAKI, N. ; OHGA, H. ; WADA, M.: FREEZE-THAW RESISTANCE OF AUTOCLAVED FLY ASH CONCRETE. CAJ REVIEW OF THE 37TH GENERAL MEETING/ TECHNICAL SESSION PP 266-269 (1984)

/O1/ OELSCHLAEGER, A.: BEITRAG ZU UNTERSUCHUNGEN VON BRAUNKOHLENFILTERASCHEN DER DDR ALS ZUMAHLSTOFFE IN DER ZEMENTINDUSTRIE. IN: BAUSTOFFINDUSTRIE 15 (1972), S. 8-12 MEET., 1964,, 1985 - ALSO IN: MATERIALS HIGHWAY RES. REC., NR.73, S. 1-12, WASHINGTON 1965

/O2/ OEZDEMIR, B.: UCUCU KUELLER KATILMIS CIMENTONUN OEZELLIKLERI. TUERKIYE CIMENTO SANAYII T.A.S. NEZ.24, ANKARA 1970

/O3/ OSTROWSKI, C.: EINFLUSS DES KALZIUMHYDROXIDS UND DES GIPSES AUF DIE PUZZOLANISCHEN EIGENSCHAFTEN VON FLUGASCHEN. IN: BAUSTOFFINDUSTRIE 19 (1976), S. 13-17, 20 (1976), S. 9-12

/O4/ OTTO, J.: FLUGASCHEN ALS BETONKOMPONENTE. IN: BAU UND BAUINDUSTRIE 23 (1970), S. 80, 82-83

/O5/ OWENS, P.L.: PULVERISED FUEL ASH - PART 1. IN: CURRENT PRACTICE SHEETS NO 54, JULY 1980 ALSO IN: CONCRETE 14 (1980), S. 35-36

/O6/ OWENS, P.L.: FLY ASH AND ITS USAGE IN CONCRETE. IN: CONCRETE 13 (1979), S. 21-26

/O7/ OWENS, P.L. ; WADDICOR, M.J.: TECHNIQUES FOR THE ASSESSMENT AND PRODUCTION CONTROL OF PULVERIZED FUEL ASH FOR USE IN CONCRETE. PROC.: INT.SYMPOS.: THE USE OF PFA IN CONCRETE DEPT. CIV. ENG. LEEDS UNIV. 1982, VOL.1, S. 51-69

/O8/ ONG, L.T. ; MUNDAY, J.G.L. ; DHIR, R.K.: MIX DESIGN FOR OPC/PFA CONCRETE. PROC.: INT. SYMPOS.: THE USE OF PFA IN CONCRETE DEPT. CIV. ENG. LEEDS UNIV. 1982, VOL.1, S. 173-180

/O9/ OATES, D.B. ; BUCKINGHAM, J.H.P.: PROPERTIES OF FLY ASH CONCRETE. PROC.: INT. SYMPOS.: THE USE OF PFA IN CONCRETE DEPT. CIV. ENG. LEEDS UNIV. 1982, VOL.1, S. 181-190

/O10/ ONG, L.T.; MUNDAY, J.G.L.; DHIR, R.K.: MOISTURE RELATED MOVEMENTS IN OPC/PFA CONCRETE. BOSTON : MATERIALS RESEARCH SOCIETY - IN: EFFECTS OF FLY-ASH INCORPORATION IN CEMENT AND CONCRETE : PROCEEDINGS SYMPOSIUM N ANNUAL MEETING (DIAMOND,S.(ED.)) BOSTON, NOVEMBER 16-18, 1981, S. 204-213

/O11/ OSBAECK, B.: EFFECTS OF GRINDING ON PROPERTIES OF FLY ASH CEMENTS. BOSTON : MATERIAL RESEARCH SOCIETY - IN: EFFECTS OF FLY-ASH INCORPORATION IN CEMENT AND CONCRETE : PROCEEDINGS SYMPOSIUM N ANNUAL MEETING (DIAMOND,S.ED.)) BOSTON, NOVEMBER 16-18, 1981, S. 280-288

/O12/ OSBORNE, G.J. ; NIXON, P.J.: A COMPARATIVE STUDY OF PFA CEMENT MADE BY BLENDING AND INTERGRINDING AND CONTAINING 5 AND 20% ASH. IN: SILICATES INDUSTRIELS 1982-1

/O13/ OWENS, P.L.: PULVERIZED FUEL ASH, PART 2 - PART 4. IN: CURRENT PRACTICE SHEETS NO 57, OCTOBER 1980; NO 75, JUNE 1982, N0 76, JULY 1982 ALSO IN: CONCRETE 14 (1980), NR. 10; 16 (1982), NR. 6, NR. 7

/O14/ OWENS, P.L.: A METHOD FOR THE SELECTION OF CONCRETE MIX PROPORTIONS INCORPORATING FLY ASH POZZOLANS. PROC.: 1ST INT. CONF. ON READY-MIXED CONCRETE, DUNDEE, OCT. 1975. EDITED BY DHIR, R.K.

/O15/ SEE/O11/

/O16/ OELMEZ, H. ET AL: THE MECHANICAL PROPERTIES OF POLYMER MODIFIED OPC/PFA PASTES. DETROIT : AMERICAN CONCRETE INSTITUTE ACI SP-79, 1983 - IN: PROCEEDINGS OF THE CANMET/ACI FIRST INTERNATIONAL CONFERENCE ON THE USE OF FLY ASH, SILICA FUME, SLAG AND OTHER MINERAL BY-PRODUCTS IN CONCRETE VOL. 1 (1983), S. 607-623

/O17/ OWENS, P.L. ; BUTTLER, F.G.: THE REACTIONS OF FLY ASH AND PORTLAND CEMENT WITH RELATION TO THE STRENGTH OF CONCRETE AS A FUNCTION OF TIME AND TEMPERATURE. PARIS : EDITION SEPTIMA, 1980 - IN: PROCEEDINGS OF THE 7TH INTERNATIONAL CONGRESS ON THE CHEMISTRY OF CEMENT, VOLUME III: COMMUNICATION, S. IV/60-VI/65

/O18/ OGAWA, K. ; UCHIWAKA, H. ; TAKEMOTO, K. ; YASUI, I.: THE MECHANISM OF THE HYDRATION IN THE SYSTEM C3S-POZZOLANA. IN: CEMENT AND CONCRETE RESEARCH 10 (1980), NR. 5, S. 683-696

/O19/ OBERHOLSTER, R.E. ; WESTRA, W.B.: THE EFFECTIVENESS OF MINERAL ADMIXTURES IN REDUCING EXPANSION DUE TO ALKALI AGGREGATES REACTION WITH MALMES-BURY GROUP AGGREGATES,PROC. FIFTH INTERNATIONAL CONF. ON ALKALI- AG-GREGATE REACTION IN CONCRETE, S. 252/31, 10 PP

/O20/ OCDE: UTILISATION DES DECHETS ET SOUS-PRODUITS EN TECHNIQUE ROUTIERE. RAPPORT P. 79-81, PARIS, 1977

/P1/ PALMIERI, M.: FLY-ASH, A POZZOLANIC MATERIAL OF GREAT USEFULLNESS. IN: CONCRETE 59 (1971), S. 20-22

/P2/ PAPROCKI, A.: ANWENDUNG VON FLUGASCHEN IM BAUWESEN. IN: TONINDUSTRIE ZEITUNG (TIZ) 102 (1978), S. 453-456

/P3/ PAPROCKI, A. ; OYRZANOWSKI, S.: HYDRAULIC ACTIVATION OF FLY ASH FOR CONCRETE. IN: BUILDING RESEARCH AND PRACTICE 3 (1975), S. 182-184

/P4/ PEARSON, A.S. ; GALLOWAY, F.R.: FLY-ASH IMPROVES CONCRETE AND LOWERS ITS COSTS. IN: CIVIL ENGINEERING 23 (1953), S. 38-41

/P5/ PETERS, H.: DIE VERWERTUNG VON FLUGASCHE IM BETON. IN: BETONSTEINZEITUNG 23 (1957), S. 27-34

/P6/ PFEIFER, D.W.: FLY ASH AGGREGATE LIGHTWEIGHT CONCRETE. IN: JOURNAL OF THE ACI 68 (1971), S. 213-216

/P7/ PHIL, K.A.: USE OF FLY ASH IN ROAD CONSTRUCTION - STATUS 1983. SOURCE UNKNOWN

/P8/ PJATSCHEW, W.A.: DIE ABHAENGIGKEIT DER FESTIGKEIT VON ASCHEN- PORTLAND-ZEMENTEN VON DER MINERALOGISCHEN ZUSAMMENSETZUNG DES KLINKERS. IN: ZEMENT 30 (1964), S. 9-10

/P9/ PLEMELJ, A.: KONSTRUKTIVER LEICHTBETON AUS FLUGASCHE. IN: BETONSTEIN ZEITUNG 29 (1963), S. 570

/P10/ POPOVICS, S.: STRENGTH RELATIONSHIPS FOR FLY ASH CONCRETE. IN: JOURNAL OF THE ACI 79 (1982), NR. 1, S. 43-49, DISCUSSION IN: JOURNAL OF THE ACI 53 (1982), NR. 6, S. 499-501

/P11/ PRICE, W.H.: POZZOLANS - A REVIEW. IN: JOURNAL OF THE ACI 46 (1975), S. 225-232

/P12/ PHILLEO, R.E.: FLY ASH IN MASS CONCRETE. PROC.: EDISON ELECTRIC INSTITUTE NATIONAL COAL ASSOC., BUREAU OF MINES SYMPOS., PITTSBURGH 1967, S. 69-79, COMPILED BY FABER, J.H. ; CAPP, P. ; SPENCER, J.D.

/P13/ PAPROCKI, A.: PRODUCTION AND UTILIZATION OF FLY ASH IN POLAND. PROC.: EDISON ELECTRIC INSTITUTE, NATIONAL COAL ASSOC., BUREAU OF MINES SYMPOS., PITTSBURGH 1967, S. 1760-186, COMPILED BY FABER,J.H.; CAPP,P.; SPENCER,J.D.

/P14/ PEARSON, A.S.: EXPERIENCE IN PRODUCTION AND UTILIZATION OF LIGHTWEIGHT AGGREGATE. PROC.: EDISON ELECTRIC INSTITUTE, NATIONAL COAL ASSOC., BUREAU OF MINES SYMPOS., PITTSBURGH 1967, S. 190-194, COMPILED BY FABER, J.H. ; CAPP, P. ; SPENCER, J.D.

/P15/ PURSGLOVE, J.JR.: FLY ASH IN THE FUTURE. PROC.: EDISON ELECTRIC INSTITUTE NATIONAL COAL ASSOC., BUREAU OF MINES SYMPOS., PITTSBURGH 1967, S. 326-329, COMPILED BY FABER, J.H. ; CAPP, P. ; SPENCER, J.D.

/P16/ PLOWMAN, C. ; CABRERA, J.G.: THE INFLUENCE OF PULVERISED FUEL ASH ON THE HYDRATION REACTIONS OF CALCIUM ALUMINATES. BOSTON : MATERIAL RESEARCH SOCIETY - IN: EFFECTS ON FLY-ASH INCORPORATION IN CEMENT AND CONCRETE : PRO- CEEDINGS SYMPOSIUM N ANNUAL MEETING (DIAMOND,S.(ED.)) BOSTON, NOVEMBER 16-18, 1981, S. 71-81

/P17/ PASQUET, A.: ESSAI D'APPROCHE METHODOLOGIQUE DE L'UTILISATION DES GRAVES TRAITES AUX LIANTS HYDRAULIQUES ET PUOZZOLANIQUES EN CON- STRUCTION ROUTIERE - CAS PARTICULIER DES GRAVES CENDRES VOLANTES - APPLICATION AU RENFORCEMENT DES CHAUSSEES. SOURCE UNKNOWN

/P18/ PASIC, Z. ; HODZIC, E. ; PASIC, E.: DEPENDENCY OF THE FLY ASH CHARACTERISTICS ON THE COMPOSITION OF COAL AND SYSTEM OF EQUIPMENT. IN: BEOGRAD (1979), S. 117-134

/P19/ PIETRZENIUK, H.-J.: VERWERTUNG VON RUECKSTAENDEN AUS KRAFTWERKEN IM BAUWESEN. IN: TIZ-FACHBERICHTE 107 (1983), NR. 4

/P20/ PIND, N.: QUANTITATIVE ELEMENTAL ANALYSIS ON SOLID MATERIALS USING ENERGY-DISPERSIVE X-RAY FLOURESCENCE SPECTROMETRY (DAN.). DISSERTA- TION, DEPARTMENT OF CHEMISTRY, UNIVERSITY OF AARHUS, DENMARK, 1983

/P21/ PALM, C.O.: EXPERIENCES WITH FINNISH FLY ASH AND SLAG (DAN.).IN: CBL SPECIAL PUBLICATION (1980) NR. 7, S. 24-25

/P22/ PAT, M.G.M. ; LANSER, P.A. ; NUITEN, P.C.: PORTLAND VLIEGASCEMENT VERSUS PORTLANDCEMENT. IN: CEMENT 34 (1982), NR. 12, S. 881-883

/P23/ PISTILLY, U.F.: AIR-VOID PARAMETERS DEVELOPED BY AIR-ENTRAINING ADMIX- TURES, AS INFLUENCED BY SOLUBLE ALKALIES FROM FLY ASH AND PORT- LAND CEMENT. IN: JOURNAL OF THE ACI (1983), NR. 3, S.217-222

/P24/ PREISS, H.P.: GRUNDLAGEN FUER DIE HERSTELLUNG VON FLIESSESTRICHEN. IN: BAUGEWERBE (1978). NR. 18, S. 171-174, NR. 19, S. 175-178

/P25/ PEABODY, T.E.: FLY ASH PRODUCTION AND UTILISATION IN AUSTRALIA. ASH UTILISATION, 1973, BUREAU OF MINES INFORMATION CIRCULAR 8640/1974, S. 106-113

/P26/ PRICE, G.C: INVESTIGATION OF CONCRETE MATERIALS FOR THE SOUTH SASKATCHEWAN RIVER DAM., PROC. ASTM VOL. 61 (1961) P 1155

/P27/ PRICE, G.C. ; PETERSON, R.: PERFORMANCE OF CONCRETE. UNIVERSITY OF TORONTO PRESS, TORONTO, 1961

/P28/ PRICE, G.C. ; PETERSON, R.: EXPERIENCE WITH CONCRETE IN SULPHATE ENVIRONMENTS IN WESTERN CANADA. TORONTO : UNIVERSITY OF TORONTO PRESS, 1968 - IN: PERFORMANCE OF CONCRETE, (SWENSON, E.G. (ED)), S. 93-112

/P29/ PRAIRIE FARM REHABILITATION ADMINISTRATION: THE DESIGN AND CONSTRUCTION OF THE GARDINER DAM AND ASSOCIATED WORKS, P 382, SUPPLY & SERVICES, CANADA, 1980

/P30/ PETTIFER, K. ; NIXON, P.J.: ALKALI METAL SULPHATE - A FACTOR COMMON TO BOTH ALKALI AGGREGATE REACTION AND SULPHATE ATTACK ON CONCRETE. IN: CEMENT AND CONCRETE RESEARCH 10 (1980), NR. 2, S. 173-181

/P31/ PEPPER, L. ; MATHER, B.: EFFECTIVENESS OF MINERAL ADMIXTURES IN PREVENTING EXCESSIVE EXPANSION OF CONCRETE DUE TO ALKALI AGGREGATE REACTION, PROC. ASTM VOL. 59 (1959), PP 1178-1202

/P32/ POWERS, T.C. ; STEINOUR, H.H.: AN INTERPRETATION OF SOME PUBLISHED RESEARCHES ON THE ALKALI-AGGREGATE REACTION. IN: JOURNAL OF THE ACI 51 (1955), PP 497-516, PP 785-812

/P33/ PAGE, C.L. ; SHORT, N.R. ; EL TARRAS, A.: DIFFUSION OF CHLORIDE IONS IN HARDENED CEMENT PASTES. IN: CEMENT AND CONCRETE RESEARCH 11 (1981), NR. 3, S. 395-406

/P34/ PARKER, T.W.: JOURNAL OF SOC. OF CHEM. INDUSTRY, 58T (1939), P 203

/R1/ RAASK, E.: CENOSPHERES IN PULVERIZED-FUEL ASH. IN: JOURNAL OF THE INSTITUTE OF FUEL 13 (1968), S. 339-344

/R2/ RAASK, E.: FUSION OF SILICATE PARTICLES IN COAL FLAMES. IN: FUEL 48 (1969), S. 366-374

/R3/ RAASK, E. ; BHASKAR, M.C.: POZZOLANIC ACTIVITY OF PULVERIZED FUEL ASH. IN: CEMENT AND CONCRETE RESEARCH 5 (1975), NR. 4, S. 363-376

/R4/ RAASK, E. ; STREET, P.J.: APPEARANCE AND POZZOLANIC ACTIVITY OF PULVERIZED FUEL ASH. FIRST INT. CONF. ON ASH TECHNOL. AND MARKETING, CENTRAL ELECTRICITY GENERATING BOARD, LONDON 1978

/R5/ RASCH, R.: FLUGASCHE, SCHMELZSCHLACKE, MUELLSCHLACKE UND DIE INDUSTRIELLE VERWERTUNG. IN: CHEMIKER ZEITUNG, CHEMISCHE APPARATUR 84 (1960), S. 564-567, S. 595-598

/R6/ RAVINA, D.: OPTIMIZED DETERMINATION OF PFA (FLY ASH) FINENESS WITH REFERENCE TO PUZZOLANIC ACTIVITY. IN: CEMENT AND CONCRETE RESEARCH 10 (1980), NR. 4, S. 573-580

/R7/ REHSI, S.S.: FLY ASH MORTAR FOR BRICK MASONRY. CENTR. BUILD. RES. INST., ROORKEE, INDIA 1971

/R8/ REHSI, S.S. ; GARG, S.K.: PROPORTIONING CONCRETE MIX CONTAINING FLY ASH. IN: J. OF THE INST. OF ENGL. (INDIA) 1964, NR.1, S. 68-76

/R9/ REHSI, S.S. ; GARG, S.K.: PRODUCTION OF CEMENT KLINKER USING FLY ASH. VI. INTER. CONGR. ON CHEM. OF CEM., SUPPL. PAP., III-3, MOSCOW 1974

/R10/ REINSDORF, S.: VERWENDUNG VON FLUGASCHEN ZUR VERBESSERUNG DER BETONTECHNOLOGIE DER LEICHTZUSCHLAGBETONE. IN: SILIKATTECHNIK 12 (1961), S. 484-487

/R11/ RICHARDSON, L. ; BAILEY, J.C.: DESIGN, CONSTRUCTION AND TESTING OF PULVERISED FUEL ASH CONCRETE STRUCTURES AT NEWMAN SPINNEY POWER STATION (PREPARED BY MIDHANDS PROJECT GROUP). CENTRAL ELECTRICITY GENERATING BOARD, RESEARCH AND DEVELOPMENT DEPARTMENT FEBRUAR 1966

/R12/ RIMPLER, W.: EIGENSCHAFTEN UND ANWENDUNG VON PUZZOLANZEMENT PUZ 225-W. IN: BAUSTOFFINDUSTRIE 12 (1969), S. 307-309

/R13/ ROEHER, H.J.: GIPS-ZEMENT-PUZZOLAN-BAUSTOFFE. SCHRIFTENREIHE DER BAUFORSCHUNG, REIHE BAUSTOFFE H.28 BAUAKADEMIE DER DEUTSCHEN DEMOKRATISCHEN REPUBLIK, BERLIN 1974

/R14/ ROMAN, G.U.: YOU CAN SAVE MONEY WITH FLY ASH. IN: COAL AGE 73 (1968), S. 60-64

/R15/ ROSE, H.J. ; RUSSEL, H.H.: FLUGASCHENVERWENDUNG IN DEN VEREINIGTEN STAATEN. IN: MITTEILUNGEN DER VGB 39 (1959), S. 335-344

/R16/ ROSENSTIEL, VON, A.P. ; GAY, A.J.: MIKROMORPHOLOGISCHE UND MIKROCHEMISCHE UNTERSUCHUNGEN AN FLUGASCHEN. 20. TAGUNG F. ELEKTRONENMIKR./

14.KOLLOQ. D. ARBEITSKREISES F. ELEKTRONENMIKR. DIREKTABBILDUNG UND ANALYSE VON OBERFLAECHEN (EBO), INNSBRUCK 1981

/R17/ ROSS, A.D.: SOME PROBLEMS IN CONCRETE CONSTRUCTION, THE CREEP OF CONCRETE CONTAINING FLY ASH. IN: CONCRETE RESEARCH 12 (1960), S. 28-29

/R18/ RUSSAK, M.A. ; TOBIN, A. ; FELDMAN, C.: DEVELOPMENT AND CHARACTERIZATION OF A CLOSED PORE INSULATION MATERIAL. IN: AMERICAN CERAMIC SOCIETY BULLETIN 55 (1976), S. 504-507

/R19/ RUSSEL, H.H.: FUTURE PROSPECTS OF FLY-ASH UTILIZATION. IN: COAL UTILISATION 10 (1956), S. 24-28

/R20/ RUSSEL, H.H.: FLY-ASH COLLECTION AND USE. IN: COAL UTILISATION 12 (1958), S. 25-27

/R21/ RUSSEL, H.H.: SUMMARY OF THE FLY-ASH DISPOSAL PROBLEM. IN: INT. AIR POLLUTION CONTR. ASS. 7 (1957), S. 46-47

/R22/ RAASK, E.: PULVERISED FUEL ASH CONSTITUENTS AND SURFACE CHARACTERISTICS IN CONCRETE APPLICATIONS. IN: PROCEEDINGS OF THE INTERNATIONAL SYMPOSIUM ON THE USE OF PFA IN CONCRETE, (J.G. CABRERA; A.R. CUSENS, EDS.), (1982), S. 5-16

/R23/ SEE/Y4/

/R24/ RAVINA, D.: EFFICIENT UTILIZATION OF COARSE AND FINE FLY ASH IN PRECAST CONCRETE BY INCORPORATING THERMAL CURING. IN: JOURNAL OF THE AMERICAN CONCRETE INSTITUTE 78 (1981), S. 194-200

/R25/ RAYMENT, P.L.: THE EFFECT OF PULVERIZED-FUEL ASH ON THE C/S MOLAR RATIO AND ALKALI CONTENT OF CALCIUM SILICATE HYDRATES IN CEMENT. IN: CEMENT AND CONCRETE RESEARCH 12 (1982), S. 133-140

/R26/ RAVINA, D.: PRODUCTION AND COLLECTION OF FLY ASH FOR USE IN CONCRETE. BOSTON : MATERIAL RESEARCH SOCIETY - IN: EFFECTS OF FLY-ASH INCORPORATION IN CEMENT AND CONCRETE : PROCEEDINGS SYMPOSIUM N ANNUAL MEETING (DIAMOND, S.(ED.)) BOSTON, NOVEMBER 16-18, 1981, S. 2-11

/R27/ RAVINA, D.: FLY ASH CONCRETE UNDER HOT WEATHER CONDITIONS. BOSTON : MATERIAL RESEARCH SOCIETY - IN: EFFECTS OF FLY-ASH INCORPORATION IN CEMENT AND CONCRETE : PROCEEDINGS SYMPOSIUM N ANNUAL MEETING (DIAMOND, S.(ED.)) BOSTON, NOVEMBER 16-18, S. 172-183

/R28/ RAMAKRISHNAN, V.;COYLE, W.;BROWN, J.;TLUSTUS, P.A.;VENKATARAMANUJAM, P.: PERFORMANCE CHARACTERISTICS OF CONCRETES CONTAINING FLY ASH. BOSTON : MATERIAL RESEARCH SOCIETY - IN: EFFECTS OF FLY-ASH INCORPORATION IN CEMENT AND CONCRETE : PROCEEDINGS SYMPOSIUM N ANNUAL MEETING (DIAMOND,S.(ED.)) BOSTON, NOVEMBER 16-18, 1981, S. 233-242

/R29/ RABA, C.F.; SMITH, R.L.; MEARING, M.: SUBBITUMINOUS FLY ASH UTILIZATION IN CONCRETE. BOSTON : MATERIAL RESEARCH SOCIETY - IN: EFFECTS OF FLY-ASH INCORPORATION IN CEMENT AND CONCRETE : PROCEEDINGS SYMPOSIUM N ANNUAL MEETING (DIAMOND,S,(ED.)) BOSTON, NOVEMBER 16-18, 1981, S. 296-305

/R30/ RONDAHL, J.: CARBONATION - EFFECT OF CEMENT TYPE (SWED.). IN: CEMENTA CM RAPPORT 82052 (1982)

/R31/ RAMIREZ, J.L.A.: ESTUDIO FISICO-QUIMICO Y TECNICO DE DIVERSOS TIPOS DE CENIZAS Y SU EMPLEO COMO MATERIAL DE CONSTRUCCION. MADRID : LABORATORIO CENTRAL DE ENSAYO DE MATERIALES DE CONSTRUCCION, PUBLICATION N. 199 (1969)

/R32/ ROZAK, W.: FLY ASH FROM COAL - A PROBLEM OF QUALITY. IN: CBL SPECIAL PUBLICATION (1980), NR. 7, S. 106-112

/R33/ RASMUSSEN, T.H.: LOW HEAT CONCRETE CONTAINING PFA, SILICA POWDER AND SUPERPLASTICIZER. IN: CBL INTERNAL REPORT NO. 28, 1981

/R34/ REICHERT, G.: HERSTELLUNG MODIFIZIERTER PORTLANDZEMENTE MIT FLUGASCHE UND HOCHOFENSCHLACKE IN DER NORWEGISCHEN ZEMENTINDUSTRIE. IN: ZEMENT-KALK-GIPS (ZKG) 35 (1982), NR. 12, S. 625-633, S. 634-640

/R35/ ROSSOUW, E. ; KRUEGER, J.: REVIEW OF SPECIFICATIONS FOR ADDITIONS FOR USE IN CONCRETE. DETROIT : AMERICAN CONCRETE INSTITUTE ACI SP-79, 1983 - IN: PROCEEDINGS OF THE CANMET/ACI FIRST INTERNATIONAL CONFERENCE ON THE USE OF FLY ASH, SILICA FUME, SLAG AND OTHER MINERAL BY-PRODUCTS IN CONCRETE VOL. 1 (1983), S. 201-220

/R36/ ROASH, E.: FLUGASCHE-CHARAKTERISTIKEN IN BEZUG AUF FOERDERUNG UND VERWERTUNG. VGB-KONFERENZ 1982: VERWERTUNG VON VERBRENNUNGSRUECKSTAENDEN IN: VGB-KRAFTWERKSTECHNIK GMBH, VGB-TB 202, S. 35-52

/R37/ RUTH, B.E. ; LARSEN, T.J.: SAVE MONEY WITH ECONOCRETE PAVEMENT SYSTEMS. IN: CONCRETE INTERNATIONAL 5 (1983), NR.5, S.23-32

/R38/ REGOURD, M. ; MORTUREUX, B. ; GAUTIER, E.: HYDRAULIC REACTIVITY OF VARIOUS POZZOLANAS. MEXICO : FACULTAD DE INGENIERIA CIVIL, UNIVERSIDAD AUTONOMA DE NUEVO LEON, 1984 IN: PROCEEDINGS OF THE 5TH INTERNATIONAL SYMPOSIUM ON CONCRETE TECHNOLOGY, S. 1-14

/R39/ REICHERT,W.W. ; CORSON,G. AND W.H.: ACTIVITIES OF THE ECONOMIC COMMISSION FOR EUROPE IN THE FIELD OF ASH UTILISATION. ASH UTILISATION, 1973, BUREAU OF MINES INFORMATION CIRCULAR 8640/1974, S.19-25

/R40/ REHSI, S.S.: STUDIES ON INDIAN FLY ASHES AND THEIR USE IN STRUCTURAL CONCRETE. ASH UTILISATION, 1973, BUREAU OF MINES INFORMATION CIRCULAR 8640/1974, S.231-245

/R41/ RUDZINSKI, L.: THE EFFECT OF FLY ASHES ON THE RHEOLOGICAL BEHAVIOUR OF CEMENT PASTES. IN: MATERIALS AND STRUCTURES 17 (1984), NR. 101

/R42/ RILEM COMMITTEE 73-SBC: FINAL REPORT : SILICEOUS BY-PRODUCTS FOR USE IN CONCRETE. IN: MATERIAUX ET CONSTRUCTION RILEM 21 (1988), NR. 121, S. 69-80

/R43/ RICHARTZ, W.: ZUSAMMENSETZUNG UND EIGENSCHAFTEN VON FLUGASCHEN. IN: ZEMENT-KALK-GIPS 37 (1984), NR. 2, S. 62-71

/R44/ ROSNER, J.C.: LET'S DESIGN FLY ASH CONCRETES, NOT COMPARE THEM. PROC. 4TH INTL. SYMP. ASH UTILIZATION, MARCH 1976, ST. LOUIS, ERDA, MERC/

SP-76/4, PP 560-572, U.S. BUREAU OF MINES, WASHINGTON, D.C., 1976

/R45/ REGOURD, M. ; HORNAIN, H. ; MORTUREUX, B.: MICROSTRUCTURE OF CONCRETE IN AGGRESSIVE ENVIRONMENTS. PHILADELPHIA : AMERICAN SOCIETY FOR TESTING AND MATERIALS, 1980 - IN: DURABILITY OF BUILDING MATERIALS AND COMPONENTS, PROCEEDINGS OF THE FIRST INTERNATIONAL CONFERENCE, ASTM STP 691, (SERE-DA, P.J. ; LITVAN, G.G.(ED)), S. 253-268

/R46/ REGOURD, M. ; HORNAIN, H. ; MORTUREUX, B.: INFLUENCE DU MODE DE CRISTALLISATION DE L'ALUMINATE TRICALCIQUE SUR LA RESISTANCE DES CIMENTS A L'EAU DE MER. IN: REVUE DES MATERIAUX DE CONSTRUCTION (1974), NR. 687, S. 69-79

/R47/ RAMACHANDRAN, V.S. ; FELDMAN, R.F. ; BEAUDOIN, J.J.: CONCRETE SCIENCE - TREATISE ON CURRENT RESEARCH. LONDON : HEYDEN & SON, 1981 IN: CONCRETE SCIENCE, S. 281-286, 306

/R48/ REGOURD, M.: PHYSICO-CHEMICAL STUDIES OF CEMENT PASTES, MORTARS, AND CONCRETES EXPOSED TO SEA WATER. DETROIT, MICHIGAN : AMERICAN CONCRETE INSTITUTE, 1980 (ACI SP-65) IN: PERFORMANCE OF CONCRETE IN MARINE ENVIRON-MENT, S. 63-82

/S1/ SAJBULATOW, S. ; KUATBAEW, K. ; RONTSCHINSKIJ, E.: DIE HERSTELLUNG VON ZIEGELN AUS KRAFTWERKSASCHE. IN: ZIEGELINDUSTRIE INTERNATIONAL 32 (1980), S. 579-581

/S2/ SAUMAN, Z.: STUDY OF REACTIONS BETWEEN CAO OR 3CAO.SIO2 AND BETA-2CAO. SIO2 AND POWER STATION FLY ASHES UNDER HYDROTHERMAL CONDITIONS. PROC.: 5TH INT. SYMPOS. ON CHEM. OF CEM., SUPPL. PAP. IV-17, TOKYO 1968

/S3/ SAUMAN, Z.: SIGNIFICANCE AND CHARACTER OF THE GLASS PHASE OF POWER STATION FLY ASHES. 11TH CONF. OF SILIC. IND., S. 461-473, BUDAPEST 1973

/S4/ SCHMIDT, H.: VERWERTUNG VON STEIN- UND BRAUNKOHLENASCHEN. IN: SPRECH-SAAL 105 (1972)

/S5/ SCHMIDT, H.: DIE HERSTELLUNG VON BETONSTEINEN AUS KRAFTWERKSAB-FAELLEN. IN: SILICATES INDUSTRIELS 41 (1976), S. 341-346

/S6/ SCHNEIDER, A.: AMERIKANISCHER PUZZOLANZEMENT. IN: BAU-MARKT 59 (1960), S. 1098-1106

/S7/ SCHOLZ, H.: DEUTSCHE STEINKOHLEN-VERBRENNUNGSRUECKSTAENDE : ARTEN UND VERWENDUNG. IN: BETONWERK UND FERTIGTEILTECHNIK 45 (1979), NR. 10, S. 595-601

/S8/ SCHOLZ, W.: EFA-FUELLER, EIN LATENTHYDRAULISCHER BETONZUSATZSTOFF NACH DIN 1045. IN: WASSER UND BODEN 32 (1980), S. 476-480

/S9/ SCHUBENZ, D.: NEUERE ERKENNTNIS FUER DIE HERSTELLUNG VON ZEMENTGEBUNDENEN TRAGSCHICHTEN. IN: BAUMASCHINE UND BAUTECHNIK 24 (1977), S. 62-76

/S10/ SCHUBERT, P. ; BERG, VOM, W.: STEINKOHLENFLUGASCHEN MIT PRUEFZEICHEN ALS BETONZUSATZSTOFF NACH DIN 1045 : EIGENSCHAFTEN VON FLUGASCHEN UND VON FLUGASCHEHALTIGEN MOERTELN UND BETONEN. IN: BETONWERK UND FERTIGTEILTECHNIK 45 (1979), NR.10, S. 602-608, NR. 11, S. 692-696

/S11/ SCHUBERT, P. ; JAEGERMANN, C.: VERSUCHE ZUR KURZZEITWARMBEHANDLUNG VON FLUGASCHEHALTIGEM BETON. IN: BETONWERK UND FERTIGTEIL-TECHNIK 43 (1977), S. 69-76

/S12/ SCHUBERT, P. ; LUEHR, H.-P.: ZUM SULFATWIDERSTAND FLUGASCHEHALTIGER MOERTEL UND BETONE. IN: BETONWERK UND FERTIGTEIL-TECHNIK 45 (1979), S. 177-182

/S13/ SCISLEWSKI, Z.: INFLUENCE DES CENDRES VOLANTES SUR LA PROTECTION DES ARMATURES DANS LE BETON. CAHIERS DU CENTRE SCIENTIFIQUE ET TECHNIQUE DU BATIMENT, NO 165 (1975)

/S14/ SERSALE, R.: STRUCTURE AND CHARACTERISATION OF POZZOLANAS AND OF FLY-ASHES. PARIS : EDITION SEPTIMA, 1980 - IN: PROCEEDINGS OF THE 7TH INTERNATIONAL CONGRESS ON THE CHEMISTRY OF CEMENT, VOLUME I: PRINCIPAL REPORTS, S. IV-1/3 - IV-1/18

/S15/ SIMONS, H.S. ; JEFFREY, J.W.: AN X-RAY STUDY OF PULVERIZED FUEL ASH. IN: JOURNAL OF APPLIED CHEMISTRY 10 (1960), S. 328-336

/S16/ SMITH, I.A.: THE DESIGN OF FLY ASH CONCRETE. PROCEEDINGS/

INSTITUTION OF CIVIL ENGINEERS (LONDON) V36 1967, S. 769-790

/S17/ SMITH, M.A.: THE ECONOMIC AND ENVIRONMENTAL BENEFITS OF INCREASED USE OF PFA AND GRANULATED SLAG. IN: RESOURCES POLICY 2 (1975), S. 154-170

/S18/ SMITH, M.A. ; HALLIWELL, F.: THE APPLICATION OF THE BS 4550 TEST FOR POZZOLANIC CEMENTS TO CEMENTS CONTAINING PULVERIZED-FUEL ASHES. IN: MAGAZINE OF CONCRETE RESEARCH 31 (1979), S. 159-170

/S19/ SNYDER, M.J.: FINAL REPORT ON FLY ASH UTILIZATION RESEARCH PROGRAM. IN: EDISON ELECTRIC INST. BULL. 12 (1966), S. 420-424

/S20/ STANGE, E. ; HAAS, H.: GASBETON AUS STEINKOHLENFLUGASCHE. IN: MITTEILUNGEN DER VGB 46 (1966), S. 145-148

/S21/ STINGLEY, W.M. ; PEYTON, R.L.: USE OF FLY ASH AS ADMIXTURE IN AN EXPERIMENTAL PAVEMENT IN KANSAS. SYMPOS. ON FLY ASH IN CONCR., 43RD ANN. MEET., 1964, HIGHWAY RES. REC., NR.73, S. 26-31, WASHINGTON 1965

/S22/ STOLNIKOV, V.V. ; KIND, V.V.: FLUGASCHE VON WAERMEKRAFTWERKEN ALS ZUSATZ ZU ZEMENT UND BETON (IN RUSS.). IN: STROITELNYE MATER. 7 (1961), S. 31-32

/S23/ STUART, K.D. ; ANDERSON, D.A. ; CADY, P.D.: COMPRESSIVE STRENGTH STUDIES ON PORTLAND CEMENT MORTARS CONTAINING FLY ASH AND SUPERPLASTICIZER. IN: CEMENT AND CONCRETE RESEARCH 10 (1980), NR. 6, S. 823-832

/S24/ SUTTON, C.A.: USE OF FLY ASH IN CONCRETE PAVEMENT CONSTRUCTED IN NEBRASKA. SYMPOS. ON FLY ASH IN CONCR., 43RD ANN. MEET., 1964, HIGHWAY RES. REC., NR.73, S. 1-12, WASHINGTON 1965

/S25/ SWAMY, R.N.: STRUCTURAL PROPERTIES OF FLY ASH CONCRETE. PROC.: INT. SYMPOS.: THE USE OF PFA IN CONCRETE, DEPT. CIV. ENG. LEEDS UNIV. 1982, VOL.1, S. 343-355

/S26/ SHEN, D.S. ; ZHANG, Y.J.: THE USE OF GROUND FLY ASH FOR THE PRODUCTION OF CEMENT OR CONCRETE. PROC.: INT. SYMPOS.: THE USE OF PFA IN CONCRETE, DEPT. CIV. ENG. LEEDS UNIV. 1982, VOL.1, S. 261-272

/S27/ SNYDER, M.J.: SPECIFICATIONS, LIMITATIONS, AND RESTRICTIONS. PROC.: EDISON ELECTRIC INSTITUTE, NATIONAL COAL ASSOC., BUREAU OF MINES SYMPOS., PITTSBURGH 1967, S. 37-45, COMPILED BY FABER, J.H.; CAPP, P.; SPENCER, J.D.

/S28/ SKAGGS, C.: PRODUCING SPECIFICATION FLY ASH. PROC.: EDISON ELECTRIC INSTITUTE, NATIONAL COAL ASSOC., BUREAU OF MINES SYMPOS., PITTSBURGH 1967, S. 52-57, COMPILBED BY FABER, J.H.; CAPP, P.; SPENCER, J.D.

/S29/ SEABRIGHT, J.: FLY ASH IN CONCRETE MANUFACTURING. PROC.: EDISON ELECTRIC INSTITUTE, NATIONAL COAL ASSOC., BUREAU OF MINES SYMPOS., PITTSBURGH 1967, S. 107-110, COMPILED BY FABER, J.H.; CAPP, P.; SPENCER, J.D.

/S30/ STOLNIKOV, V.V.: FLY ASH IN CEMENTS AND CONCRETES. PROC.: EDISON ELECTRIC INSTITUTE, NATIONAL COAL ASSOC., BUREAU OF MINES SYMPOS., PITTSBURGH 1967, S. 187-189, COMPILED BY FABER, J.H.; CAPP, P.; SPENCER, J.D.

/S31/ SHAFER, H.E. JR.: STATUS REPORT ON BRICKS FROM FLY ASH. PROC.: EDISON ELECTRIC INSTITUTE, NATIONAL COAL ASSOC., BUREAU OF MINES SYMPOS., PITTSBURGH 1967, S. 195-203, COMPILED BY FABER, J.H.; CAPP, P.; SPENCER, J.D.

/S32/ STEWART, R.F. ; FARRIOR, W.F. JR.: NUCLEAR MEASUREMENT OF CARBON IN FLY ASH. PROC.: EDISON ELECTRIC INSTITUTE, NATIONAL COAL ASSOC., BUREAU OF MINES SYMPOS., PITTSBURGH 1967, S. 262-270, COMPILED BY FABER, J.H.; CAPP, P.; SPENCER, J.D.

/S33/ SCHEETZ, B.E. ; STRICKLER, D.W. ; GRUTZECK, M.W. ; ROY, D.M.: PHYSICAL AND CHEMICAL BEHAVIOUR OF SELECTIVELY ETCHED FLY ASHES. BOSTON : MATERIAL RESEARCH SOCIETY - IN: EFFECT OF FLY ASH INCORPORATION IN CEMENT AND CONCRETE : PROCEEDINGS SYMPOSIUM N ANNUAL MEETING (DIAMOND,S.(ED.)) BOSTON, NOVEMBER 16-18, S. 24-33

/S34/ SEDNER, H.: KRAFTWERKS-DESIGN "MADE IN DENMARK". IN: DK2-3/82, S. 8-13

/S35/ SANMAMED DE, L.U.F.: UTILIZACION DE CENIZAS VOLANTES EN LA FABRICACION DE CEMENTOS. LICENCIADO EN CIENCIAS QUIMCIAS, UNIVERSIDAD DE SANTIAGO DE COMPOSTELA

/S36/ SMITH, M.A.: REVIEW OF STANDARD SPECIFICATIONS FOR FLY ASH FOR USE IN CONCRETE. GARSTON WATFORD : BUILDING RESEARCH ESTABLISHMENT ; BUILDING RESEARCH STATION, 1975 - IN: BUILDING RESEARCH ESTABLISHMENT CURRENT PAPER, NR. CP 8/75 (1975)

/S37/ SEE/S7/

/S38/ SCHULTZ, H. ; JENSEN, S.A. ; NEPPER-CHRISTENSEN, P.: FLY ASH AND CONCRETE. IN: BETONTEKNIK 1/03/1978, (DENMARK)

/S39/ SOERENSEN, E.V.: PROPERTIES OF CONCRETE WITH FLY-ASH CEMENT OR FLY-ASH ADDITION. INTERNATIONAL CONFERENCE ON SLAGS AND BLENDED CEMENTS, MONS, BELGIUM, SEPT. 1981

/S40/ SMITH, I.A.: A MODERN APPROACH TO FLY ASH IN CONCRETE. PROC. 1ST INT. CONF. ON READY MIXED CONCRETE, DUNDEE, OCT. 1975 EDITED BY DHIR, R.K.

/S41/ STOLTENBERG HANSSON, E. ; SOEPLER, B.: HERSTELLUNG MODIFIZIERTER PORTLANDZEMENTE MIT FLUGASCHE UND HOCHOFENSCHLACKE IN DER NORWEGISCHEN ZEMENTINDUSTRIE, TEIL 1: FORSCHUNG-ENTWICKLUNG-QUALITAETS-BETRACHTUNG-SYSTEMPLANUNG. IN: ZEMENT-KALK-GIPS 35 (1982), NR. 12, S. 625-633

/S42/ SORETZ, ST.: BEITRAG ZUM FROSTWIDERSTAND VON FLUGASCHEZEMENTEN. IN: MATERIAL UND TECHNIK 1980, NO. 3

/S43/ STOLTENBERG HANSSON, E.: DEVELOPMENT OF NORWEGIAN BLENDED CEMENTS (DAN.). IN: CBL SPECIAL PUBLICATION (1980), NR. 7, S. 21-23

/S44/ SOERENSEN, E.V.: INVESTIGATIONS AT THE CEMENT- AND CONCRETE LABORATORY CONCERNING THE USE OF FLY ASH (DAN.). IN: CBL SPECIAL PUBLICATION (1980), NR. 7, S. 90-95

/S45/ SEE/S39/

/S46/ SCHUBENZ, D.: HYDRAULISCH GEBUNDENE FLUGASCHE FUER TRAGSCHICHTEN VON VERKEHRSFLAECHEN - TECHNOLOGIE, TECHNIK, ERFAHRUNGEN -. VGB-KONFERENZ 1982: VERWERTUNG VOV VERBRENNUNGSRUECKSTAENDEN. IN: VGB-KRAFTWERKSTECHNIK GMBH, VGB-TB 202, S. 171-183

/S47/ SCHORR, K.: FROST-TAUSALZ-WIDERSTAND VON ZEMENTSTEIN AUS VERSCHIEDENEN ZEMENTARTEN UND MIT UNTERSCHIEDLICHEN BEIMENGUNGEN VON FLUGASCHEN. IN: BETONWERK UND FERTIGTEIL-TECHNIK 49 (1983), NR. 1, S. 16-21, NR. 2, S. 96-100

/S48/ STURRUP, V.R. ; HOOTON, R.D. ; CLENDENNING, T.G.: DURABILITY OF FLY ASH CONCRETE. DETROIT : AMERICAN CONCRETE INSTITUTE SP-79, 1983 - IN : PROCEEDINGS OF THE CANMET/ACI FIRST INTERNATIONAL CONFERENCE ON THE USE OF FLY ASH, SILICA FUME, SLAG AND OTHER MINERAL BY-PRODUCTS IN CONCRETE VOL. 1 (1983), S. 47-69

/S49/ SAMARIN, A. ; MUNN, R.L. ; ASHBY, J.B.: THE USE OF FLY ASH IN CONCRETE - AUSTRALIAN EXPERIENCE. DETROIT : AMERICAN CONCRETE INSTITUTE ACI SP-79, 1983 - IN: PROCEEDINGS OF THE CANMET/ACI FIRST INTERNATIONAL CONFERENCE ON THE USE OF FLY ASH, SILICA FUME, SLAG AND OTHER MINERAL BY-PRODUCTS IN CONCRETE VOL. 1 (1983), S. 143-172

/S50/ SULLENTRUP, M.G. ; BALDWIN, J.W.: HIGH LIME FLY ASH AS A CEMENTING AGENT. DETROIT : AMERICAN CONCRETE INSTITUTE ACI SP-79, 1983 - IN: PROCEEDINGS OF THE CANMET/ACI FIRST INTERNATIONAL CONFERENCE ON THE USE OF FLY ASH, SILICA FUME, SLAG AND OTHER MINERAL BY-PRODUCTS IN CONCRETE VOL. 1 (1983), S. 321-331

/S51/ SWAMY, R.N. ; ALI, S.A.R. ; THEODORAKOPOULOS, D.D.: ENGINEERING PROPERTIES OF CONCRETE COMPOSITE MATERIALS INCORPORATING FLY ASH AND STEEL FIBRES. DETROIT : AMERICAN CONCRETE INSTITUTE ACI SP-79, 1983 - IN: PROCEEDINGS OF THE CANMET/ACI FIRST INTERNATIONAL CONFERENCE ON THE USE OF FLY ASH, SILICA FUME, SLAG AND OTHER MINERAL BY-PRODUCTS IN CONCRETE VOL. 1 (1983), S. 559-588

/S52/ SCHMIDT, W.: AUFBEREITUNG UND WEITERVERWERTUNG VON FLUGASCHE EINSCHLIESSLICH SINTERUNG ZU LEICHTBAUZUSCHLAGSTOFFEN. VGB-KONFERENZ 1982: VERWERTUNG VON VERBRENNUNGSRUECKSTAENDEN. IN: VGB-KRAFTWERKSTECHNIK GMBH, VGB- TB 202, S. 155-161

/S53/ SCHOLZ, W. ; SCHOLZ, H.: DAUERHAFTIGKEIT VON BETON. IN: TIEFBAU, INGENIEURBAU, STRASSENBAU 26 (1984), NR. 7, S. 369-374

/S54/ SWAMY, R.N. ; LAMBERT, G.H.: MIX DESIGN AND PROPERTIES OF CONCRETE MADE FROM PFA COARSE AGGREGATES AND SAND. IN: THE INTERNATIONAL JOURNAL OF CEMENT COMPOSITES AND LIGHTWEIGHT CONCRETE 5 (1983), NR.4

/S55/ STERN, R.M. ; FURST, A.: METALLIC MUTAGENS IN FLY ASH. COPENHAGEN : THE DANISH WELDING INSTITUTE, 1982

/S56/ SLOANKER, J.F. ; LEONARD, J.W.: REVIEW OF CURRENT RESEARCH ON COAL ASH IN THE UNITED STATES. ASH UTILISATION, 1973, BUREAU OF MINES INFORMATION CIRCULAR 8640/1974, S.26-32

/S57/ SELMECZI, J.G. ; KNIGHT, R.G.: PROPERTIES OF POWERPLANT WASTE SLUDGES. ASH UTILISATION, 1973, BUREAU OF MINES INFORMATION CIRCULAR 8640/1974, S.123-138

/S58/ SMITH, P.H.: LARGE TONNAGE USES OF PFA IN ENGLAND AND OTHER EUROPEAN COUNTRIES. ASH UTILISATION, 1973, BUREAU OF MINES INFORMATION CIRCULAR 8640/1974, S.139-147

/S59/ STEOPOE: SUR LA DETERMINATION DE LA ACTIVITE HYDRAULIQUE DES POUZZOLANES. IN: REVUE DES MATERIAUX (1956), NR 492, S. 210-212

/S6O/ SMITH, R.L. ET AL: THE UTILIZATION OF CLASS C FLY ASH IN CONCRETE, THE CHALLENGE OF CHANGE. SIXTH INTERNATIONAL ASH UTILIZATION SYMPOSIUM PROCEEDINGS. RENO, NEVADA, 1982, PP 164-189

/S61/ SWAMY, R.N. ; ALI, S.A.R. ; THEODORAKOPOULOS, D.D.: EARLY STRENGTH FLY ASH CONCRETE FOR STRUCTURAL APPLICATIONS. IN: JOURNAL OF THE ACI 80 (1983), NR. 5, S. 414-423

/S62/ SAYAL, S.K. ; KATARIA, S.S.: EFFECT OF FLY ASHES ON DEVELOPMENT AND OPTIMIZATION OF BLENDED CEMENT AND CONCRETES, WORLD CEM. TECHNOL., 12 (1981), PP 461-468

/S63/ SAMARIN, A. ; RYAN, W.G.L.: EXPERIENCE IN USE OF ADMIXTURES IN CONCRETE CONTAINING CEMENT AND FLY ASH, PROC. WORKSHOP USE OF CHEMICAL ADMIXTURES IN CONCRETE CONTAINING CEMENT AND FLY ASH, DEC. 1975, UNIVERSITY OF NEW SOUTH WALES, PP 91-112, NEW SOUTH WALES, 1975

/S64/ SEMMELINK, C.J.: FIELD SURVEY ON THE EXTENT OF CRACKING AND OTHER DETAILS OF CONCRETE STRUCTURES SHOWING DETERIORATION DUE TO ALKALI-AGGREGATE REACTION IN THE SOUTH WESTERN CAPE PROVINCE, PROC. FIFTH INTERNATIONAL CONF. ON ALKALI-AGGREGATE REACTION IN CONCRETE, S.252/19, 12 PP

/S65/ SWENSON, E.G. ; GILLOTT, J.E.: CHARACTERISTICS OF KINGSTON CARBONATE ROCK REACTIONS, HIGHWAY RESEARCH BOARD, BULLETIN NO. 275, PP 18-31

/S66/ SWENSON, E.G. ; GILLOT, J.E.: ALKALI REACTIVITY OF DOLOMITIC LIMESTONE AGGREGATE. IN: MAGAZIN OF CONCRETE RESEARCH 19 (1969), NR. 59, S. 95-104

/S67/ STANTON, T.E.: STUDIES ON THE USE OF POZZOLANS FOR COUNTERACTING EXCESSIVE CONCRETE EXPANSION RESULTING FROM REACTION BETWEEN AGGREGATES AND ALKALIS IN CEMENT, ASTM STP 99 (1949), PP 178-301

/S68/ SCHIESSL, P.: CORROSION OF REINFORCEMENT, DURABILITY OF CONCRETE STRUCTURES. IN: CEB-RILEM INTERNATIONAL WORKSHOP, 1983, (ROSLAM, S.(ED)), S. 73-93

/S69/ STANTON, T.E: EXPANSION OF CONCRETE THROUGH REACTION BETWEEN CEMENT AND AGGREGATE, PROC. ASCE, VOL. 66 (1940), PP 1781-1812

/S70/ SHERWOOD, P.T.: THE USE OF WASTE AND LOW-GRADED MATERIALS IN ROAD CONSTRUCTION. 3. PULVERIZED ASH, DEP. OF ENVIRONMENT. TRRL REPORT NO. LR 686, CROWTHORNE, 1976

/S71/ SUTHERLAND, H.B. ET AL: ENGINEERING AND RELATED PROPERTIES OF PULVERIZED ASH. IN: JOURNAL INST. HIGHW. ENG. 5 (1968), NR. 6, PP 1-16

/T1/ TAKEMOTO, K. ; UCHIKAWA, H.: HYDRATION OF POZZOLANIC CEMENT. PARIS : EDITION SEPTIMA, 1980 - IN: PROCEEDINGS OF THE 7TH INTERNATIONAL CONGRESS ON THE CHEMISTRY OF CEMENT, VOLUME I: PRINCIPAL REPORTS, S. IV.2/1-IV.2/29

/T2/ TANNER, E.: UNTERBRINGUNG UND VERWERTUNG DER ASCHE BEI GROSS- KESSEL-ANLAGEN. IN: MITTEILUNGEN DER VGB 38 (1955), S. 773-784

/T3/ TERRIER, P. ; MOREAU, M.: RECHERCHE SUR LE MECANISME DE L' ACTION POUZZOLANIQUE DES CENDRES VOLANTES DANS LE CIMENT. IN: REVUE DES MATERIAUX DES CONSTRUCTION ET DE TRAVAUX PUBLICS NO 613 (1966), S. 379-396 UND NO 614 (1966), S. 440-451

/T4/ SEE/W2/

/T5/ TIMMS, A.G. ; GRIEB, W.E.: USE OF FLY ASH IN CONCRETE. IN: PUBLIC ROADS 29 (1957), S. 142-150

/T6/ TRAN-THANH-PHAT: LA DURETE DES BETONS DE CIMENTS AUX CENDRES. CENTRE D' ETUDES DE L' INDUSTRIE DES LIANTS HYDRAULIQUES. PUBL. NO 217, PARIS 1971

/T7/ TSUKAYAMA, R. ; ABE, H. ; NAGATAKI, S.: LONG-TERM EXPERIMENTS ON THE NEUTRALISATION OF CONCRETE MIXED WITH FLY ASH AND THE CORROSION OF REINFORCEMENT. 7TH INT. CONGR. ON CHEM. OF CEM. PARIS 1980, PROC.: VOL. III, S. IV-30-35 PARIS, 1981

/T8/ TEOREANU, I. ; NICOLESCU, L.D.: THE PROPERTIES OF POWER STATION FLY ASH CONCRETE. PROC.: INT. SYMPOS.: THE USE OF PFA IN CONCRETE, DEPT. CIV. ENG. LEEDS UNIV. 1982, VOL.1 S. 231-241

/T9/ TILLINGHAST, J.A.: FLY ASH AND THE ELECTRIC UTILITY INDUSTRY. PROC.: EDISON ELECTRIC INSTITUTE, NATIONAL COAL ASSOC., BUREAU OF MINES SYMPOS., PITTSBURGH 1967, S. 316-320, COMPILED BY FABER, J.H.; CAPP, P.; SPENCER, J.D.

/T10/ TOGNON, G. ; URSELLA, P.: COMBINED LIME AND SPECIFIC SURFACE AREA OF THE HYDRATION PRODUCTS OF LIME-POZZOLANA AND LIME-FLY ASH MIXES. BOSTON : MATERIAL RESEARCH SOCIETY - IN: EFFECTS OF FLY-ASH INCORPORATION IN CEMENT AND CONCRETE : PROCEEDINGS SYMPOSIUM N ANNUAL MEETING (DIAMOND, S. (ED.)) BOSTON, NOVEMBER, 1981, S. 145-154

/T11/ TUCKER, J.F.E.: BRENNEN VON ZIEGELN AUS BRENNSTOFFHALTIGEM ROHMATERIAL IM TUNNELOFEN. IN: ZIEGELINDUSTRIE INTERNATIONAL (1980), S. 9-11

/T12/ TURNSEK, V.: FLY ASH USED IN BUILDING CONSTRUCTION. IN: SAVETOVANJE GRADEVINSKIH STRUCNJAKA, LJUBLJANA REFERAT 12 (1961)

/T13/ THEISSING, E.M.: VLIEGAS IN BETON. IN: CEMENT (1982), S. 215-218

/T14/ TAEUBERT, U.: RESTSTOFFE AUS KOHLEKRAFTWERKEN FUER DIE BAUTECHNIK. IN: VGB KRAFTWERKSTECHNIK 62 (1982), NR. 11, S. 977-978

/T15/ TAEUBERT, U.: RESTSTOFFE AUS KOHLEKRAFTWERKEN FUER DIE BAUTECHNIK - VERWENDUNG VON FLUGSTAUB IM STRASSENBAU -. VGB-KONFERENZ 1982: VERWERTUNG VON VERBRENNUNGSRUECKSTAENDEN. IN: VGB-KRAFTWERKETECHNIK GMBH, VGB-TB 202, S. 184-207

/T16/ TAEUBERT, U.: FLUGASCHE ALS NEUER BAUSTOFF IM STRASSENBAU - TEIL 1. IN: STRASSEN- UND TIEFBAU 37 (1983), NR. 5, S. 24-27

/T17/ THORVALDSON, T.: CHEMICAL ASPECTS OF THE DURABILITY OF CEMENT PRODUCTS. 3RD. INTERN. SYMPOSIUM CHEM. CEMENT, PP 436-484, LONDON, 1952

/T18/ THORVALDSON, T. ; VIGFUSSON, V.A. ; WOLOCHOW, D.: THE ACTION OF SULFATES ON THE COMPONENTS OF PORTLAND CEMENT, TRANS. ROY. SOC. CANADA, 21, 1927, SEC. III, P 295

/T19/ TUTHILL, L.H.: RESISTANCE TO CHEMICAL ATTACK, SIGNIFICANCE OF TESTS AND PROPERTIES OF CONCRETE AND CONCRETEMAKING MATERIALS, ASTM SPECIAL TECHN. PUBL. 169B (1978), PP 369-387

/T20/ TYNDALL, C.J. ; MUNN, R.L.: DESIGN AND TESTING OF CONCRETE MIXES, BAYSWATER POWER STATION FINAL REPORT FOR THE ELECT. COMM OF NEW SOUTH WALES, QUOTATION SPEC, NO W247, OCTOBER 1981

/T21/ TRITTHART, J.: BEWEHRUNGSKORROSION - ZUR FRAGE DES CHLORIDBINDEVER-MOEGENS VON ZEMENT. IN: ZEMENT-KALK-GIPS 37 (1984), NR. 4, S. 200-204

/T22/ THORNTON, S.I. ; PARKER, D.G.: FLY ASH AS FILL AND BASE MATERIAL IN ARKANSAS HIGHWAYS. FINAL REPORT H.R. PROJECT 43, ARKANSAS STATE H. DEPT., 1975

/U1/ UCHIKAWA, H. ; UCHIDA, S. ; OGAWA, K.: INFLUENCE OF THE PROPERTIES OF FLY ASH ON THE FLUIDITY AND STRUCTURE OF FLY ASH CEMENT PASTE. PROC.: INT. SYMPOS.: THE USE OF PFA IN CONCRETE, DEPT. CIV.. ENG. LEEDS UNIV. 1982, VOL 1, S. 830-894

/U2/ UCHIKAWA, H. ; UCHIDA, S.: INFLUENCE OF POZZOLANA ON THE HYDRATION OF C3A. PARIS : EDITIONS SEPTIMA, 1980 - IN: 7TH INTERNATIONAL CONGRESS ON THE CHEMISTRY OF CEMENT, VOL. III (1980), S. IV.24-IV.29

/V1/ VAQUIER, A. ; CARLES-GIBERGUES, A.: SUR L' IMPORTANCE DES SULFATES DANS LE CARACTERE POUZZOLANIQUE D'UNE CENDRE VOLANTE SILICOALUMINEUSE DE CENTRALE THERMIQUE. IN: REVUE DES MATERIAUX DES CONSTRUCTION ET DE TRAVAUX PUBLICS NO 662 (1970), S. 331-337

/V2/ VENUAT, M.: CIMENT AUX CENDRES VOLANTES - INFLUENCE DE LA PROPORTION DE CENDRE SUR LES PROPRIETES DES CIMENTS. IN: CENTRE D' ETUDE ET DES RECHERCHES DE L' IND. DES LIANTS HYDRAUL., PUBL. TECH. NO 133, ALSO IN: REVUE DES MATERIAUX DES CONSTRUCTION NO 565, 566, 567 (1962), S.349FF

/V3/ VENUAT, M.: CIMENTS AUX CENDRES VOLANTES - INFLUENCE DE LA FINESSE DES CONSTITUANTS SUR LES PROPRIETES DES CIMENTS. IN: CENTRE D' ETUDE ET DES RECHERCHES DE L' IND. DES LIANTS HYDRAUL., PUBL. TECH. NO 165, ALSO IN: REVUE DES MATERIAUX DES CONSTR. NO 595 (1965), S. 208-212, NO 596 (1965), S. 235-250

/V4/ VENUAT, M.: LES CIMENTS CONTENANT DU LATIER OU DES CENDRES VOLANTES. IN: REVUE DES MATERIAUX DES CONSTRUCTION ET DE TRAVAUX PUBLICS NO 692 (1975), S. 30-35

/V5/ VENUAT, M. ; ALEXANDRE, J.: LE L'ETUDE DU COMPORTEMENT RHEOLOGIQUE DE QUELQUES CENDRES VOLANTES. PARIS : CENTRE D'ETUDES ET DE RECHERCHES DE L'INDUSTRIE DES LIANTS HYDRAULIQUES, 1966 - IN: REIHE : PUBLICATION TECHNIQUE, NR. 177

/V6/ VOYATZAKIS, E. ; SIPITANOS, K.M. ; CHRISTAKI-PAPAGEORGIOU, A.: ETUDE SUR LE CENDRES VOLANTES DES LIGNITES GRECS - UTILISATION DANS LE CIMENT - REGION DE PTOLEMAEIDE. IN: REVUE DES MATERIAUX DES CONSTRUCTION ET DE TRAVAUX PUBLICS, NO 703 (1976), S. 341-34

/V7/ VANHANEN, A.: POWER PLANT ASHES AND THEIR UTILIZATION, PART 3: USE IN AUTOCLAVECURED PRODUCTS (FIN.). TECHNICAL RESEARCH CENTRE OF FINLAND, CONCRETE AND SILICATE LABORATORY, REPORT 63

/V8/ VANHANEN, A.: POWER PLANT ASHES AND THEIR UTILIZATION, PART 4: USE AS CONSTITUENT OF MORTAR (FIN.). TECHNICAL RESEARCH CENTRE OF FINLAND, CONCRETE AND SILICATE LABORATORY, REPORT 64

/V9/ VANHANEN, A.: POWER PLANT ASHES AND THEIR UTILIZATION PART 9: OTHER WAYS OF UTILIZATION (FIN.). TECHNICAL RESEARCH CENTRE OF FINLAND, CONCRETE AND SILICATE LABORATORY, REPORT 67

/V10/ VERHASSELT, A.: CARACTERISATION DE LA POUZZOLANICITE DE CENDRES VOLANTES. IN: COMPTES RENDUS DU 7EME CONGRES INTERNATIONAL DE LA CHIMIE DES CIMENTS VOLUME III - COMMUNICATIONS (SUITE), PP.IV - 116 A 121, PARIS, 1980

/V11/ VIRTANEN, J.: FREEZE-THAW RESISTANCE OF CONCRETE CONTAINING BLAST-FURNACE SLAG, FLY ASH OR CONDENSED SILICA FUME. PROC.: THE USE OF FLY ASH, SILICA FUME, SLAG AND OTHER MINERAL BY-PRODUCTS IN CONCRETE; MONTEBELLO, 1983 ACI PUBLICATION SP-79, VOL. II, PP. 923-942

/V12/ VAMOS, G.: TENDENZEN DER FLUGASCHEVERWERTUNG IN UNGARN. VGB-KONFERENZ 1982: VERWERTUNG VON VERBRENNUNGSRUECKSTAENDEN. IN: VGB-KRAFTWERKSTECHNIK GMBH, VGB-TB 202, S. 67-72

/V13/ VINCENT, R.D. ; MATEOS, M. ; DAVIDSON, D.T.: VARIATION IN POZZOLANIC BEHAVIOR OF FLY ASHES. PHILADELPHIA : AMERICAN SOCIETY FOR TESTING AND MATERIALS, 1961 - IN: PROCEEDINGS OF THE AMERICAN SOCIETY FOR TESTING AND MATERIALS 61 (1961), S. 1094-1116

/V14/ VENUAT, M.: FLY ASH AND ITS USE IN CEMENT AND CONCRETE. BUILDING RESEARCH ESTABLISHMENT, DEPARTMENT OF THE ENVIRONMENT, LIBRARY TRANSLATION 1841, 1974, 14P

/V15/ VIVIAN, H.E.: STUDIES IN CEMENT-AGGREGATE REACTION, CSIRO BULLETIN NR. 256 (1950), PP 21-230

/V16/ VAZQUEZ, E. ; ALONSO, E.E.: FLY ASH STABILIZATION OF DECOMPOSED GRANITE. PROC. X. INT. CONF. SOIL MECH. AND FOUND. ENG. VOL. 2, P 391-397, STOCKHOLM, 1981

/V17/ VISKOCHIL, R.K. ET AL: EFFECT OF DENSITY ON STRENGTH OF LIME-FLY ASH STABILIZED SOIL. HIGHWAY RESEARCH BOARD BULLETIN (1957), NR. 183

/V18/ VAZQUEZ, E. ; ROVIRA, J. ; BASTUS, E. ; YAGUE, E.: APPLICATIONS OF FLY ASH FROM LIGNITE TO CONSTRUCTION MATERIALS DIFFERENT FROM CONCRETE. 1ST INT. CONF. ON THE USE OF FLY ASH..., MONTEBELLO, 1983

/W1/ WATT, J.D. ; THORNE, D.J.: COMPOSITION AND POZZOLANIC PROPERTIES OF PULVERISED FUEL ASHES. I. COMPOSITION OF FLY ASHES FROM SOME BRITISH POWER STATIONS AND PROPERTIES OF THEIR COMPONENT PARTICLES. IN: JOURNAL OF APPLIED CHEMISTRY 15 (1965), S. 585-595

/W2/ WATT, J.D. ; THORNE, D.J.: COMPOSITION AND POZZOLANIC PROPERTIES OF PULVERISED FUEL ASHES. II. POZZOLANIC PROPERTIES OF FLY ASHES, AS DETERMINED BY CRUSHING STRENGTH TESTS ON LIME MORTARS. IN: JOURNAL OF APPLIED CHEMISTRY 15 (1965), S. 595-604

/W3/ WATT, J.D. ; THORNE, D.J.: THE COMPOSITION AND POZZOLANIC PROPERTIES OF PULVERISED FUEL ASHES. III. POZZOLANIC PROPERTIES OF FLY ASHES AS DETER-

MINED BY CHEMICAL METHODS. IN: JOURNAL OF APPLIED CHEMISTRY 16 (1966), S. 33-39

/W4/ WIERIG, H.-J.: UNTERSUCHUNGEN UEBER DIE FROSTBESTAENDIGKEIT VON ZEMENTMOERTEL MIT FLUGASCHEZUSAETZEN. IN: BETON 20 (1970), NR. 6, S.233-246

/W5/ WOGRIN, A.: VERWENDBARKEIT DER OESTERREICHISCHEN FLUGASCHE ALS TEILWEISES ZEMENTERSATZMITTEL. IN: OESTERREICHISCHE INGENIEUR ZEITSCHRIFT 3 (1960), S. 403-408

/W6/ WOOLGAR, G. ; OATES, D.B.: FLY ASH AND THE READY-MIXED CONCRETE PRODUCER. IN: CONCRETE INTERNATIONAL (1979), S. 34-40

/W7/ WILLIAMS, J.T. ; OWENS, P.L.: THE IMPLICATIONS OF A SELECTED GRADE OF UNITED KINGDOM PULVERIZED FUEL ASH ON THE ENGINEERING DESIGN AND USE IN STRUCTURAL CONCRETE. PROC.: INT. SYMPOS.: THE USE OF PFA IN CONCRETE, DEPT. CIV. ENG. LEEDS UNIV. 1982, VOL.1, S. 301-311

/W8/ WOOLLEY, G.R.: CONSTRUCTION OF STRUCTURES AT DRAX POWER STATION USING PFA CONCRETE. PROC.: INT. SYMPOS.: THE USE OF PFA IN CONCRETE, DEPT. CIV. ENG. LEEDS UNIV. 1982, VOL.1, S. 313-321

/W9/ WEIGLER, H.: BETON EIN DAUERHAFTER BAUSTOFF - ANFORDERUNGEN UND MOEGLICHKEITEN. IN: DAS BAUZENTRUM (1981), S. 9-12

/W10/ WESCHE, K. ; SCHUBERT, P.: ZUMAHLSTOFFZEMENTE UND BETONZUSATZSTOFFE. IN: BETONTECHNIK 3 (1982), NR. 3, S. 80-84

/W11/ WESCHE, K. ; BERG, VOM, W.: PROPERTIES OF FLY ASH USED IN GERMANY. BOSTON: MATERIAL RESEARCH SOCIETY - IN: EFFECTS OF FLY-ASH INCORPORATON IN CEMENT AND CONCRETE : PRODEEDINGS SYMPOSIUM N ANNUAL MEETING (DIAMOND,S.(ED.)) BOSTON, NOVEMBER 16-18, 1981, S. 45-52

/W12/ WESCHE, K. ; SCHUBERT, P.: PROPERTIES OF MORTARS AND CONCRETE MADE WITH FLY ASH. BOSTON : MATERIAL RESEARCH SOCIETY - IN: EFFECTS OF FLY-ASH INCORPORATION IN CEMENT AND CONCRETE : PROCEEDINGS SYMPOSIUM N ANNUAL MEETING (DIAMOND,S.(ED.)) BOSTON, NOVEMBER 16-18, 1981 S. 225-232

/W13/ WISCHERS, G. ; RICHARTZ, W.: EINFLUSS DER BESTANDTEILE UND DER GRANULOMETRIE DES ZEMENTS AUF DAS GEFUEGE DES ZEMENTSTEINS. DUESSELDORF : BETON- VERLAG, 1984 - IN: BETONTECHNISCHE BERICHTE 1982/83, S. 61-93 ALSO IN: BETON 32 (1982), NR. 9, S. 337-339, NR. 10, S. 379-385

/W14/ WARRIS, B.: STRENGTH OF CONCRETE CONTAINING SECONDARY CEMENTING MATERIALS. DETROIT : AMERICAN CONCRETE INSTITUTE ACI SP-79, 1983 - IN: PROCEEDINGS OF THE CANMET/ACI FIRST INTERNATIONAL CONFERENCE ON THE USE OF FLY ASH, SILICA FUME, SLAG AND OTHER MINERAL BY-PRODUCTS IN CONCRETE VOL. 1 (1983), S. 539-557

/W15/ WISCHERS, G.: FLUGASCHE ALS PUZZOLANISCHER BESTANDTEIL DES ZEMENTS. VGB-KONFERENZ 1982: VERWERTUNG VON VERBRENNUNGSRUECKSTAENDEN. IN: VGB-KRAFTWERKSTECHNIK GMBH, VGB-TB 202, S. 86-100

/W16/ WARRIS, B.: THE CONTRIBUTION OF CONCRETE TO REINFORCEMENT PROTECTION AGAINST CORROSION AND RESISTANCE TO WEATHERING. ERMCO/1983, BRITISH READY-MIXED CONCRETE ASSOCIATION, LONDON: PROGRESS IN CEMENTS TO IMPROVE THE COMPETITIVENESS OF CONCRETE

/W17/ WATSON PEDLAR, J.: CENOSPHERES. ASH UTILISATION, 1973, BUREAU OF MINES INFORMATION CIRCULAR 8640/1974, S.33-43

/W18/ WILSON, C.F.: A PRACTICAL APPROACH TO PRODUCING PUMPABLE CONCRETE. ASH UTILISATION, 1973, BUREAU OF MINES INFORMATION CIRCULAR 8640/1974, S.94-106

/W19/ WHITTEMORE, R.C.: AN EVALUATION OF THE ADSORPTIVE PROPERTIES OF FLY ASH WITH REFERENCE TO A PULP AND PAPER MILL WASTE EFFLUENT. ASH UTILISATION, 1973, BUREAU OF MINES INFORMATION CIRCULAR 8640/1974, S.296-317

/W20/ WESCHE, K.: BAUSTOFFE FUER TRAGENDE BAUTEILE. BAND 2: BETON. 2. AUFL. WIESBADEN : BAUVERLAG, 1981 -

/W21/ WARD, M.A.: DURABILITY OF CONCRETES CONTAINING FLY ASH. PRESENTED AT ACI SEMINAR ON UTILIZATION OF FLY ASH, CALGARY, NOV. 11,12, 1981

/W22/ WOOD, H.: DURABILITY OF CONCRETE CONSTRUCTION, ACI MONOGRAPH (1968), NR. 4, PP 57-81

/W23/ WALKER, H.N.: CHAPTER 41: CHEMICAL REACTIONS OF CARBONATE AGGREGATES IN CEMENT PASTE, SIGNIFICANCE OF TESTS AND PROPERTIES OF CONCRETE MAKING MATERIALS, ASTM STP 168B (1978), PP 722-743

/X1/ ASTM C-311-85: METHODS OF SAMPLING AND TESTING FLY ASH OR NATURAL POZZOLANS FOR USE AS A MINERAL ADMIXTURE IN PORTLAND CEMENT CONCRETE.

/X2/ ASTM C-593-85: SPECIFICATIONS FOR FLY ASH AND OTHER POZZOLANS FOR USE WITH LIME.

/X3/ ASTM C-618-85: SPECIFICATION FOR FLY ASH AND RAW OR CALCINED NATURAL POZZOLAN FOR USE AS A MINERAL ADMIXTURE IN PORTLAND CEMENT CONCRETE.

/X4/ BS 3892-1982: PART1: SPECIFICATION FOR PULVERIZED-FUEL ASH FOR USE AS A CEMENTITIOUS COMPONENT IN STRUCTURAL CONCRETE.

/X5/ DEUTSCHES AMT FUER MESSWESEN UND WARENPRUEFUNG: FILTERASCHE ALS ZUSATZSTOFF FUER SCHWERBETON UND LEICHTZUSCHLAGSTOFFBETON. DAMWVORSCHRIFT WARENPRUEFUNG (1961)

/X6/ RICHTLINIE FUER DIE DURCHFUEHRUNG DER UEBERWACHUNG DER HERSTELLUNG VON STEINKOHLENFLUGASCHE ALS BETONZUSATZSTOFF NACH DIN 1045 (UEBERWACHUNGS- RICHTLINIE). IN: MITTEILUNGEN DES INSTITUTES FUER BAUTECHNIK 11 (1980), S.39-43

/X7/ RICHTLINIE FUER DIE ERTEILUNG VON PRUEFZEICHEN FUER STEINKOHLEN- FLUGASCHE ALS BETONZUSATZSTOFF NACH DIN 1045 (PRUEFZEICHENRICHTLINIE). IN: MITTEILUNGEN DES INSTITUTES FUER BAUTECHNIK 11 (1980), S.43-45

/X8/ TS 639: PORTLAND CIMENTOSU KLINKERINE VE PORTLAND CIMENTOSU BETONUNA KARISTIRILAK UCUCU KUELLER. TEMMUZ 1968

/X9/ TS 640: UCUCU KUELLUE CIMENTO. TEMMUZ 1968

/X10/ SEE/X4/

/X11/ OENORM B 3319, AUSGABE MAI 1962: FLUGASCHE ALS HYDRAULISCHE ZUMAHLKOMPONENTE FUER DIE ZEMENTERZEUGUNG.

/X12/ SEE/X1/

/X13/ JIS A 6201 - 1977: JAPANESE INDUSTRIAL FLY ASH STANDARD.

/X14/ OENORM B 3319, AUSGABE OKTOBER 1981: ZUMAHLFLUGASCHE FUER DIE ZEMENTERZEUGUNG. (S./X15/)

/X15/ OENORM B 3320, AUSGABE MAERZ 1982: ZUSATZFLUGASCHE FUER DIE BETON- ERZEUGUNG.

/X16/ DS 411: SUPPLEMENT TO THE DANISH CODE OF PRACTICE FOR THE STRUCTURAL USE OF CONCRETE (DS 411) CONCERNING THE USE OF POZZOLANS (PARTIAL TRANSLATION IN ENGL.). AALBORG PORTLAND, 1979

/X17/ GUIDELINE FOR FEDERAL PROCUREMENT OF CEMENT AND CONCRETE CONTAINING FLY ASH. IN: ENVIROMENTAL PROTECTION AGENCY, RULES AND REGULATIONS, WASHINGTON, 40 CFR PART 249, FEDERAL REGISTER, VOL. 48, NO.20, 28.1.1983

/X18/ BRITISH STANDARDS INSTITUTION: CONCRETE: SPECIFICATIONS, PRODUCTION AND CONSTRUCTION. THE STRUCTURAL USE OF CONCRETE, PART 1, SECTION 6, TECHNICAL COMMITTEE CSB/39

/X19/ STANDARDS ASSOCIATION OF AUSTRALIA: FLY ASH FOR USE IN CONCRETE. AUSTRALIAN STANDARDS 1129 AND 1130-1971

/X20/ INDIAN STANDARD - 3812-1966: SPECIFICATION FOR FLY ASH: PART I: FOR USE AS POZZOLANA. PART II: FOR USE AS ADMIXTURE FOR CONCRETE. INDIAN STANDARDS INSTITUTION, NEW DELHI

/X21/ NEDERLANDS NORMALISATIE INSTITUUT: TOEPASSING VAN PORTLANDVLIEGASCEMENT EN VLIEGAS IN BETON. NORMCOMMISSIE 35307 "CEMENT"

/X22/ ACI COMMITTEE 212: ADMIXTURES FOR CONCRETE AND GUIDE FOR USE OF ADMIXTURES IN CONCRETE. DETROIT : AMERICAN CONCRETE INSTITUTE, 1981

/X23/ ASTM C227-87: STANDARD TEST METHOD FOR POTENTIAL ALKALI REACTIVITY OF CEMENT-AGGREGATE COMBINATIONS (MORTAR-BAR METHOD). IN: ANNUAL BOOK OF ASTM STANDARDS (1990), VOL. 04.02, S. 124-128

/X24/ ASTM C157-89: STANDARD TEST METHOD FOR LENGTH CHANGE OF HARDENED HYDRAULIC-CEMENT MORTAR AND CONCRETE. IN: ANNUAL BOOK OF ASTM STANDARDS (1990), VOL. 04.02, S. 99-104

/X25/ ASTM C586-69 (REAPPROVED 1986): STANDARD TEST METHOD FOR POTENTIAL ALKALI REACTIVITY OF CARBONATE ROCKS FOR CONCRETE AGGREGATES (ROCK CYLINDER METHOD). IN: ANNUAL BOOK OF ASTM STANDARDS (1990), VOL. 04.02, S. 282-285

/X26/ ASTM C 289-87: STANDARD TEST METHOD FOR POTENTIAL REACTIVITY OF AGGREGATES (CHEMICAL METHOD). IN: ANNUAL BOOK OF ASTM STANDARDS (1990), VOL. 04.02, S. 159-165

/X27/ ASTM C441-89: STANDARD TEST METHOD FOR EFFECTIVENESS OF MINERAL ADMIXTURES OR GROUND BLAST-FURNACE SLAG IN PREVENTING EXCESSIVE EXPANSION OF CONCRETE DUE TO THE ALKALI-SILICA REACTION. IN: ANNUAL BOOK OF ASTM STANDARDS (1990), VOL. 04.02, S. 224-226

/X28/ DIN 4030 11.69. BEURTEILUNG BETONANGREIFENDER WAESSER, BOEDEN UND GASE
DIN 4030 TEIL 1 ENTWURF 12.89. BEURTEILUNG BETONANGREIFENDER WAESSER, BOEDEN UND GASE; GRUNDLAGEN UND GRENZWERTE
DIN 4030 TEIL 2 ENTWURF 12.89. -; ENTNAHME UND ANALYSE VON WASSER- UND BODENPROBEN

/X29/ ACI COMMITTEE 201, 1984: GUIDE TO DURABLE CONCRETE. IN: ACI MANUAL OF CONCRETE PRACTICE, PP 201.2R-24 - 2O1.2R-32

/X30/ JIS A 6204: CHEMICAL ADMIXTURE FOR CONCRETE

/X31/ ASTM C666-84: STANDARD TEST METHOD FOR RESISTANCE OF CONCRETE TO RAPID FREEZING AND THAWING. IN: ASTM ANNUAL BOOK OF ASTM STANDARDS (1990), VOL. 04.02, S. 314-319

/X32/ ASTM C671-86: STANDARD TEST METHOD FOR CRITICAL DILATION OF CONCRETE SPECIMENS SUBJECTED TO FREEZING. IN: ANNUAL BOOK OF ASTM STANDARDS (1990), VOL 04.02, S. 327-331

/X33/ ASTM C672-84: STANDARD TEST METHOD FOR SCALING RESISTANCE OF CONCRETE SURFACES EXPOSED TO DEICING CHEMICALS. IN: ANNUAL BOOK OF ASTM STANDARDS (1990), VOL. 04.02, S. 332-334

/X34/ ASTM C457-82: MICROSCOPICAL DETERMINATION OF AIR-VOID CONTENT AND PARAMETERS OF THE AIR-VOID SYSTEM IN HARDENED CONRETE. IN: ANNUAL BOOK OF ASTM STANDARDS (199O), VOL. 04.02, S. 227-237

/X35/ TGL 28104/17 12.1973: PRUEFUNG VON ZEMENTEN; CHEMISCHE PRUEFUNG; BESTIMMUNG DER ALKALIEN

/X36/ TGL 32363/03 09.1976: PRUEFUNG FESTER BRENNSTOFFE; BESTIMMUNG DER CHEMISCHEN ZUSAMMENSETZUNG VON ASCHE; PHOTOMETRISCH-TITRIMETRISCHES VERFAHREN

/X37/ TGL 23157/12.1970: PROBENAHME UND PROBENVORBEREITUNG VON FESTEN SCHUETTFAEHIGEN, PASTENFOERMIGEN UND FLUESSIGEN STOFFEN - ALLGEMEINE FESTLEGUNGEN

/Y1/ YUAN, R.L. ; COOK, J.E.: TIME-DEPENDENT DEFORMATION OF HIGH STRENGTH FLY ASH CONCRETE. PROC.: INT. SYMPOS.: THE USE OF PFA IN CONCRETE, DEPT. CIV. ENG. LEEDS UNIV. 1982, VOL.1, S. 255-260

/Y2/ YAMATO, T. ; SUJITA, H.: SHRINKAGE AND CREEP OF MASS CONCRETE CONTAINING FLY ASH. DETROIT : AMERICAN CONCRETE INSTITUTE, ACI SP-79, 1983 - IN: PROCEEDINGS OF THE CANMET/ACI FIRST INTERNATIONAL CONFERENCE ON THE USE OF FLY ASH, SILICA FUME, SLAG AND OTHER MINERAL BY-PRODUCTS IN CONCRETE VOL. 1 (1983), S. 87-102

/Y3/ YUAN, R.L. ; COOK, J.E.: STUDY OF A CLASS C FLY ASH CONCRETE. DETROIT: AMERICAN CONCRETE INSTITUTE SP-79, 1983 - IN: PROCEEDINGS OF THE CANMET/ACI FIRST INTERNATIONAL CONFERENCE ON THE USE OF FLY ASH, SILICA FUME, SLAG AND OTHER MINERAL BY-PRODUCTS IN CONCRETE VOL. 1, S. 307-319

/Y4/ YUAN, RUN-ZHAN ; ZHU, JIE-AN ; ZHANG, LI-YUN: COMPOSITION AND STRUCTURE OF FLY ASHES AND THEIR PUZZOLANIC REACTIVITY. LEEDS : DEPARTMENT OF CIVIL ENGINEERING, 1982 - IN: PROCEEDINGS OF THE INTERNATIONAL SYMPOSIUM ON THE USE OF PFA IN CONCRETE, S. 61-69

/Z1/ ZMACHINSKY, A.E. ; LYASHKEVICH, I.M. ; CHERNAYA, L.G.: PRODUCTION OF GYPSUM CONCRETE WITH PFA INGREDIENTS. PROC.: INT. SYMPOS.: THE USE OF PFA IN CONCRETE, DEPT. CIV. ENG. LEEDS UNIV. 1982, VOL.1, S. 273-276

/Z2/ ZIMBELMANN, R.: VERWENDUNG VON FLUGASCHE ZUR BETONHERSTELLUNG. IN: BETONWERK UND FERTIGTEIL-TECHNIK (1983), NR. 11, S. 705-709

/Z3/ ZIVICA, V.: THE CORROSION OF MORTARS DUE TO THE ACTION OF MGSO4 SOLUTIONS. IN: RILEM, INT. SYMPOSIUM; DURABILITY OF CONCRETE - 1969 FINAL REPORT, PART II, C72-C82

/Z4/ ZIVICA, V.: UEBER DIE KORROSIVE WIRKUNG VON CALCIUMNITRAT-LOESUNGEN AUF ERHAERTETE ZEMENTPASTE. IN: ZEMENT-KALK-GIPS (1971), NR.4, S. 175-179

/Z5/ ZALTZMAN, R.: LET'S GET RID OF "THE DUMP". ASH UTILISATION, 1973, BUREAU OF MINES INFORMATION CIRCULAR 8640/1974, S.280-295

/Z6/ ZIMMER, F.V.: FLY ASH A BITUMINOUS FILLER. BUREAU OF MINES INF. CIRCULAR, 197

APPENDIX

RILEM RECOMMENDATIONS TC FAB-67 USE OF FLY ASH IN BUILDING

Fly ash in concrete – Test methods

FAB 1: Test methods for determining the properties of fly ash

FAB 2: Test methods for determining the properties of fly ash in concrete

Materials and Structures/Matériaux et Constructions, 1989, **22**, 299–308

RILEM RECOMMENDATIONS
RECOMMANDATIONS DE LA RILEM
TC FAB-67 USE OF FLY ASH IN BUILDING
CT 67-FAB UTILISATION DES CENDRES VOLANTES DANS LA CONSTRUCTION

Fly ash in concrete – Test methods

K. WESCHE, *Chairman*
Aachen, Germany
I. L. ALONSO
Madrid, Spain
I. BIJEN
Maastricht, Netherlands
P. SCHUBERT
Aachen, Germany
W. VOM BERG
Aachen, Germany
R. RANKERS (Co-Worker)
Aachen, Germany

The Recommendations presented here were first issued as drafts for comment in July 1989. They have been finalised after taking into consideration the comments submitted.

FAB 1: Test methods for determining the properties of fly ash

CONTENTS

1. SCOPE

This test guideline relates to the properties of fly ashes for use in cement, mortar and concrete. It does not establish specifications, which are to be drawn from the national and international standards.

2. DEFINITION

Fly ash according to this Recommendation is a fine powder of mainly spherical, glassy particles having pozzolanic properties and consisting essentially of SiO_2 and Al_2O_3. Fly ash is obtained by electrostatic or mechanical precipitation of dust-like particles from the flue gases of furnaces fired with pulverized coal.

3. SAMPLE SIZE

The sample should be representative for the test purpose. The taking of a sample of at least 4 kg for complete testing is recommended. From this sample a laboratory sample of at least 1 kg is obtained by subdividing, such as quartering.

4. CHEMICAL COMPOSITION

4.1 General

The Recommendation describes the reference procedures and, in certain cases, an alternative method which can be considered as giving equivalent results. If other methods are used it is necessary to show that they give results equivalent to those given by the reference methods. In the case of a dispute, only the reference procedures are used.

The chemical composition and the moisture content shall be expressed as proportions by mass of dry ash.

4.2 Preparation of samples

The laboratory sample is to be divided into a fraction of approximately 100 g. The moisture content is then to be determined according to section 4.3. The dried sample is sieved on a 90-μm sieve (63 μm for determination of free calcium oxide) according to ISO 565 and the residue fined down by grinding until the complete sample passes the mesh. The sample is then homogenized, dried to constant mass at 105 ± 5°C and subsequently kept under airtight conditions above a drying agent in a desiccator.

4.3 Moisture content

Procedure

Spread approximately 10 g (weighed to an accuracy of 0.1 mg (m_1)) of one fraction of the original sample of fly ash in a flat dish and dry to constant mass in a well-ventilated oven at 105 ± 5°C. After cooling in a desiccator, reweigh the sample (m_2).

Expression of results

The following equation is used to calculate the moisture content h_m related to the dried fly ash:

$$h_m = \frac{m_1 - m_2}{m_2} \times 100 \text{ [\% by mass]}$$

where
m_1 = original mass of wet fly ash in g,
m_2 = final mass of dried fly ash in g.

4.4 Loss on ignition

Procedure

Place approximately 1 g (weighed to an accuracy of 0.1 mg) of the dried, fined fly ash (m_3) in a porcelain crucible previously raised to red heat and heat in an oven at 975 ± 25°C for 60 min. After cooling in a desiccator, reweigh the sample (m_4).

Expression of results

The following equation is used to calculate the loss on ignition (LOI) of dried fly ash:

$$\text{LOI} = \frac{m_3 - m_4}{m_3} \times 100 \text{ [\% by mass]}$$

where
m_3: original mass of dried fly ash,
m_4: final mass of calcined fly ash.

4.5 Silicon Oxide (SiO_2)

4.5.1 Total Silicon Oxide

Procedure

Grind approximately 0.5 g (weighed to an accuracy of 0.1 mg) of the dried, fined fly ash (m_5) in an agate mortar with 6 to 10 g of decomposition mixture comprising 1 part anhydrous sodium carbonate (Na_2CO_3) to 1.3 parts anhydrous potassium carbonate (K_2CO_3). Heat the mixture in a 50 ml platinum crucible. Melt the mixture, slowly at first, and then to red heat (approximately 1100°C), until the melt flows evenly. During cooling, use tongs to agitate the platinum crucible so that the melt solidifies in a thin layer on the inner surface of the platinum crucible. After complete cooling, place the platinum crucible together with its contents in a 500 ml porcelain casserole and cover with water. Heat the casserole. Remove the platinum crucible from the casserole and spray with dilute hydrochloric acid (HCl 1 + 3). Unite the washing water with the contents of the porcelain casserole and carefully add 20 ml of concentrated hydrochloric acid.

Evaporate the contents of the porcelain casserole to dryness and leave in the heating cabinet for two hours at 135°C. Digest the evaporate in 50 ml concentrated hydrochloric acid at 80°C, dilute the mixture with 150 ml hot water, heat and filter through a medium-textured filter. Wash out the filter sediment three times with hot dilute hydrochloric acid (HCl 1 + 1) and once with hot water. Evaporate the filtrate and treat the evaporate in the same way. Please note that any cloudiness in the filtrate due to titanium (IV) oxyhydrate is to be dissolved by addition of concentrated hydrochloric acid and heating.

Incinerate the two filters with the contaminated silicic acid in a platinum crucible and heat the residue to constant mass at 1100 ± 50°C. After cooling in a desiccator, weigh the platinum crucible with the residue to an accuracy of 0.1 mg (m_6). Moisten the residue in the platinum crucible with water and fume with 5 drops of concentrated sulphuric acid and 5 to 10 ml hydrofluoric acid. Heat the crucible with the fumed residue at 1100 ± 50°C and weigh to an accuracy of 0.1 mg (m_7) after cooling in a desiccator. Decompose the fumed residue in the platinum crucible with 2 g of the mixture (1 part anhydrous sodium carbonate (Na_2CO_3) and 1.3 parts anhydrous potassium carbonate (K_2CO_3)), and dissolve the melt in dilute hydrochloric acid (HCl 1 + 1) under heat. Unite the filtrate from the silicon oxide separation with the fumed residue solution and fill with water in a 500 ml volumetric flask (final solution).

Expression of results

The following equation is used to determine the silicon oxide content in % by mass of the LOI-free fly ash:

$$\text{tot } SiO_2 = \frac{m_6 - m_7}{m_5} \times 100$$

where

m_5: original mass in g of the dried fined fly ash,
m_6: mass of the platinum crucible including residue in g after igniting,
m_7: mass of the platinum crucible including residue in g after fuming and igniting.

4.5.2 Soluble Silicon Oxide

Procedure

Soluble silicon oxide content is obtained indirectly by determining the silicon oxide contents of the dried fly ash and of the residue insoluble in hydrochloric acid and potassium hydroxide. The method for determining the residue insoluble in hydrochloric acid and potassium hydroxide is described in EN 196 Part 2, Section 10. The silicon oxide content is determined according to Section 4.5.1 of this guideline.

Expression of results

The following equation is used to calculate the soluble silicon oxide content in % by mass of the fly ash:

$$\text{sol } SiO_2 = \text{tot } SiO_2 - \frac{IR}{100} \times SiO_2{}^{IR}$$

where

tot SiO_2: silicon oxide content in % by mass of the fly ash,
IR: residue insoluble in hydrochloric acid and potassium hydroxide in % by mass of the dried fly ash,
$SiO_2{}^{IR}$: silicon oxide content in % by mass of the residue insoluble in hydrochloric acid and potassium hydroxide.

4.6 Ferric Oxide (Fe_2O_3)

Procedure

Ferric oxide content is determined by complexometric titration, using the method described in EN 196, Part 2, Section 13.10. A fraction of the final solution prepared according to Section 4.5.1 is employed for analysis.

Expression of results

The following equation is used to calculate the ferric oxide content in % by mass of the fly ash:

$$Fe_2O_3 = \frac{159.692 \times C_1 \times V_{T_1} \times f_1 \times 500}{2 \times m_5 \times V_{P_1}} \times 100$$

where

C_1: concentration of the EDTA solution in mol dm^{-3},
V_{T_1}: volume of the EDTA solution in ml used for titration,
f_1: factor of the EDTA solution,
m_5: original mass in g of dried, fined fly ash (*vide* Section 4.2),
V_{P_1}: fraction of the final solution (in ml) used to determine ferric oxide content.

4.7 Aluminium Oxide (Al_2O_3)

Procedure

Aluminium oxide content is determined by complexometric titration, using the method described in EN 196, Part 2, Section 13.11. The fraction of the final solution titrated according to section 4.6 is used for analysis.

Expression of results

The following equation is used to calculate the aluminium oxide content in % by mass of the fly ash:

$$Al_2O_3 = \frac{101.961 \times C_1 \times V_{T_2} \times f_1 \times 500}{2 \times m_5 \times 1000 \times V_{P_1}} \times 100$$

where

C_1: concentration of the EDTA solution in mol dm^{-3},

V_{T_2}: volume of the EDTA solution in ml used for titration,
f_1: factor of the EDTA solution,
m_5: original mass in g of dried, fined fly ash (*vide* Section 4.2),
V_{P_1}: fraction of the final solution in ml used to determine the aluminium oxide content.

4.8 Calcium Oxide (CaO)

Procedure

Calcium oxide content is determined by complexometric titration, using the method described in EN 196, Part 2, Section 13.12 (reference method) or 13.14 (alternative method). A fraction of the final solution prepared according to Section 4.5.1 is employed for analysis.

Expression of results

The following equation is used to calculate the calcium oxide content in % by mass of the fly ash:

$$CaO = \frac{56.08 \times C_2 \times 500 \times V_{T_3} \times f_2}{m_5 \times 1000 \times V_{P_2}} \times 100$$

where
C_2: concentration of the EGTA or EDTA solution in mol dm^{-3},
V_{T_3}: volume of the EGTA or EDTA in ml used for the titration,
f_2: factor of the EGTA or EDTA solution,
m_5: original mass in g of the dried, fined fly ash (*vide* Section 4.2),
V_{P_2}: fraction of the final solution in ml used to determine the calcium oxide content.

4.9 Magnesium Oxide (MgO)

Procedure

Magnesium oxide content is determined by complexometric titration, using the method described in EN 196, Part 2, Section 13.13 (reference method) or 13.15 (alternative method). A fraction of the master solution prepared according to Section 4.5.1 is employed for analysis.

Expression of results

The following equations are used to calculate the calcium oxide content in % by mass of the fly ash:
- For determination according to EN 196, Part 2, Section 13.13:

$$MgO = \frac{40.311 \times C_3 \times V_{T_4} \times f_3 \times 500}{m_5 \times V_{P_3} \times 1000} \times 100$$

where
C_3: Concentration of the DCTA solution in mol dm^{-3},
V_{T_4}: volume of the DCTA solution in ml used for the titration,
f_3: factor of the DCTA solution,
m_5: original mass in g of dried, fined fly ash (*vide* Section 4.2),
V_{P_3}: fraction of the final solution in ml used to determine the magnesium oxide content.

- For determination according to EN 196, Part 2, Section 13.15:

$$MgO = \frac{40.311 \times C_2 \times (V_{T_5} - V_{T_3}) \times f_1 \times 500}{m_5 \times V_{P_2}} \times 100$$

where
C_2: concentration of the EDTA solution in ml,
V_{T_5}: volume of the EDTA solution in ml used for the titration of the CaO and MgO content,
V_{T_3}: volume of the EDTA solution in ml used for the titration of the CaO-content (*vide* Section 4.8),
f_1: factor of the EDTA solution,
m_5: original mass in g of dried fly ash,
V_{P_2}: fraction of the final solution in ml used to determine the calcium and magnesium oxide content.

4.10 Sulphuric Anhydride (SO_3)

Procedure

Sulphate content is determined by gravimetric analysis, using the method described in EN 196, Part 2, Section 8. 1 g (weighed to an accuracy of 0.1 mg) of the dried, fined fly ash is employed for analysis.

Expression of results

The following equation is used to calculate the sulphate content of the dried fly ash in % by mass:

$$SO_3 = \frac{m_9 \times 0.343}{m_8} \times 100$$

where
m_9: final mass in g of $BaSO_4$,
m_8: original mass in g of dried, fined fly ash.

4.11 Chloride (Cl^-)

Procedure

Chloride content is determined by volumetric analysis, using the method described in EN 196, Part 21, Section 4. 5 g (weighed to an accuracy of 0.1 mg) of the dried, fined fly ash are employed for analysis.

Expression of results

The following equation is used to calculate the chloride content in % by mass of the dried fly ash:

$$Cl^- = \frac{35.453 \times C_4 \times (V_{T_{7}} - V_{T_{7}}) \times f_4}{m_{10} \times 1000} \times 100$$

where

C_4: concentration of the NH_4SCN solution in mol dm^{-3},
V_{T_6}: volume of the NH_4SCN solution in ml used for the titration of the blank value,
V_{T_7}: volume of the NH_4SCN solution in ml used for the titration of the test solution,
m_{10}: original mass in g of dried, fined fly ash,
f_4: factor of the NH_4SCN solution.

4.12 Free Calcium Oxide (free CaO), including $Ca(OH)_2$

Procedure

Place 1 to 1.5 g (weighed to an accuracy of 0.1 mg) of dried, fined fly ash in a 250 ml Erlenmeyer flask together with a mixture of 12 ml ethyl acetoacetate and 80 ml isobutanol. Fit a spiral reflux condenser and heat for 1 h, agitating the mixture. Use a dry pipe filled with sodium hydroxide on an inorganic carrier throughout the test to shield the reflux condenser against penetration of atmospheric CO_2. Filter the warm mixture via a filter crucible. Wash the residue with 50 ml iso-propanol. If the filtrate is clouded the procedure must be repeated. Add a few drops of bromphenol blue (0.1 g bromphenol blue in 100 ml ethanol) to the filtrate and titrate the mixture with 0.1 N hydrochloric acid until it turns yellow.

Expression of results

The following equation is used to calculate the free calcium oxide content in % by mass of the dried fly ash:

$$\text{free CaO} = \frac{28.04 \times C_5 \times V_{T_8} \times f_5}{m_{11} \times 1000} \times 100$$

where

C_5: concentration of the HCl solution in mol dm^{-3},
V_{T_8}: volume of the HCl solution used for titration in ml,
f_5: factor of the HCl solution,
m_{11}: original mass in g of the dried, fined fly ash.

4.13 Total Alkali Oxides (Na_2O, K_2O)

Procedure

Total alkali content is determined by flame photometry or atomic absorption spectroscopy using the method described in EN 196, Part 21, Section 7 or 8. The fly ash is decomposed according to the method described in EN 196, Part 21, Section 7.5.2. 0.2 g (weighed to an accuracy of 0.1 mg) of the dried, fined fly ash is used for analysis.

Expression of results

The alkali content of the dried fly ash is assessed by comparison with reference curves plotted as described in EN 196, Part 21, Section 7.4 or 8.4.

4.14 Ammonium (NH_4^+)

Procedure

Add 30 ml of sodium hydroxide solution, 30% (m/m) to approximately 50 g (weighed to an accuracy of 0.1 mg) of the dried, fined fly ash in a sealed Kjehdahl steam distillation apparatus. Distil the ammonium as ammonia and capture in a receiver of 50 ml 0.005 molar sulphuric acid. Titrate the receiver with a 0.01 molar sodium hydroxide solution against bromophenol blue until the colour changes from blue to yellow. A blank sample is similarly distilled and titrated using 50 ml water without ammonium.

Expression of results

The following equation is used to calculate the ammonium content in % by mass of the dried fly ash:

$$NH_4^+ = \frac{18.04 \times C_6 \times (V_{T_9} - V_{T_{10}}) \times f_6}{m_{12} \times 1000} \times 100$$

where

C_6: concentration of the NaOH solution in mol dm^{-3},
V_{T_9}: volume of the NaOH solution in ml used for the titration of the blank value,
$V_{T_{10}}$: volume of the NaOH solution in ml used for the titration of the test solution,
f_6: factor of the NaOH solution,
m_{12}: original mass in g of the dried, fined fly ash.

5. GLASS CONTENT

Visual assessment with an optical microscope fails to determine the glass content of a fly ash satisfactorily. Good results may be achieved with the aid of X-ray diffraction spectroscopy. With this method, the crystalline constituents (e.g. quartz, hematite) of the fly ash are quantified and the glass content calculated. Tests have shown that the glass content can be correlated with the content of constituents soluble in hydrochloric acid and potassium hydroxide, providing an alternative method of determining glass content. The method used to determine the insoluble residue is described in EN 196, Part 2, Section 10 (cf. Section 4.5.2 of this guideline).

Calculation

The following equation is used to calculate the content in % by mass of the LOI-free fly ash of constituents soluble in hydrochloric acid and potassium hydroxide (S):

$$S = 100 - IR$$

where

S: residue in % by mass soluble in hydrochloric acid and potassium hydroxide (in relation to the dried fly ash),

IR: residue in % by mass insoluble in hydrochloric acid and potassium hydroxide (in relation to the dried fly ash).

6. PARTICLE DENSITY

6.1 Scope

The particle density of fly ash is defined as the mass per unit volume of the solids received after drying according to Section 6.2. Ground fly ash may have a higher density.

6.2 Preparation of the sample

The initial sample is quartered and sub-divided until approximately 200 g of the material is obtained. This is spread in a shallow container and dried in a well-ventilated oven at 105 ± 5°C to constant mass. It is then cooled in a desiccator.

6.3 Determining particle density

The particle density of the fly ash is determined by displacement of liquid in a pycnometer bottle of at least 25 ml capacity or in a Le Chatelier flask (*vide* ASTM C 188). A non-reactive liquid such as n-heptane, naphtha or redistilled kerosene (paraffin oil) is used for the displacement measurement. The particle density determination test is to be performed three times with an accuracy of 0.01 g cm^{-3}.

6.3.1 Method of Determination Using a Pycnometer Bottle

Place 10 g (proportionately a larger quantity if a larger bottle is used) of the sample, prepared as described in Section 6.2 and free of lumps, in a 25 ml pycnometer bottle. Half fill the bottle with the selected liquid, place in a desiccator and evacuate with a vacuum pump until all air has been removed. Fill the bottle with the liquid and maintain at a constant temperature of between 15°C and 25°C. The chosen temperature must not vary by more than 0.5 K during the test.

6.3.2 Method of Determination using a Le Chatelier Flask

Fill the Le Chatelier flask to the appropriate mark with the selected liquid. Immerse the flask in a constant-temperature water bath and take the first reading when a constant temperature in the flask has been achieved. Tip approximately 50 g (weighed to an accuracy of 0.05 g) of fly ash into the flask. Replace the stopper and roll the flask or spin it horizontally to remove all adhering air. Take the final reading after the flask has been immersed in the water bath long enough to exclude temperature variations in excess of 0.2 K between the initial and final reading.

7. FINENESS

7.1 General

The fineness of fly ash can be determined by wet sieving (Section 7.2), dry sieving (Section 7.3) or with the Blaine air permeability apparatus (Section 7.4).

7.2 Fineness by wet sieving

7.2.1 Scope

This test method describes the determination of fly ash fineness by wet sieving on a 45 μm sieve (ISO 565). Fineness is expressed as the percentage (m/m) retained on the sieve.

7.2.2 Apparatus (Fig. 1)

Sieve. The sieve frame shall be constructed of durable material not susceptible to corrosion or distortion by oven heat. The frame is essentially a tube of 50 mm nominal diameter and measuring 75 mm from the top of the frame to the sieve cloth, with facilities for removing and replacing the cloth. The 45 μm stainless steel sieve cloth shall comply with the requirements of ISO 565 and ISO 3310/1 and be free of visible irregularities in mesh size when inspected visually as described in ISO 3310/1.
Spray nozzle. The spray nozzle (Fig. 2) shall be constructed of metal not susceptible to corrosion by water, with an inside diameter of 17.5 mm. The spray nozzle has a central hole drilled parallel to the longitudinal axis, an intermediate row of eight holes drilled 6 mm centre-to-centre at an angle of 5° to the longitudinal axis and an outer row of eight holes drilled 11 mm centre-to-centre at an angle of 10° to the longitudinal axis. All holes shall be 0.5 mm in diameter.
Pressure gauge. The pressure gauge employed must have a minimum diameter of 80 mm and a maximum capacity of 160 kPa, graduated at maximum intervals of 5 kPa. The accuracy of the gauge shall be ±2 kPa.
Oven. A well-ventilated drying oven.
Balance. A balance capable of weighing to the nearest 1 mg.

7.2.3 Checking the test sieve

A reference sample* with a known proportion of material coarser than the specified mesh size is recommended for checking the specified sieve. The material shall be

* Reference material is not yet available and a proposal should be made for an internationally accepted material/institution.

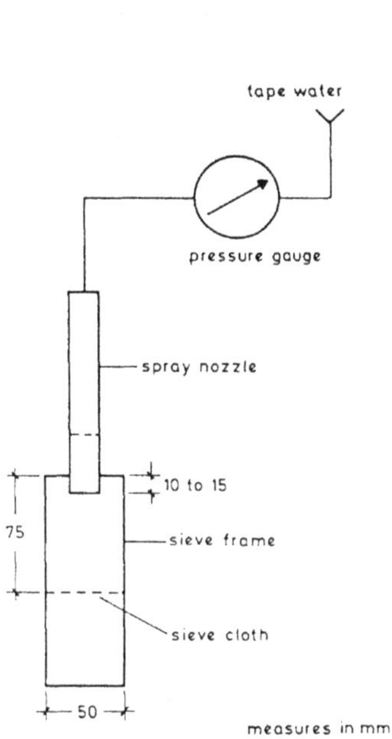

Fig. 1 Apparatus.

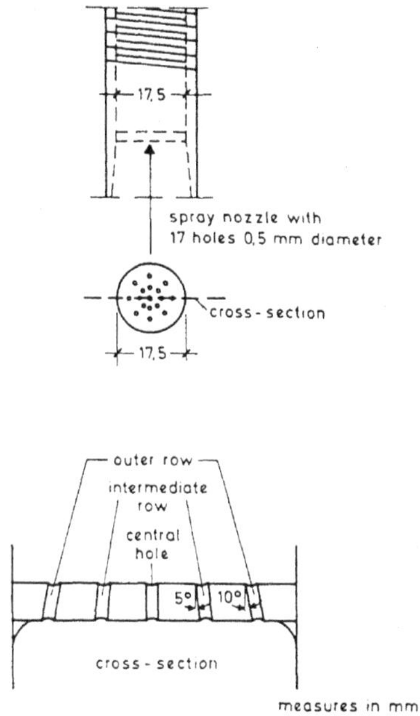

Fig.2. Spray nozzle.

stored in sealed, airtight containers, to preclude changes in its properties due to deposition or absorption from the atmosphere. Containers shall be marked with the sieve residue of the reference material.

Test sieves shall be checked when new and at intervals not exceeding 100 tests. The sieve cloth is first inspected visually as described in ISO 3310/1. Any sieve with an imperfect or damaged sieve cloth must be rejected. The sieve shall be cleaned after each 5 tests. The fineness of the reference material is determined as described in Section 7.2.4. The correction factor *f* for the test sieve is calculated as follows:

$$f = r_r/r_t$$

where

r_r: proportion in % by mass of the reference material retained by the test sieve;

r_t: known 45 μm sieve residue in % by mass of the reference material.

The test is performed twice and the mean value of *f* taken as the correction factor.

7.2.4 *Method of determination*

The method of determination is as follows. Dry the sample in the oven at 105 ± 5°C. Transfer approximately 1 g (weighed to the nearest 1 mg) of the oven-dry sample to a clean, dry sieve. Wet the sample thoroughly with a gentle flow of water, using the spray nozzle. Remove the sieve from the spray nozzle and adjust the water pressure to 80 ± 5 kPa. Place the sieve in position under the nozzle and wash for 60 ± 10 s, keeping the lower end of the nozzle between 10 mm and 15 mm below the top of the sieve frame, and swirling the sieve horizontally at about 1 revolution per second. Remove the sieve from its position under the nozzle, rinse with approximately 50 ml alcohol or distilled water and blot up residual moisture from the underside of the sieve cloth. Dry the sieve and the residue in the ventilated oven at 105 ± 5°C. Cool the sieve and the residue in a desiccator and weigh the residue to the nearest 1 mg.

7.2.5 *Calculation*

The following equation is used to calculate the fineness of the sample to the nearest 0.1%:

$$r = \frac{f \times m_s}{m_o} \times 100$$

r: 45μm sieve residue in % by mass,

f sieve correction factor (*vide* Section 7.2.3),

m_s: sample residue in g,

m_o: original mass of sample, in g.

7.2.6 Result

The test is to be performed twice and the mean value of *r* taken as the fineness of the sample.

7.3 Fineness by dry sieving

7.3.1 Scope

This method describes the determination of fly ash fineness by dry sieving on a test sieve according to ISO 565. Fly ash fineness as determined by dry sieving is expressed as the percentage (m/m) retained on the test sieve used.

7.3.2 Apparatus and method

Sieving is to be carried out in accordance with ISO 2591, using test sieves with woven wire cloth complying with ISO 565. The diameter of the sieve frame shall be 200 mm and a weighted fly ash sample of approximately 20 g should be used. If the nominal size of the test sieve is greater than or equal to 63 μm, the test may be performed by hand sieving according to EN 196, Part 6,* Section 3. With nominal test sieve sizes below 63 μm; a sieving machine should be used e.g. airjet sieving machine.‡ In accordance with ISO 2591, the sieving process should be terminated when the quantity passing through the sieve in 1 min is less than 0.1% of the charge.

*Currently in draft form.
‡The method works by underpressure and constantly purging the sieve meshes by air current.

7.4 Fineness with the Blaine air permeability apparatus

7.4.1 Scope

This test method covers the determination of fineness of fly ash, using the Blaine air permeability apparatus. The fineness of fly ash in terms of the specific surface is determined by measuring the time taken for a fixed quantity of air to flow through a compacted fly ash bed of specified dimensions and porosity. Under standardized conditions, the specific surface of the fly ash is proportional to $\sqrt{t}$ where t is the time in s for a given quantity of air to flow through the compacted fly ash bed. The number and size range of individual pores in the specified bed are governed by the fly ash particle size distribution, which also determines the time for the specified air flow. The method is comparative rather than absolute, and a reference sample with a known specific surface is therefore required to calibrate the apparatus.†

7.4.2 Apparatus and Method

The test is to be performed according to EN 196, Part 6*, Section 4, with the following supplements: (1) Fly ash is to be used as the reference material† and (2) Because of the differences in particle shape and particle size distribution for fly ash as opposed to cement, the specific porosity of the fly ash bed in the Blaine apparatus usually differs from $\epsilon = 0.500$; e is to be calculated therefore according to EN 196, Part 6*, Section 4.8.

†Reference material is not yet available and a proposal should be made for an internationally accepted material/institution.

FAB 2: Test methods for determining the properties of fly ash in concrete

CONTENTS

1. SCOPE

This test guideline covers the properties of fly ash as an additive for mortar and concrete. It does not establish specifications, which are to be drawn from national and international standards. The guideline confines itself to describing test methods significant for the assessment of fly ash suitability and not already covered by other international standards or RILEM Recommendations for cement, mortar and concrete.

2. DEFINITION

Fly ash is a fine powder of mainly spherical, glassy particles having pozzolanic properties and consisting essentially of SiO_2 and Al_2O_3. Fly ash is obtained by electrostatic or mechanical precipitation of dust-like particles from the flue gases of furnaces fired with pulverized coal.

3. SAMPLE SIZE

For a complete test according to the following sections at least 5 kg of fly ash are required. Please refer to the relevant sections for the size of sample needed for individual tests.

4. CEMENT

It should be noted that the activity of fly ash depends not only on its own properties, but on the physical and chemical properties of the cement employed, even within the same cement type. The fly ash should therefore be tested with the cement intended to be used in practice in mortar and concrete. Unless otherwise specified, an 'ordinary PC' should be used to test basic activity.

5. GENERAL REMARKS ON THE COMPOSITION OF MIXES

The water requirement of fly ash according to Section 7 may be determined by comparing control mixes and test mixes of paste, mortar or concrete. Pozzolanic activity according to Section 8 may be determined by comparing control mixes and test mixes of mortar or concrete. Portland cement should be used in control mixes, and it is suggested that at least 20% (m/m) of the cement should be replaced by fly ash in the test mixes. Paste used to determine the water requirement of fly ash is to be prepared according to ISO/DIS 9597,* Sections 1 to 5. Mortar composition should correspond to ISO/DIS 679,* except that the water content should be adjusted to achieve equal consistency (*vide* Sections 7.3 and 8.2). Concrete composition should correspond to the RILEM Reference Concrete† with a cement or cement plus fly ash content of 300 kg m^3. Aggregate gradings should be within the range shown in Fig. 1.

6. SOUNDNESS

The soundness is expressed as the expansion tested by the method described in EN 196/Part 3, Section 7.

7. WATER REQUIREMENT

7.1 General

The water requirement of fly ash may be tested on paste, mortar or concrete. Results cannot be transferred from one type of mix to another. In order to determine the water requirement, a portland cement control mix and a portland cement/fly ash test mix are prepared, with the respective water contents necessary to yield nominally equal consistency numbers. The water content of the test mix expressed as a percentage of the water content of the control mix is termed the 'water requirement of fly ash in the mix'.

7.2 Water requirement of fly ash in paste

The water requirement is to be determined in accordance with ISO/DIS 9597,* Section 5, using a portland cement control paste and a portland cement/fly ash test paste. The water content of the test paste expressed as a percentage of the water content of the control paste is termed the 'water requirement of the fly ash in paste'.

7.3 Water requirement of fly ash in mortar

A portland cement control mortar and a portland cement/fly ash test mortar are prepared in accordance with ISO/DIS 679.* The water content of the mixes must be adjusted to achieve a flow of 160 ± 10 mm when tested 10 minutes after addition of mixing water to the mix, in accordance with RILEM Recommendation MR-11. The water content of the test mortar expressed as a percentage of the water content of the control mortar is termed the 'water requirement of the fly ash in mortar'.

7.4 Water requirement of fly ash in concrete

A portland cement control concrete and a portland cement/fly ash test concrete are prepared in accordance with the RILEM Reference Concrete as described in Section 5. The water content of the concrete must be adjusted to achieve a slump of 60 ± 10 mm according to ISO 4109 or a flow diameter of 400 ± 30 mm according to ISO 9812* when tested 10 minutes after the addition of mixing water to the mix. The water content of the test concrete expressed as a percentage of the water content of the control concrete is termed the 'water requirement of the fly ash in concrete'.

8. ACTIVITY RELATED TO STRENGTH

8.1 General

The method tests the effects of fly ash as a substitute for cement on the compressive strength of mortar and concrete. The activity is characterized by the ratio of the compressive strengths of the mortars or concretes with fly ash to that of the control mortars or concretes without fly ash. Specimens are stored in water at either +20°C or +40°C. Storage at +40°C is intended to estimate the activity of the fly ash over a longer period.

8.2 Composition of the mortars or concretes

The mortars or concretes with and without fly ash may be prepared with the same consistency or the same water

*Currently a draft is in preparation.
†RILEM COMMITTEE 14-CPC
Reference Concrete
Materials and Structures **12** (1979), No. 68, pp. 140–141.

content. Preparation and Composition shall be in accordance with sections 7.3 and 7.4 and the cited standards and recommendations.

8.3 Preparation and curing of the specimens

Mortar specimens are prisms, each measuring 40 mm × 40 mm × 160 mm prepared according to ISO/DIS 679. Concrete specimens are 150 mm cubes prepared according to ISO 2736/2. Three specimens are to be prepared for each test date. Immediately after production, the moulds with the specimens, covered with a glass plate, are to be cured in a moist atmosphere for 24 h at 20 ± 2°C and ≥95% relative humidity.

Subsequent curing is alternatively.

(i) In water at +20°C. The specimens are removed from the moulds and cured up to testing in a suitable container in water at a temperature of 20 ± 2°C.

(ii) In water at +40°C. The specimens are removed from the moulds and cured up to testing in a suitable container in water at 40 ± 2°C. 2 hours before testing, the specimens are to be cured in water at 20 ± 2°C.

In both cases the distance between the surfaces of the specimens and the water surface must be at least 50 mm. The distance between the specimens and the bottom of the container must be at least 20 mm.

8.4 Determining compressive strength

Compressive strength is usually determined at an age of 28 days. Tests on mortar are carried out according to ISO/DIS 679* and on concrete according to ISO 4012.

8.5 Activity index

The Activity Index r_a is the ratio

$$r_a = \frac{f_f}{f_o} \times 100 \text{ (as \%)}$$

*Currently a draft is in preparation.

and is to be calculated from the compressive strength values determined for the mortars or concretes with fly ash (index f) and without fly ash (index o) under identical curing conditions. The age at testing and the form of curing are to be quoted.

RELATED INTERNATIONAL STANDARDS

ASTM C 188 1984. Test method for density of hydraulic cement.
EN 196 Part 2 1987. Methods of testing cement; chemical analysis of cement.
EN 196 Part 3 1987. Methods of testing cement; determination of setting time and soundness.
EN 196 Part 6 1987 (Draft). Methods for testing cement; Determination of fineness.
EN 196 Part 21 1987. Methods for testing cement; Determination of the chloride carbon dioxide and alkali content of cement.
ISO 565 1983. Test sieves — Woven metal wire cloth, perforated plate and electroformed sheet — Nominal sizes of opening.
ISO/DIS 679 1987. Methods of testing cements — determination of strength.
ISO 2591 1973. Test sieving.
ISO 2736/2 1986. Concrete tests — Test specimens — Part 2: Making and curing of test specimens for strength test.
ISO 3310/1 1982. Test sieves — Technical requirements and testing. Part 1: Test sieves of metal wire cloth.
ISO 4012 1978. Concrete; determination of compressive strength of test specimens.
ISO 4109 1980. Fresh concrete; determination of the consistency; slump test.
ISO/DIS 9597 1987. Methods of testing cements; Determination of setting time and soundness.
ISO/DP 9812. Fresh concrete; determination of consistency–flow test
RILEM Recommendation MR1–21 1982 (E). MR 11: Determination of mortar consistency using the flow table.

INDEX

For Product Safety Concerns and Information please contact our EU representative GPSR@taylorandfrancis.com
Taylor & Francis Verlag GmbH, Kaufingerstraße 24, 80331 München, Germany

www.ingramcontent.com/pod-product-compliance
Ingram Content Group UK Ltd.
Pitfield, Milton Keynes, MK11 3LW, UK
UKHW010959180926
475701UK00004B/448
9780367863968